Software Patterns, Knowledge Maps, and Domain Analysis

Mohamed E. Fayad • Huascar A. Sanchez
Srikanth G.K. Hegde • Anshu Basia • Ashka Vakil

CRC Press
Taylor & Francis Group
Boca Raton London New York

CRC Press is an imprint of the
Taylor & Francis Group, an **informa** business
AN AUERBACH BOOK

CRC Press
Taylor & Francis Group
6000 Broken Sound Parkway NW, Suite 300
Boca Raton, FL 33487-2742

Printed on acid-free paper
Version Date: 20141006

International Standard Book Number-13: 978-1-4665-7143-3 (Hardback)

Visit the Taylor & Francis Web site at
http://www.taylorandfrancis.com

and the CRC Press Web site at
http://www.crcpress.com

Dedication

To the land of the Delta—Egypt, and the land of Dreams—The United States of America, and the love of science and humanity that they share... Pouring out a flood of knowledge for centuries and centuries between the Nile and the Mississippi... Between the millstones of my heart, Egypt and the United States of America...

Mohamed E. Fayad

To my wife, Claudia, and my two beautiful daughters, Isabella and Camilla.

Huascar A. Sanchez

To my wife, Kumuda.

Srikanth G. K. Hegde

To my family, for their encouragement and support.

Ashka Vakil

Contents

SECTION III Knowledge Map Capabilities

Preface

This book delineates a new creation process and provides an understanding of software pattern languages and true domain analysis based on the fundamental concepts of software stability. It also introduces a well-defined paradigm for creating pattern languages, software patterns, and better software development methodology that leads to highly reusable artifacts and high-quality, cost-effective systems. Each chapter of the book concludes with an open research issue, review questions, exercises, and projects.

The main goal of this book is to define *knowledge maps* as the groundwork for an insightful classification of the software patterns governing or administering a particular discipline. Knowledge maps are the enduring mirrors of experience and best practices of what a discipline is, why it is so necessary, and how developers can exploit it.

This book addresses various issues related to stable software patterns, knowledge maps, and domain analysis and eventually analyzes different paradigms and factors that result in the creation of stable software systems that are reusable and extremely cost-effective to produce. It is written for use as an advanced textbook for software course developers, students of software development, researchers, and academicians.

WHY THIS BOOK?

Software analysis and design patterns are known to play a vital role in enhancing the quality and merit of a software product. In addition, they are also known to lessen the final cost of software products as well as reduce their life cycle. Despite the immense usefulness of these patterns, we may still need to sort out a series of critical problems that usually occur in today's domain of contemporary analysis and design patterns, such as instability, absence of right abstraction levels, and improper/insufficient documentation of procedures and processes. All these factors, in combination, might significantly bring down factors like reusability, repeatability, stability, robustness, and overall effectiveness of finished software products.

Hence, software developers, pattern makers, and programmers may need to focus their professional attention on using different patterns in combination, to solve myriad problems that might pose numerous challenges, while developing a product, and to provide practical solutions to those problems to bring effective resolutions, which eventually lead to the development of a robust and stable product with sustained life cycle and durability.

While conducting detailed research on these issues, we analyzed how numerous drawbacks of current software approaches that deal with software patterns, especially in software pattern compositions, traceability, generality, and so on, hindered the quality of built systems in one way or another (e.g., design trade-offs, loss of generality). In order to overcome these drawbacks, we have provided a standard way for conceiving, building, and deploying systems by using a topology of software patterns. This topology is known as knowledge maps. The knowledge map will serve as the road map or supporting technique to guide software practitioners as they delve into the rationale, business rules, and context of application of a set of problem domains and come up with a high-quality software system.

The essence of knowledge maps is twofold: *a clear methodology* and a *precise visual representation*. For the methodology approach, we have provided a set of guidelines, heuristics, and quality factors that will simplify the process of creating knowledge maps, along with their realization and documentation. However, for visual representation, we have provided the visual gadgets or symbols that convey how the knowledge maps and their enclosed elements look, and in what manner they interact with other enclosed elements or other knowledge maps. Together, both methodology and visual representation serve as the road map for building systems from software patterns in a cost-effective manner. In addition to this, this road map will also allow the creation of synergies between managers and technical staff, especially when creating systems in terms of goals and capabilities. As a result, these synergies will provide the ways and means for reducing existing communication gaps between the managerial and technical staff.

In essence, this book provides readers with a detailed view of the art and practice of creating meaningful knowledge maps that help software developers build software products from stable, enduring, and cost-effective software patterns.

WHOM IS THIS BOOK FOR?

Software students who read this book will gain a basic, as well as advanced, understanding of principles and issues related to the creation of stable and robust software patterns, meaningful knowledge maps, and their domain analysis. While using knowledge maps, we can expect great team dynamics between managers and technical staffs. They are capable of creating an environment where the initial clashing of ideas that might occur because of one's own beliefs and experience is immediately detected and recognized for immediate action and identification of suitable solutions. This environment will also allow managers and technical staff to focus on the merit of the problem and not on the irrelevant and trifles, for example, implementation details. At the same time, it will also create a common language for communicating ideas between managers and the technical staff.

Students, software developers, software designers, and technical managers with a basic background in software development and engineering will find information contained in this book easy to understand. Although some of the material in this book relates to advanced programming, readers (both beginning and advanced) can easily understand its essence and get the big picture of creating knowledge maps and robust software patterns.

This book could be of great help for a large community of computing and modeling academics, students, software technologists, software methodologists, software pattern communities, component developers, software reuse communities, and software professionals (analysts, designers, architects, programmers, testers, maintainers, and developers) who are involved in the management, research, and development of methodologies and software patterns. Industry agents, who work on any technology project and want to improve the project's reliability and cost-effectiveness, will also benefit hugely by reading this book.

We also anticipate and assume that the concepts presented in this book will greatly affect the development of new software systems and application frameworks for the next two or three decades. This book will be very valuable for database designers, knowledge management and development professionals, and knowledge ontology scientists. We expect this book to be a leading choice for many graduate courses on software engineering, system engineering, software modeling, knowledge modeling, domain analysis, requirement engineering, software architectures, software design, and programmers.

HOW TO USE THIS BOOK

This book is designed to allow readers to master the basics of knowledge maps from their theoretical aspects to practical application. To allow easy reading and better understanding of individual topics, this book is divided into 14 chapters, each of which deals with separate aspects of knowledge maps.

CHAPTER CONTENTS IN DETAIL

This book is stratified and segregated as follows. Chapter 1 provides an overview of the contents of the entire book and sets the stage for its proper development. Chapter 2 examines the methodology for forming knowledge maps in a cost-effective manner. Chapter 3 explores the term *goals* and its importance in the formation of knowledge maps, whereas Chapters 4 and 5 provide a set of complete documentation of two stable analysis patterns: discovery and knowledge. A goal without a trace of capability is not a goal; therefore, in Chapter 6, we will give readers a detailed description of the capabilities of the element of knowledge maps and their role in building them. Chapters 7 and 8 provide complete documentation of two stable design patterns: AnyMap and AnyContext. Chapter 9 provides additional details and describes what knowledge maps and their system of patterns are and their role in the understanding and mastering of any discipline of interest. Chapter 10 concentrates and focuses on the formation of development scenarios, especially in the identification of context-specific classes, and how they are hooked into the core formed by goals and capabilities. Chapter 11 provides insight and a summary of the ways and manner in which knowledge maps are deployed. Chapter 12 provides detailed descriptions on knowledge map engines and how software protagonists can start initial work on the formation of a stable engine. Chapter 13 provides information on CRC cards and their relationship with knowledge maps. Chapter 14 focuses on the book's concluding remarks, where we provide a summary of what we have performed throughout the book and what we will do as a future task.

Mohamed E. Fayad
San Jose State University

Huascar A. Sanchez
University of California Santa Cruz

Srikanth G.K. Hegde
Freelance Writer

Anshu Basia
Quisk

Ashka Vakil
SAP

Acknowledgments

This book would not have been completed without the help of many great people; I *thank them all.* I am honored to work with my friend and coauthor of this book, Srikanth G. K. Hegde, and three of my best students and coauthors of this book, Huascar A. Sanchez, Anshu Basia, and Ashka Vakil. This was a great and fun project because of your tremendous help and extensive patience. I also thank all of my student assistants, Hema Veeraragavathatham, Vishnu Sai Reddy Gangireddy, Mansi Joshi, Siddharth Jindal, and Pavan Pavuluri, for their work on the figures and diagrams. Thanks to my dear friend and colleague Professor Supratik Mukhopadhyay for his contribution of two sidebars, on formality and goal-oriented development. Special thanks to my San Jose State University students—Chintan Shah, Hardik Shah, Viral Sonawala, Ashutosh Kulkarni, Sapna Suku, Shashi Bhushan Kedilaya , Swetha Seshadri; the Magnum team—Santosh Kumar Gottipamula, Vamseedhar Vuppu, Dhiwakar Mani, Nirav Kumar Patel, Ashka Vakil, Lalitha Venkataramani, Padmavathi Chaganti, Suju Koshy, Lois Desplat, Fayad Hussain, Ashira Khera, Ali Parandian, Mary Elaine David, Anu Ganesan, and Joselyn Tapas; and Abhishek Maloo, Rahul Panjrath, Ruchin Kabra, Fan Ieong, Priya Lobo, and Mrunali Mohane for helping me create the sample requirements in Appendix D.

Special thanks to my wife, Raefa, my lovely daughters Rodina and Rawan and my son Ahmad for their great patience and understanding. Special thanks also to Srikanth's wife, Kumuda Srikanth, for help with reviewing some of the chapters. Special thanks to all my friends all over the world for their encouragement and long discussions about the topics and the issues in this book. Thanks to all my students and coauthors of many articles related to this topic, in particular, Haitham Hamza, Ahmed Mahdy, Shasha Wu, Peeyush Tugnawat, and Shivanshu Singh; to my friends Davide Brugali and Ahmed Yousif for their encouragement during this project; and to the *Communications of the ACM* staff—my friends Diana Crawford, the executive editor, Thomas E. Lambert, the managing editor, and Andrew Rosenbloom, the senior editor.

On behalf of the authors of this book, I acknowledge and thank all of those who have had a part in the production of this book. First, and foremost, we owe our families a huge debt of gratitude for being so patient while we put their world in a whirl by injecting this writing activity into their already busy lives. We also thank the various reviewers and editors who have helped in so many ways to get the book together. We thank our associates who offered their advice and wisdom in defining the content of the book. We owe special thanks to those who have worked on the various projects covered in the case studies and examples.

Finally, we acknowledge and thank the work of some of the people who helped us in this effort: John Wyzalek, acquisition editor, Jill J. Jurgensen, senior project coordinator, Keyle Meyer, project editor, and Rebecca Rothschild, the marketing manager at CRC Press, Taylor & Francis Group, LLC, for their excellent and quality support and work done to produce this book; a special note of acknowledgment and thanks to Indumathi S., project management executive at Lumina Datamatics Ltd., whose team did a tremendous job proofreading and copy editing all the chapters in detail, including the elegant and focused way in which Indumathi took care of the day-to-day handling of this book; and special thanks to all the people in marketing and design and the support staff at CRC Press, Taylor & Francis Group, LLC, and Lumina Datamatics Ltd.

Authors

Dr. Mohamed E. Fayad is a full professor of computer engineering at San Jose State University from 2002 to the present. He was a J.D. Edwards Professor, Computer Science and Engineering, at the University of Nebraska, Lincoln, from 1999 to 2002; an associate professor at the computer science and computer engineering faculty at the University of Nevada, from 1995 to 1999; and an editor-in-chief for IEEE Computer Society Press—Computer Science and Engineering Practice Press, from 1995 to 1997. He has 15+ years of industrial experience. Dr. Fayad is an IEEE distinguished speaker, an associate editor, editorial advisor, a columnist for the *Communications of the ACM* (Thinking Objectively), and a columnist for the *Al-Ahram* newspaper (two million subscribers); he was a general chair of the IEEE/Arab Computer Society International Conference on Computer Systems and Applications (AICCSA 2001), Beirut, Lebanon, June 26–29, 2001; he is the founder of the Arab Computer Society (ACS), serving as its president from April 2004 to April 2007.

Dr. Fayad is a known and well-recognized authority in the domain of theory and the applications of software engineering. Dr. Fayad was a guest editor on 12 theme issues: *CACM's OO Experiences*, October 1995; *IEEE Computer's Managing OO Software Development Projects*, September 1996; *CACM's Software Patterns*, October 1996; *CACM's OO Application Frameworks*, October 1997; *ACM Computing Surveys—OO Application Frameworks*, March 2000; *IEEE Software—Software Engineering in-the-Small*, September/October 2000; *International Journal on Software Practice and Experiences*, July 2001; *IEEE Transaction on Robotics and Automation—Object-Oriented Methods for Distributed Control Architecture*, October 2002; *Annals of Software Engineering Journal—OO Web-Based Software Engineering*, October 2002; *Journal of Systems and Software, Software Architectures and Mobility*, July 2010; and *Pattern Languages: Addressing the Challenges, Wiley Software: Practice and Experience*, March–April 2012.

Dr. Fayad has published more than 218 high-quality papers, which include profound and well-cited reports (more than 50) in reputed journals, 84 articles in refereed conferences, more than 20 well-received and cited journal columns, 16 blogged columns; 12 well-cited theme issues in prestigious journals and flagship magazines; 24 different workshops in respected conferences; and over 125 tutorials, seminars, and short presentations in 30+ different countries, such as Hong Kong (thrice), Canada (12 times), Bahrain (twice), Saudi Arabia (4 times), Egypt (30 times), Lebanon (twice), UAE (twice), Qatar (twice), Portugal (twice), Finland (twice), United Kingdom (thrice), Holland (thrice), Germany (4 times), Mexico (once), Argentina (thrice), Chile (once), Peru (once), Spain (once), and Brazil (once). Dr. Fayad is founder of 7 online journals, NASA Red Team Review of QRAS and NSF-USA Research Delegations' Workshops to Argentina and Chile, and 4 authoritative books, of which three are translated into different languages, such as Chinese; over 5 books are currently in progress. Dr. Fayad is filing for 8 valuable and innovative patents and has developed over 800 stable software patterns. Dr. Fayad earned an MS and a PhD in computer science from the University of Minnesota at Minneapolis. His research title was OO Software Engineering: Problems and Perspectives. He is the lead author of several classic works: *Transition to OO Software Development*, August 1998; *Building Application*

Frameworks, September 1999; *Implementing Application Frameworks*, September 1999; *Domain-Specific Application Frameworks*, October 1999. Dr. Fayad's books in progress include *Stable Analysis Patterns*, *Stable Design Patterns*, *Unified Software Architectures*, *Service and Production Engines*, *Moviemaking: UML and Knowledge Map in Action*, *UML in Action*, *Unified Software Engine*, and *Knowledge Map: True Domain Analysis Approach*.

Huascar A. Sanchez is a PhD candidate in the University of California Santa Cruz's Computer Science Department. His research interests include software engineering, specifically source code curation, an approach to discovering, cleaning, and refining online code snippets upon which to build programs. Sanchez has earned an MS in software engineering from San Jose State University.

Srikanth G. K. Hegde is a professional Internet security consultant and a freelance writer with a master's degree to his credit. His areas of interest include Internet security, networking, social media marketing, antivirus software, adware, spyware removal and its management, Internet safety, network security policy, and broadband and Internet/security protocols. In addition, his domains of interest also include preparing articles, whitepapers, and status reports on diverse industries, businesses, global events, finance, and business management. Furthermore, he has published numerous articles on software patterns (analysis and design), pattern development, patterns composition, and knowledge maps, in association with Dr. Prof. M. E. Fayad. Srikanth is also an experienced freelance writer with more than 15 years of experience in writing books, articles, columns, critiques, and various other e-publications.

Anshu Basia is a software engineer with extensive experience and management skills. Her specialization includes analyzing, designing, and developing complex enterprise solutions in agile environments. Anshu is highly proficient in Java, Spring, Struts, HTML, JavaScript, Rest Web services, SQL, and a multitude of other technologies used in modern applications.

Currently, Anshu works for Quisk, a global technology company that partners with financial institutions and others to digitize cash and provide safe, simple, and secure financial services and cashless transactions for anyone with a mobile phone number. Prior to joining Quisk, Anshu worked as a software engineer at A2Z (subsidiary of Amazon.com) now known as Amazon Music. Anshu completed her second masters in software engineering with a focus on enterprise software technologies from San Jose State University, graduating with highest honors. Anshu's first master's degree is in computer applications from Banasthali Vidyapith, India.

Ashka Vakil is an accomplished software engineer who specializes in building highly complex enterprise applications. Ashka has 8+ years of extensive experience in architecture, design, and agile development. She is an expert in mobile application and cloud application development. Ashka is highly proficient in Java, HTML, JavaScript, web services, SQL, and a multitude of other technologies used in modern web applications.

Currently, Ashka works for SAP, a German multinational software corporation that makes enterprise software to manage business operations and customer relations. As a senior software engineer, she is responsible for building custom enterprise-grade mobile applications for SAP customers. Prior to joining SAP, Ashka worked as a software engineer at Tata Consultancy Services, the largest India-based IT services company. Ashka holds a master's degree in software engineering with a focus on enterprise software technologies from San Jose State University, graduating with highest honors.

Section I

Introduction

A knowledge map is a topology of patterns that is driven by the principles of software stability concepts (Fayad 2002a, 2002b; Fayad and Altman 2001). In this section, we will provide its structure, mantra, and the rationale-driven language use to discover and visualize elemental pieces of knowledge (patterns), how to organize them, and how to relate them to formulate an accurate solution in contexts, which shares the same core knowledge (rationale or goals, and capabilities).

Building a knowledge map (Sanchez 2006) for a determined discipline involves myriad skills, knowledge, and steps beyond the identification of the tangible artifacts that are bound to a specific context of applicability. It also requires a systematic capture and full understanding of the domain, where our solution would be laid down and expanded. That includes describing the problem not from its tangible side, but focusing more on its conceptual side, describing underlying affairs with respect to the problem, and describing the elements required to fulfill them. Section I is made up of two chapters and five sidebars.

Chapter 1 is titled "An Overview of Knowledge Maps," and it introduces the key concepts and technologies of knowledge maps, such as software stability model, the definitions of enduring business themes or goals, and the definition of business objects or capabilities. It also discusses the existing problems with traditional pattern languages and software patterns, enumerates the objectives of knowledge maps approach briefly, defines the software stability concepts, shows the representation of knowledge maps, and compares the essential differences between traditional pattern languages and knowledge maps. This chapter concludes with a summary and numerous open research issues. This chapter also provides a number of review questions, exercises, and projects.

Chapter 2 is titled "Abstraction: Knowledge Maps, Stability, and Patterns," and it discusses knowledge maps level of abstraction, charts knowledge maps elements to software stability concepts, and patterns world; it also illustrates the software stability model steps like goals, capabilities, and knowledge map.

Sidebar 1.1 is titled "Traditional Pattern Languages," and it provides a brief introduction for traditional pattern languages, as a structured method of describing better design practices within a field of expertise or domain. A pattern language consists of a cascade or *hierarchy of parts*, linked together by *patterns* that solve generic recurring problems associated with the parts. Each pattern has a definite title and collectively the titles form a language for design (http://www.designmatrix.com/pl/anatomy.html).

Sidebar 1.2 is titled "Hooks or Extension Points." Hooks are the important and critical extension points that are used as a means to extend, augment, activate, modify, replace, and add new functionality (Fayad, Schmidt, and Johnson 1999; Froehlich et al. 1997), to adapt, customize, personalize, trace, and/or integrate knowledge by application developers, and to design and produce brand new applications from the core knowledge or knowledge map (Shtivastava 2005).

Sidebar 1.3 is titled "Hook Engine." Hook Engine is a special web-based engine, and hook facility is an enduring tool that facilitates knowledge map hooks. The engine or tool maintains a rich repository of the hook templates and existing software patterns defined specially for the core knowledge of the applications driven from the entire knowledge map. The engine or tool also supports addition, modification, and deletion of hook templates in the given repository. Each hook template specifies a list of changes or editions needed for the core knowledge classes known as business objects.

Sidebar 2.1 is titled "Formal Methods and Formal Languages." Popular knowledge representation techniques can include various monotonic and nonmonotonic logics

(Barwise 2006), such as description logics (Baader et al. 2003) and default logics (Besnard 1989). Goals can be specified as intentional knowledge in a knowledge base. Capabilities can be added to provide the extensional definitions. Goals can be reified by automatically *connecting* the intentional goals to the extensional knowledge, by using the deductive reasoning capabilities of the underlying logical framework (Boddu et al. 2004). Reifying goals usually results in ontology.

Sidebar 2.2 is titled "The Definition of Ontology." Ontology is a specification of a conceptualization (Chandrasekaran et al. 1999; Gruber 1993, 1995). The word *ontology* seems to generate a lot of controversies in discussions about artificial intelligence. It has a long history and tradition in philosophy, in which it refers to the subject of existence. It is also often confused with epistemology, which is about knowledge and knowing.

1 An Overview of Knowledge Maps

He who knows not and knows not he knows not: he is a fool—shun him. He who knows not and knows he knows not: he is simple—teach him. He who knows and knows not he knows: he is asleep—wake him. He who knows and knows he knows: he is wise—follow him.

Old Arabian Proverb

1.1 INTRODUCTION: KEY CONCEPTS—SOFTWARE STABLE MODELS, KNOWLEDGE MAPS, PATTERN LANGUAGE, GOALS, CAPABILITIES (ENDURING BUSINESS THEMES + BUSINESS OBJECTS)

Right now, a number of factors, for example, overall increase or bulkiness in software size, complexity, hefty costs incurred in design and development, and an increase in the need for more insightful and practical techniques, exist that require total software development time and complexity to be reduced. A number of companies and corporate firms are now attempting to design and develop their diverse software products and applications in lesser amount of time and with lower cost, all the while maintaining and preserving a very high quality in the products designed and created. In fact, in the 2000s, novel and innovative concepts of software patterns emerged before us, as the magical potions to achieve these underlying goals and thereby creating very high quality developed software products (Gamma et al. 1995; Schmidt, Fayad, and Johnson 1996). However, we are yet to realize the potentiality of using these patterns in developing robust systems.

When the meaningful theory of software patterns emerged some years back, its proponents foresaw and visualized a huge and unlimited potential for developing and conveying flexible, practical, useful, and quality software solutions. The main objective of developing such solutions was to embed and include given software products with an uncanny ability to adapt to new needs and requirements, with ease and without any visible or serious side effects (i.e., bugs), via useful software patterns. Software patterns are successful solutions to recurring software problems within a context (Coplien 1996; Schmidt, Fayad, and Johnson 1996). Although there have been a number of successful stories quoted for using software patterns (e.g., analysis and design) (Buschmann 1996; Fowler 1997), we still do not know how one can weave and hem several software patterns together to build a stable system of patterns. These systems of patterns are simply a set of related patterns, insightfully and intelligently woven, that later communicate a measure of architectural knowledge and styles for a set of high-level problems in particular contexts. Along with the appearance of software patterns, the concept of pattern languages (as defined in the Sidebar 1.1) also arose to attempt to ease the weaving of software patterns and form a system of patterns. Pattern languages are simply a collection of interrelated patterns (Schmidt, Fayad, and Johnson 1996). One can combine these patterns in any way and manner to design and create new environments and ambiences, where practitioners and developers can solve context-specific problems with few problems. More precisely, the concept of pattern languages has spilled over into the software engineering domain to describe prior experiences and the processes that arise from them (patterns) in a simple and straightforward language, where one can skillfully weave and combine patterns in any way to solve a particular problem. Yet, this process is still ad hoc in nature and very casual, and it is not simple and straightforward enough to ease and speed the software development process up.

This book poses you three main questions. First, how can we classify, develop, and utilize analysis and design patterns together toward problem resolution? Second, what are the *behind-the-scenes* language and scripts that guide the sewing of patterns together as a whole? Third, how can we overcome and face a range of unique challenges other than pattern composition problems (patterns traceability) that can hinder and obstruct the development of a system of patterns? The inability to answer these subtle questions detrimentally impacts the understanding of how to put patterns in practice and will therefore make the use of software patterns more complex than it should.

1.2 THE MOTIVATION

The main motivation for writing this book is to answer the aforementioned questions in a meaningful manner, and to synthesize and crystallize the foundations for patterns classification, composition, traceability, and deployment, with the sole purpose of building stable systems by use of patterns in a systematic manner. We are able to fulfill this motivation by providing or suggesting two important ideas. First, we will offer a set of quality factors that will evaluate the definition, application, and solution accuracy of software patterns. Second, we will also provide a new and different representation of pattern languages. This new representation is called *knowledge maps*, or knowledge core sets that describe a topology of software patterns.

The main goal of this book is to define *knowledge maps* as the groundwork for an insightful classification of the software patterns governing a particular discipline. Knowledge maps are the enduring mirrors of experience and best practices of what a discipline is. The most important driving and motivating force of knowledge maps is the innovative approach of software stability concepts (Fayad 2002a, 2002b; Fayad and Altman 2001; Mahdy and Fayad 2002). This unique approach allows us to classify software patterns within knowledge maps and according to their application rationale and nature (e.g., analysis, design).

We will also demonstrate how knowledge maps overcome those problems experienced in traditional pattern languages, by surveying a number of examples within the genre of pattern languages, and later analyzing detected commonalities and drawbacks in these forms, by maintaining the reference to the instances surveyed. The surveyed pattern languages will undergo detailed comparisons with knowledge maps by using previously defined quality factors and parameters.

1.3 THE PROBLEM

Building high-quality systems is not an easy task, nor is it a work carried out in a moment, especially when several factors can undermine and hinder their success, such as cost, time, and lack of systematic approaches. The promise of using software patterns in software development to deal with these aforementioned factors or obstacles has led and made software developers to strongly affirm their belief in the power of pattern languages as the sole means for constructing complex systems in a constrained environment.

Software patterns, along with pattern languages, have attracted software developers for more than a decade. In fact, they have visualized software patterns and pattern languages as promising and emerging techniques that can ease and speed up their software development processes (Appleton 1997; Coplien 1996; Gamma et al. 1995; Schmidt, Fayad, and Johnson 1996). However, developing a set of robust software patterns and pattern languages is yet to reach the expected level of ease it should have when dealing with determined software problems, such as pattern composition and stability. Instead, they end up in constructing models that lack some important and essential qualities that diminish the quality of the system rather than improve it (Wu, Hamza, and Fayad 2003).

Our calculated and calibrated response to the aforementioned critical issues is the introduction of *knowledge maps* or *stable pattern language* as a standard means to classify, organize, weave, and deploy knowledge core sets or a group of patterns according to their rationales. These knowledge core sets consist of software patterns that are pertinent and important to particular domains. To classify these software patterns in accordance with their rationale and create knowledge maps, we will also use software stability concepts as the main and leading approach. The succeeding sections of this book will provide detailed descriptions of this approach.

1.4 THE OBJECTIVES

This research effort also aims to achieve a knowledge synthesis for building systems by using patterns, that is, creating *knowledge maps*. We will plan, intend, and project to highlight and emphasize, through an extensive study, how current approaches that are using pattern languages to build systems strive in providing a systematic and cost-effective manner to weave patterns together and create immensely complex systems. To confront this unique problem, we will also propose a new and distinct representation of pattern languages, called *knowledge maps*, the realization of which is mainly driven by the software stability concepts approach.

Throughout the course of this book, we will debate and confront several important issues and topics related to pattern classification, composition, traceability, deployment, and development to support our concept of knowledge maps. The main objective of this book is to provide patterns researchers, framework developers, and application developers a stable means and mode for answering critical questions or queries, such as how one can weave together similar and different kinds of patterns, what the relationships between analysis patterns and design patterns really are, and what those behind-the-curtain guidelines for sewing these patterns together really are.

The next section will describe the approach that drives the knowledge map realization, providing the required semantics, knowledge organization, organization, and understanding.

1.5 OVERVIEW OF SOFTWARE STABILITY CONCEPTS

Software stability concepts segregate or classify the classes of any system into three main layers of understanding (Fayad 2002a, 2002b; Fayad and Altman 2001): the enduring business themes (EBTs) layer, the business objects (BOs) layer, and the industrial objects (IOs) layer. It is possible to assign and tag each class to a particular layer based on its nature and level of tangibility.

EBTs represent the specification classes within a problem's understanding, whereas the nature of EBTs is entirely conceptual, which means that their structure is internally and externally very stable or durable (Fayad and Altman 2001). BOs are semitangible artifacts that are internally stable and externally adaptable, via a number of extension points called *hooks*, or existing or traditional patterns, such as gang of four patterns (Gamma et al. 1995). Hooks are extension points used as a means to extend, enhance, or augment knowledge of BOs by application developers, to produce new applications from the core knowledge (EBTs + BOs), by activating, modifying, replacing, and/or adding new functionality to the core knowledge. Hooks also provide other critical services like adaptability, customization and personalization, integration, and configuration, as discussed in Sidebars 1.2 and 1.3. They also represent the business rules or process abstractions that are necessary to carry out a determined EBT—they are, in other words, the workhorses of the EBTs. The last artifact is the IO. IOs are the context-specific classes that attach themselves to the core formed by EBTs and BOs. The nature of IOs is entire tangible, which means that they are both internally and externally unstable. They always keep changing proportionally with the occurrence of new business requirements (Fayad, Hamza, and Sanchez 2005).

Software stability concepts also provide practical foundations for domain-neutral core sets or stable patterns. These domain-neutral core sets are not bound to any application-specific concerns by any means. Instead, they remain the same and almost constant whenever they appear, regardless of the application context. Figure 1.1 represents a concise view of software stability concepts (Fayad 2002a, 2002b; Fayad and Altman 2001; Hamza and Fayad 2004, pp. 197–208).

1.6 OVERVIEW OF KNOWLEDGE MAPS

A knowledge map or stable pattern language is a topology of patterns driven by the essential principles of software stability concepts (Fayad 2002a, 2002b; Fayad and Altman 2001). It also consists of knowledge core sets or stable patterns that host the pertinent features and functionality of a particular domain. In addition, one can also utilize it to build other foundation sets or knowledge maps of other domains.

Building a knowledge map for a determined and set discipline involves usage of numerous skills and knowledge and a number of steps beyond the identification of tangible artifacts bound to a specific context of applicability. It also requires systematic capture and full understanding of the domain where we are planning to deploy and expand the proposed solution. This also includes describing the problem in detail, not from its tangible side, but focusing more on its conceptual side, describing the underlying affairs with respect to the problem, and using the elements required to fulfill them.

The ultimate representation of knowledge maps is driven mainly by the significant mantra *divide and conquer*, which is applied throughout the structure of knowledge maps, as shown

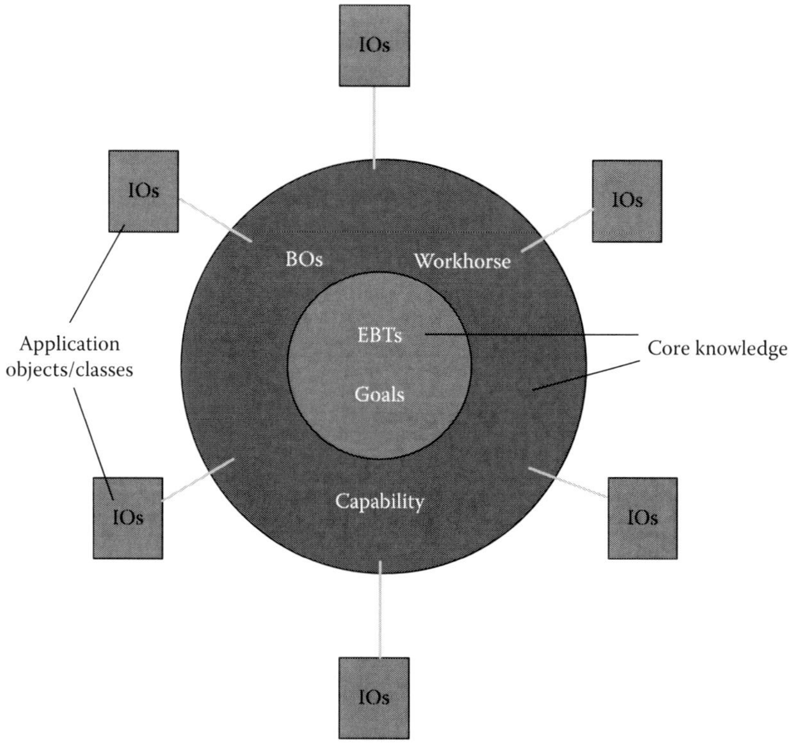

FIGURE 1.1 The software stability concepts approach.

in Figure 1.2. For instance, knowledge maps are the products of partitioning a domain into different levels of granularity, so that we can manage and administer each level with considerable ease. In addition, knowledge maps are stratified into the following five main concerns:

- Analysis concerns (goals)
- Design concerns (capabilities)
- Knowledge concerns (goals and capabilities together)
- Development concerns (development scenarios)
- Deployment concerns (deployment scenarios)

Each one of the knowledge concerns found on a knowledge map is stratified and graded into three layers of software stability, EBTs, BOs, and IOs, whereas EBTs and BOs together are called the *core knowledge* of any domain.

1.7 PATTERN LANGUAGES VERSUS KNOWLEDGE MAPS: A BRIEF COMPARISON

The novel concept of pattern languages (Appleton 1997; Buschmann 1996; Fincher 1999; Salingaros 2000) is spilling over into the software engineering field to describe experiences or best practices of software development, by using a coherent language that can be used for both talking and describing about a particular problem and creating new environments from the patterns it conveys. This special language works specifically by connecting a collection

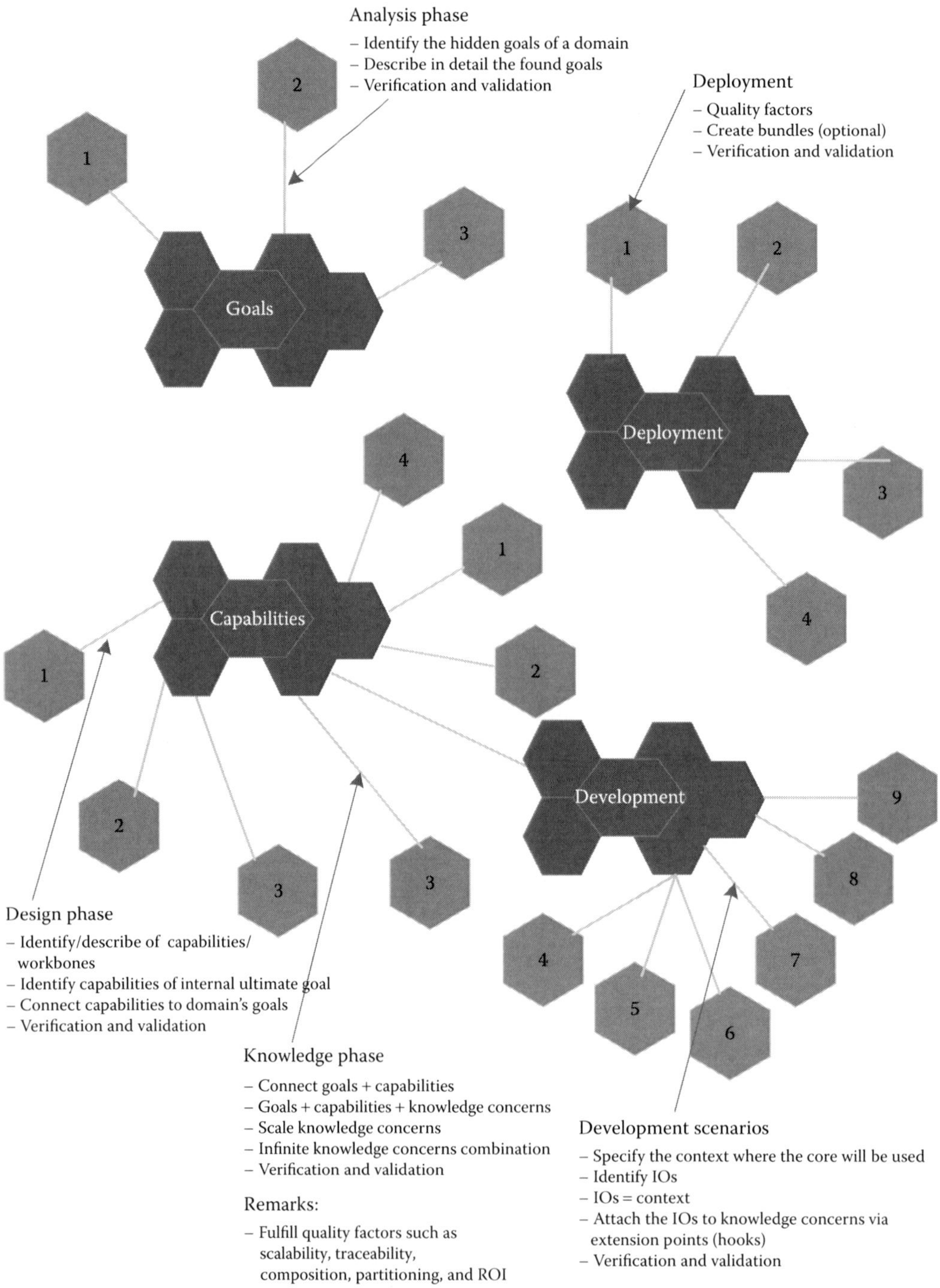

FIGURE 1.2 The representation of knowledge maps.

of patterns, as if they were in a neatly narrated story. Each of the patterns in this collection is an insightful way to handle or solve a set of recurrent problems in a particular context (Appleton 1997; Fincher 1999; Schmidt, Fayad, and Johnson 1996). As a whole, they also make visible both the knowledge that is pertinent or relevant to a particular domain and the solutions for a set of recurrent problems within this particular domain.

TABLE 1.1
Knowledge Maps versus Pattern Languages

Knowledge Maps	Pattern Languages
Knowledge maps provide a systematic approach that ensures focused software solutions	Pattern languages lack indicators/guidelines to determine within-context software solutions
They classify patterns according to their rationale, that is, EBTs, BOs, and IOs	They lack indicators that determine the rationale of their enclosed patterns
They provide full traceability of their enclosed patterns	Traceability is lost, especially when dealing with deeper levels of pattern language's implementation
They provide full generality of their enclosed patterns	They do not guarantee full generality of their enclosed patterns
They provide enduring solutions. Maintenance is minimal	They are hard to maintain and they struggle in providing enduring solutions
They are quite easy to understand and use, when dealing with determined software problems. Everything is based on goals, capabilities, etc.	They are hard to use and understand, when solving a determined software problem
They distinguish between direct and remote knowledge	They do not distinguish between associate (direct) and remote knowledge

Any existing approaches and pattern language representations, not driven by software stability concepts, will refer to as traditional approaches and traditional pattern languages, and the ones driven by software stability concepts will refer to as knowledge maps through out this book. Throughout the section, we will also try to briefly compare and contrast both traditional pattern languages and knowledge maps. This study will efficiently determine the path taken in this book and offer the benefits of using knowledge maps to ease and speed up the software development process.

The comparison between these two pattern language representations is illustrated in Table 1.1, a brief description of the generated outcomes.

The next section describes the systematic approach for implementing knowledge maps.

1.8 THE SOLUTION

The following sections provide a brief explanation of the distinct pieces of our proposed solution. It starts with the detailed description of the methodology to be used for building knowledge maps. Then, it will proceed with the employed research methodology that will support the completion of this book.

1.9 KNOWLEDGE MAPS METHODOLOGY OR CONCURRENT SOFTWARE DEVELOPMENT MODEL

From a software perspective, several requirements of prime importance must be fulfilled before, during, and after investigating any problem. These important things are as follows:

- Choosing the right approach for understanding the problem, by using a systematic and precise problem analysis process

- Creating a suitable design process to support analysis outcomes and foresee or visualize future project changes
- Providing a classification of the distinct building blocks that conform to the analysis and design outcomes, which will then form knowledge and the best practices
- Using validation and verification techniques to evidence your work value and the integration accuracy of building blocks
- Creating and making development cost and time reduction
- Encouraging a unique *welcome-change* attitude when the developers are working with constantly evolving situations

In this book, we will also present a suitable approach to address all the aforementioned challenges and questions. This special approach consists of the unification of two promising approaches: the *pattern languages* approach and the *software stability concepts* approach (Fayad 2002a, 2002b; Fayad and Altman 2001; Mahdy and Fayad 2002; Salingaros 2000; Wu, Hamza, and Fayad 2003). This unified approach relies heavily on a rationale-driven view to discover and visualize stable knowledge core sets (patterns) within a particular domain and methods to organize and relate them to formulate an accurate solution for a myriad of contexts that share the same core knowledge (rationale or goals and capabilities).

From a bird's-eye perspective, we could suggest that the overall process of creating a knowledge map involves five main steps, as shown in Figure 1.3: *Goals* or *classification*, *capabilities/properties of a particular discipline*, *knowledge maps formation*, *development scenarios*, and finally, *solution deployment*. However, later in the book, we will describe a detailed process about how to design and formulate knowledge maps in a systematic and organized manner. For each one of these main steps involved in the creation of knowledge foundations sets (knowledge map), we will help readers discover a set of distinctive patterns that they can interconnect to form accurate solutions that can satisfy a rationale of the domain in question.

1. The first main step, *analysis/goals*, is concerned with surfacing the implicit goals hidden within a particular discipline. These goals are the EBTs (Fayad and Altman 2001; Hamza and Fayad 2004, pp. 197–208). This process requires the capture and full understanding of the context, where one can use solutions. This process also incorporates describing the goals not from their tangible side, but focusing more on their conceptual side. This process may imply the necessity and need to delve or deliberate into the internal structure of the goals, flush out, and obtain any hidden insight or knowledge core sets and rules that aid the problem's resolution. The outcome is stable analysis patterns (SAPs) (Hamza 2002; Hamza and Fayad 2002).
2. The second main step, *design/capabilities*, emphasizes the discovery of the recipes and potions required to fulfill the stated goals of a particular domain. These recipes are the BOs (Fayad 2002a, 2002b; Fayad and Altman 2001). Without these recipes or stable patterns, only a vague understanding (almost none) of the domain's goals will be achieved. As with goals, the accurate and correct understanding of capabilities may require a deep analysis of the elements that build them. That is, capabilities may contain a second level of abstraction or internal structure. When such a capacity occurs, we will label them as *Pattern-BO*. However, this second level of abstraction is still not made public. Therefore, the Pattern-BO will be represented as a single unit of interest using only its first level

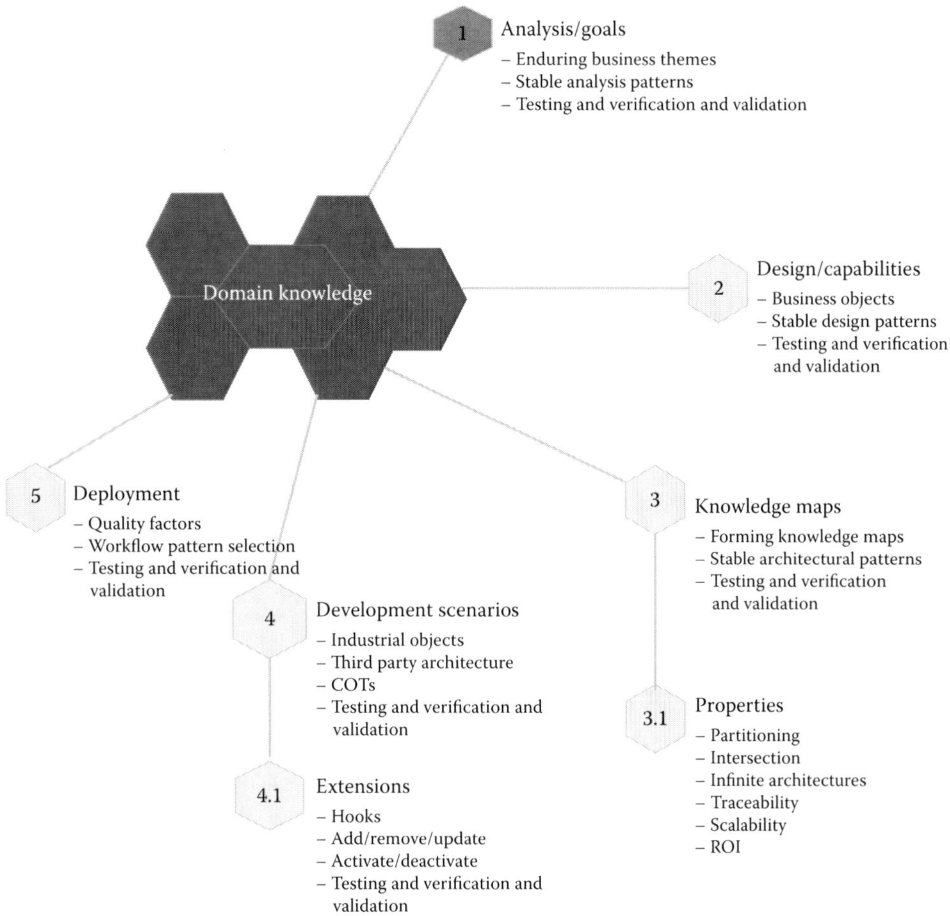

FIGURE 1.3 Methodology of knowledge maps.

of abstraction. The outcome of this step is the stable design patterns (SDPs) (Chen, Hamza, and Fayad 2005, pp. 592–596; Fayad and Kilaru 2005, pp. 108–115).

3. The third main step is forming the *knowledge maps*. Intuition and experiences from practitioners (i.e., analysts and designers) will support the formation of knowledge maps. First, practitioners must know the environment wherein the problem is happening or occurring. Second, practitioners must examine the overall goals and capabilities required to describe that environment (i.e., the solution within the context). Third, after acknowledging the existing environment, goals, and capabilities, practitioners must then create synergies between these elements and form knowledge concerns or stable architectural patterns that will handle the given problem of interest. In short, the main objective of this step is to compose knowledge core sets from goals and capabilities. One can realize this composed knowledge via the distinct routes/paths taken during the synergy between two or more patterns or one goal and other capabilities. Each one of the complete routes taken will satisfy a distinct need in a particular domain. This step produces a number of different stable architectures that include SAPs and SDPs, and its outcome is the stable architectural patterns (Fayad 2015a, 2015b, 2015c).

4. The fourth step, *development scenarios*, provides the essential qualities of standards to our software solution, such as scalability, traceability, maintainability, stability, and return on investment, due to the inherent qualities of the inherited software stability concepts. This particular step is concerned with how the knowledge core sets will be adapted to specific contexts, based upon the utilization of tangible artifacts or IOs. Such an adaptation is possible with extension points called *hooks* that will attach context-specific classes (IOs) to the core. By achieving this, unprecedented flexibility and ease will arise, enabling businesses and firms to add, remove, update, and extend functionalities from their systems on a real-time basis.
5. The last main step, *deployment*, deals not only with how a particular solution and its enclosed knowledge core sets would be deployed in particular domain, but also with the representation of the artifacts or domain-specific patterns that will aid the deployment process. This includes EBTs, Pattern-EBTs, BOs, and Pattern-BOs, and their formed context (IOs) (Hamza and Fayad 2002).

The knowledge maps methodology or concurrent software development method offers many advantages:

1. Applications created by using this methodology are quite adaptable and amenable to ever-changing needs and requirements. This is because the core is quite robust and stable and is unaffected in any way by the changes or modifications in the requirements. As a result, the application built by using knowledge maps can be modified very easily to satisfy the changing business needs.
2. Applications can also be scaled easily with minimal efforts. Because it is possible to plug the application-specific IOs to the stable knowledge map core, one can easily extend and adapt a specified application.
3. Using a knowledge map, an infinite number of diverse applications can be created within a fraction of time and with minimum effort. The knowledge map serves as a building block for the applications. IOs have to be just plugged to the knowledge map with the help of hooks and existing design patterns.
4. Because the knowledge map methodology is synonymous with concurrent development, all the phases of software life-cycle analysis, design, implementation, and testing can be carried out simultaneously. This eventually results in our being able to verify the work at every step, rather than waiting until the testing phase. In addition, a thorough and complete understanding of the problem is possible. Again, changes or modifications to the design are possible at any time, as all phases are carried out concurrently.

1.10 WHY KNOWLEDGE MAPS?

1.10.1 Research Methodology Undertaken

The development and synthesis of this book will be on a systematic and organized manner. We have already invested considerable time and effort in collecting and organizing fundamental information with respect to pattern languages, pattern organization, and collections. This strategy also includes acquiring a number of books that reflect and provide up-to-date, readily available software patterns/pattern language techniques—usage and addressed problems, production and deployment processes commonly used in industry—and current limitations and future trends related to this area of study.

This strategy also includes accessing good online libraries, looking for proceedings of software patterns conferences, white papers, published research papers, and scientific journals. Additionally, the trial of currently available commercial software, such as Eclipse and Rational XDE, which use patterns to support software development, will also be an essential part of our strategic and knowledge input. In doing so, we will also explore the required knowledge hidden behind their usage and capabilities currently offered. Such a task is complicated, complex, and Herculean, because most of these commercial and noncommercial systems never disclose or announce their internal structure or source code.

After collecting and understanding information and details regarding software patterns, pattern languages, their underlying techniques, and ad hoc classifications, we will introduce our new approach called *knowledge maps*. We will describe the structure, semantics, quality factors, and properties of knowledge maps, including the extremely promising benefits to software development activities.

After completing these essential steps, we will also proceed with the utilization of the offered benefits and semantics of knowledge maps, by quoting real-life examples and experiences. These examples will show the benefits of knowledge maps in terms of pattern discovery, organization, classification, development, and deployment.

1.10.2 Research Verification and Validation

We will verify and validate our work with the help of two factors: using a set of *applicability scenarios* that imply common software patterns/pattern languages usage and *expected outcomes and results*. Nevertheless, we will also provide considerable enhancement in their organization and visualization, via a software stable model (SSM) (Mahdy and Fayad 2002), which is actually the visual representation of software stability concepts. In addition, we will validate each information source to see and examine if it accurately addresses what it is actually disclosing and divulging.

We can consider the determined numbers of scenarios per pattern based on proof of validity requirements of current patterns: by providing three or more applications or scenarios to prove pattern accuracy and correctness. Thus, practitioners will be able to visualize the underlying rationale and its associated capabilities and perfectly aid and assist context-specific aspects to rapidly realize and form a suitable solution.

1.10.3 The Stratification of This Book

The book provides the following to its readers:

1. Knowledge synthesis for building systems using knowledge maps, along with the knowledge core sets that form its main structure
2. An initial framework on knowledge maps with some of their significant qualities, like scalability and adaptability, including a discussion of source code
3. A representation of knowledge maps to help readers understand the path of execution for all the stable patterns provided, as well as their objectives and the part of the knowledge infrastructure they actually represent

4. Documentation of some of the stable patterns, via the use of a noteworthy documentation template
5. Two significant case studies illustrating the actual applicability of the framework of knowledge maps
6. Stable patterns implementation heuristics to simplify the software development process of stable patterns

This book is stratified and segregated as follows. This chapter provides an overview of the contents of the entire book and sets the stage for its proper development. Chapter 2 examines the methodology for forming knowledge maps in a cost-effective manner. Chapter 3 explores the term *goals* and its importance in the formation of knowledge maps, whereas Chapters 4 and 5 provide a set of complete documentation of two SAPs: knowledge and discovery. A goal without even a trace of capability is not a goal; therefore, in Chapter 6, we will give readers a detailed description on the capabilities of the element of knowledge maps and their role in building knowledge maps. Chapters 7 through 10 provide a complete documentation of four SDPs: AnyMap, AnyContext, AnyAgreement, and AnyPartition. Chapter 11 provides additional details and describes what a knowledge map is and its role in the understanding and mastering of any discipline of interest. Chapter 12 focuses on the formation of development scenarios, especially in the identification of context-specific classes, and how they are hooked into the core formed by goals and capabilities. Chapter 13 provides an insight into and summary about the ways and manner in which knowledge maps are deployed, whereas Chapters 14 and 15 provide detailed descriptions of the two critical case studies. Chapter 16 focuses on the book's concluding remarks, where we will provide a summary of what we have performed throughout the book and what we will do as a future task. This also includes a description of the book's contributions.

SUMMARY

This section will represent each of the contributions expected from this book. These contributions will list themselves according to the merit of the problem they address and encounter. We will also deal with the following aspects.

The main objectives of Chapter 1 were to introduce knowledge maps or the topology of stable patterns, as the means for developing software systems in a cost-effective manner, to show its perceived superiority over traditional pattern languages and to specify how the rest of the book will flow. This includes a brief and concise introduction of the content that will appear in each chapter. The chapter also provides a brief description of the structure and properties of knowledge maps along with their benefits, challenges, and constraints.

OPEN RESEARCH ISSUES

There is nothing more difficult to take in hand, more perilous to conduct, or more uncertain in its success, than to take the lead in the introduction of a new order of things (Machiavelli 1913).

The above-mentioned quote is very true in the context of knowledge maps, as this concept is the newest development in the field of software engineering. Moreover, every new invention has to undergo the test of passage of time, so knowledge maps will also have to answer a number of questions before being accepted.

1. *Potential of knowledge maps.* Knowledge maps have immense potential to change, transform, or modify the way in which software development is currently perceived and felt. By using a knowledge map, it is possible to generate stable applications in double-quick time. However, the concept of knowledge map is still in its infant stages of development and considerable work needs to be carried out for knowledge maps to replace the existing traditional methodologies. As a result, one of the issues that need immediate attention is how to implement knowledge maps to achieve a stable core. Another open issue is of verifiability. How to be sure that the knowledge map is stable and satisfies the need at hand? In other words, how to build knowledge maps that are correct by construction, that is, are stable, and meet the customer's requirements? How can we carry out testing of the knowledge map? Knowledge maps will definitely face some form of competition or even challenge, because traditional approaches have been in vogue for quite some time.
2. *Concurrent-oriented software development versus existing software development models.* Contrast concurrent-oriented software development with other existing software development models and methodologies, such as incremental development model, spiral model, aspect-oriented programming, and iterative process, based on quantitative criteria, such as time, cost, and number of recurrence (applications), and qualitative criteria, such as scalability, reusability, flexibility, accuracy, completeness, applicability, and maintainability.
3. *The utilization of concurrent-oriented software development as dynamic analysis.* Using a knowledge map, an infinite number of diverse applications can be created within a fraction of time. The knowledge map serves as the main building block for the applications. One needs to plug the IOs to the knowledge map with the help of hooks and existing design patterns. Therefore, we can easily generate sophisticated applications very quickly and perform dynamic analysis of each of the generated applications on top of the same core. This will ultimately lead to comparative studies and real-time data about dynamic analysis and allow the developers and users of the applications to give concrete results based on real running systems or applications.
4. *Goals for requirements formation and true problem understanding.* How to use goals for requirements formation and true problem understanding? Goals that correspond to SAPs and impose an accurate list of requirements which are based on problem formation contribute to a true problem understanding. The main idea behind the goals or SAPs is to analyze the overall problem under question, in terms of its EBTs and the BOs, mainly with the objective of increased stability and broader reuse. By deeply analyzing the problem in terms of its EBTs and the BOs, the consequent pattern will form the core knowledge of the problem. The ultimate goal of this new concept is enduring stability. Accordingly, these stable patterns could be easily comprehended and reused to model the same underlying problem, under any given situation and context. Data must be collected in relation to how

accurately the problem is spaced and how much is understood by all the people involved in software development and management.

5. *Business objects or capabilities—Software design base or ultimate solution space.* By applying stability model concepts to design patterns, we hereby propose the new concept of SDPs, or BOs. The important idea behind SDPs is to design an enduring solution to the problem under consideration, in terms of its EBTs and the BOs, with the main goal of increased stability and broader reuse. By developing the problem solution in terms of its EBTs and the BOs, the resulting pattern could easily be reused to solve the same problem under any given context and domain. Data sets must be collected in relation to how accurate and complete the solution space is to all the people involved in software development and management.
6. *EBTs + BOs = software architectures or mapping any software architecture to model-driven architectures.* The rapid growth of emerging technology coupled with tightened or constricted software development time and production cost constraints has imposed and exerted tremendous pressure on and an intense desire for software enterprises and firms to design and create new and innovative designs to respond to a rapidly changing business environment. Enterprises must heavily invest in building stable architectures that are readily adapted in many different ways to meet the new challenges and risks. These kinds of architectures are called *architectures on demand*, as they can be adapted accordingly to meet the future requirements and changes in the system. The primary focus of this issue is to show how software stability concepts are used to develop on-demand architectures. This issue also focuses on three key aspects: (1) EBTs or business goals and transformations, which we call SAPs; (2) BOs or business process design, which we call SDPs; and (3) IOs or application objects. Both EBTs and BOs form a stable core and thus provide architectures on demand for any domain. We will call these architectures as *stable architectural patterns.* Data must be collected in relation to how often and how many architectures on demand can be generated per knowledge map. EBTs and BOs are *stable software patterns*, and a combination of EBTs and BOs forms the *core knowledge* for a given domain. The core knowledge for any domain is called a *stable architectural pattern* that you can extend and adapt through the application of hooks. The quality of stable architectural patterns creates competitive advantages through differentiation and productivity. It will also integrate partners in order to increase adaptive capabilities.
7. *Pitfalls of traditional pattern languages.* Software patterns, along with traditional pattern languages, have attracted software developers for more than a decade or so. In fact, developers have visualized software patterns and existing pattern languages as promising techniques that simplify and speed up their software development process (Appleton 1997; Coplien 1996; Gamma et al. 1995; Schmidt, Fayad, and Johnson 1996). However, developing a set of robust software patterns and traditional pattern languages is yet to reach expected ease—as it should have—when dealing with determined software problems, such as pattern composition and stability; instead, developers construct models that lack some essential qualities that diminish the quality of the system rather than improve it (Wu, Hamza, and Fayad 2003).

Our innovative response to the aforementioned critical issues needs further advanced research and discussions on the pitfalls of traditional pattern languages; we provide solutions for each one of these pitfalls.

8. *EBTs + BOs = Unified engine for any domain.* This method leads to a very highly reusable and unified software engine (USE) technology for developing service and/or production systems, which are called *service engines* and *production engines.* USEs for any domain are an open research issue and topic, because building such engines is not an easy exercise, specifically, when several conflicting factors can undermine or impede their success, such as cost, time, and lack of systematic approaches. The main difference between software developments (business as usual), application and enterprise frameworks, and the USEs also needs further research and development in a comprehensive manner.

REVIEW QUESTIONS

1. What is a knowledge map?
2. What is a traditional pattern language?
3. What are the major differences between knowledge maps and pattern languages?
4. Knowledge map = stable pattern language. Explain.
5. Knowledge map methodology is equivalent to concurrent software development model. Explain.
6. Into how many types can you classify the classes of a system by using software stability concept? Name them.
7. Compare EBTs, BOs, and IOs.
8. Knowledge map is based on the concept of ________________.
9. Is the following statement true or false? Software stability concept results in stable patterns.
10. Is the following statement true or false? Core knowledge of any domain is represented by EBTs and IOs.
11. Into how many concerns can you stratify knowledge maps? Describe each concern briefly.
12. __________________ is the mantra used to create knowledge maps.
13. What are hooks?
14. What are the advantages of using hooks?
15. Match the following:
    a. Goals – Development scenarios
    b. Capabilities – Knowledge phase
    c. IOs – Analysis phase
    d. EBTs + BOs – Deployment
    e. Quality factors – Design phase
16. Is the following statement true or false? Traditional pattern languages and traditional approaches are driven by software stability concepts.
17. Describe knowledge map methodology in brief.
18. List the advantages of knowledge map methodology.
19. What is meant by Patterns–BO?

20. Specify against each of the traits listed below, whether they belong to knowledge map or pattern languages:
    a. Lack of systematic approach
    b. Result in classification of patterns according to their rationale
    c. Can be traced
    d. Difficult to maintain
    e. Hard to use
    f. Does not distinguish between direct and remote knowledge
21. Is the following statement true or false? BOs are usually conceptual.
22. Is the following statement true or false? IOs are application independent.
23. Is the following statement true or false? EBTs represent the ultimate goal of the system.
24. Is the following statement true or false? IOs are stable over a period of time.
25. Is the following statement true or false? BOs are externally stable and internally adaptable.
26. Is the following statement true or false? EBTs are stable.
27. Is the following statement true or false? BOs represent the capabilities needed to satisfy the goal of the system.
28. Is the following statement true or false? BOs are always tangible.
29. Infinite number of applications can be built by the use of knowledge map. Is this statement correct? Explain.
30. Stable analysis pattern is represented by ________________________.
31. Is the following statement true or false? BOs represent stable architectural patterns.
32. ______________________________ can be formed by using knowledge maps.
33. Is the following statement true or false? IOs result in third-party architectures.

EXERCISES

1. Name, list, categorize, and describe all the patterns of the traditional patterns language of sample requirements D1, titled "Ocean Resources Management System." (see Appendix D)
2. Name, list, categorize, and describe all the patterns of the traditional patterns language of sample requirements D2, titled "Dengue Fever Prevention and Outbreak Management System." (see Appendix D)
3. Name, list, categorize, and describe all the patterns of the traditional patterns language of sample requirements D3, titled "Organizing Cricket World Cup." (see Appendix D)
4. Name, list, categorize, and describe all the patterns of the traditional patterns language of sample requirements D4, titled "Pollution Management." (see Appendix D)

PROJECTS

1. Show the relationships between all the categories that are based on domain names of all of the patterns, within the pattern language of Exercise 1. List all the patterns per category.
2. Show the relationships between all the categories that are based on domain names of all of the patterns, within the pattern language of Exercise 2. List all the patterns per category.

3. Show the relationships between all the categories that are based on domain names of all of the patterns, within the pattern language of Exercise 3. List all the patterns per category.
4. Show the relationships between all the categories that are based on domain names of all of the patterns, within the pattern language of Exercise 4. List all the patterns per category.
5. Patterns that appear in the above-mentioned four projects are *common patterns*. List all the common patterns, specify the pattern type (analysis, design, process, etc.), and describe them. Document three of the common patterns using Appendix A and the pattern documentation template.

SIDEBAR 1.1 Traditional Pattern Languages

According to Wikipedia, the free online encyclopedia (Alexander 1977, 1979):

A *pattern language* is a structured method of describing better design practices within a field of expertise or domain.

It is essentially characterized by the following:

1. Noticing and naming common problems in a field of interest
2. Describing the key characteristics of effective solutions for meeting some stated goals
3. Helping a designer migrate from one problem to another in a logical way
4. Allowing different paths through the design processes

Christopher Alexander, an architect and author, coined the term *pattern language*. He used it to refer to common problems of civil and architectural designs, from how cities and towns should be laid out, to where windows should be placed in a room. The main idea was initially popularized in his book *A Pattern Language* (Alexander 1977).

A Pattern Language consists of a cascade or *hierarchy of parts*, all linked together by *patterns* that solve generic recurring problems associated with the parts. Each pattern has a definite title and collectively the titles form a language for design (Hamza and Fayad 2002). In a pattern language, individual patterns are not isolated. The structure of the language is composed of the links from larger patterns to smaller patterns, together creating a network. Thus, for a single pattern to work completely, it must not only be followed through by implementing the smaller patterns that complete it, it must, if possible, be connected to certain larger patterns (Coplien and Schmidt 1995; Khadpe 2005).

REFERENCES

Alexander, C. *A Pattern Language: Towns, Buildings, Construction*. New York, NY: Oxford University Press, 1977.

Alexander, C. *The Timeless Way of Building*. New York, NY: Oxford University Press, 1979.

Coplien, J. O., and D. C. Schmidt (eds.). *Pattern Languages of Program Design*. Addison-Wesley, 1995.

Hamza, H., and M. E. Fayad. "A Pattern Language for Building Stable Analysis Patterns." Paper presented at the 9th Pattern Languages of Programs Conference, Monticello, IL, September 8–12, 2002.

Khadpe, P. "Pattern Language for Data Mining." Master's Thesis Report, San Jose State University, San Jose, CA, May 2005.

SIDEBAR 1.2 Hooks or Extension Points

Hooks are the important and critical extension points that are used as a means to extend, augment, activate, modify, replace, and add new functionality (Fayad, Schmidt, and Johnson; Froehlich et al. 1999), to adapt, customize, personalize, trace, and/or integrate knowledge by application

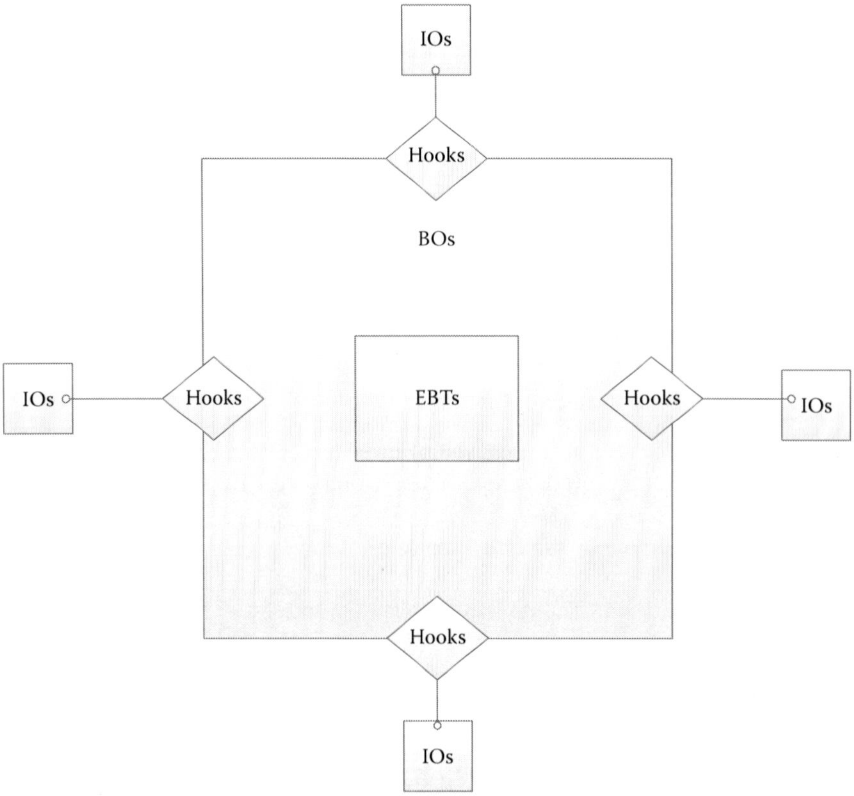

FIGURE 1.4 Hooks in SSM architecture.

developers, and to design and produce brand new applications from the core knowledge or knowledge map (Shtivastava 2005). Systems developed based on the software stable model (SSM) are highly stable and robust, and they could easily accommodate new needs and requirements. This is possible as software stability stratifies classes of the system into three layers (Figure 1.4).

EBTs. These contain classes that present the enduring and basic knowledge of the underlying industry or business domain. These are extremely stable and enduring, as they represent the goals of the system to be developed (Shtivastava 2005).

BOs. These contain classes that map the EBTs of the system into more concrete objects, which serve as capabilities required to achieve the goals of the system. These BOs implement generic functionality, which will be needed by all the applications of the domain (Shtivastava 2005).

IOs. These contain classes that map the BOs of the system into application-specific components. The IOs implement application-specific functionality (Shtivastava 2005).

A pattern designed using SSM system could be further extended, enhanced, or customized to build multiple applications. If the changes or corrections are directly performed on the BOs, then the core will no longer remain stable, while the rules of SSM would be intentionally violated. Hooks are used for this sole purpose. They take the load of modifying the BOs to achieve application-specific behaviors through the same core, without actually making any significant changes in the core classes.

The extensions and customizations are specified through hook templates. Hooks interface between the IOs and BOs (Figure 1.4). They provide a flexible mapping between BOs and IOs. IOs for different applications could be easily associated with BOs through hooks. Figure 1.4 depicts the hooks with the three layers of the software stability model.

In addition to stable analysis and design patterns, more than a 1000 existing patterns, such as the Gang of Four, Java design patterns, and Siemens group patterns, are implemented as essential parts of the hook facility. The hook is described and implemented through several templates: a base and several specialized templates, such as customization, adaptation, and integration templates. Each hook description is written and codified in a specific format made up of base template, and as many as you wish of specialized template and/or existing software patterns. Here, the application developers will be able at change or tweak the given application at ease and within the shortest possible time through the application of hook or create a new application (Fayad, Schmidt, and Johnson 1999; Froehlich et al. 1997). The hook template(s) serve as an enduring guide to the application developers by using the knowledge map. The hook template helps and assists in organizing the available information in a precise and less ambiguous way (Shtivastava 2005).

REFERENCES

Fayad, M. E., D. C. Schmidt, and R. E. Johnson. *Building Application Frameworks: Object-Oriented Foundations of Framework Design.* New York, NY: Wiley, 1999.

Froehlich, G., H. J. Hoover, L. Liu, and P. Sorenson. "Hooking into Object-Oriented Application Frameworks." Proceedings of the International Conference on Software Engineering, Boston, MA, May 1997, 491–501.

Shtivastava, P. The Hook Facility, MS Project Report, San Jose State University, San Jose, CA, May 2005.

SIDEBAR 1.3 Hook Engine

Hook engine is a special web-based engine, while hook facility is an enduring tool that facilitates knowledge map hooks. The engine or tool maintains a rich repository of the hook templates and existing software patterns, defined specially for the core knowledge of the applications driven from the entire knowledge map. The engine or tool also supports addition, modification, and deletion of hook templates in the given repository. Each hook template specifies a list of changes or editions needed for the core knowledge classes known as BOs. These perceived changes or modifications are specified by using hook grammar that has specific statements to modify/replace existing behavior and add new behavior to adapt, change, customize, integrate, trace, and configure the core BOs. There are essential statements to add new properties and operations, override, extend, and copy methods, as shown in the base hook template in Figure 1.5.

The tool also understands hook grammar syntax, which uses Enhanced Backus–Naur Form (EBNF) syntax (Niklaus 1982; Peter 1960; Scowen 1993). It parses each change statement individually and automates the process of applying the changes specified in the statement. Additional statements have been integrated into the hook grammar to address the need of keeping a flexible mapping between BOs and IOs. This enhancement has added more value and additional functionalities to the hook concept and increased/enhanced its applicability. By using the engine or tool and application-specific functionality, one can easily extend and enhance the applications to the core knowledge and it is possible to develop new applications in no time. The engine or tool has an intuitive and useful user interface that makes it very easy and straightforward to use. The tool requires minimal configuration to be put to use. Figure 1.5 shows the details of the base template of the hook on the right side and many options on the left side of the screen.

Change statements. Figure 1.6 shows the change section of the template in a separate tab.

Creating new hook template. Clicking on Create button brings up an empty form to help create a new hook template. The Save button should be clicked to save the new template as shown in Figure 1.7.

Editing existing templates. One can make changes to the template in the repository by selecting it from the list and clicking on the Edit button. The Save button helps you to save the ensuing changes as shown in Figure 1.8.

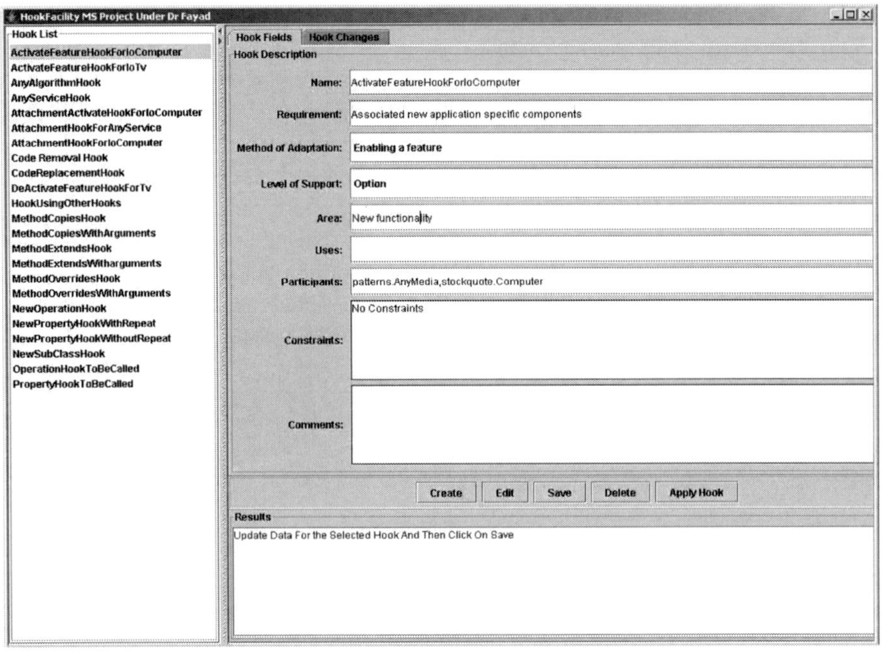

FIGURE 1.5 The hook base template.

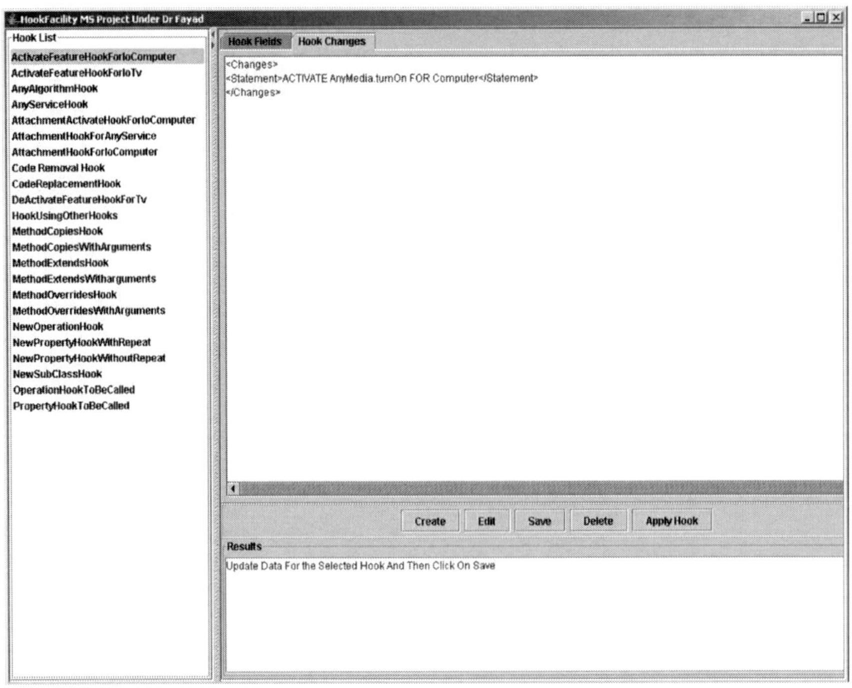

FIGURE 1.6 The change statement screen.

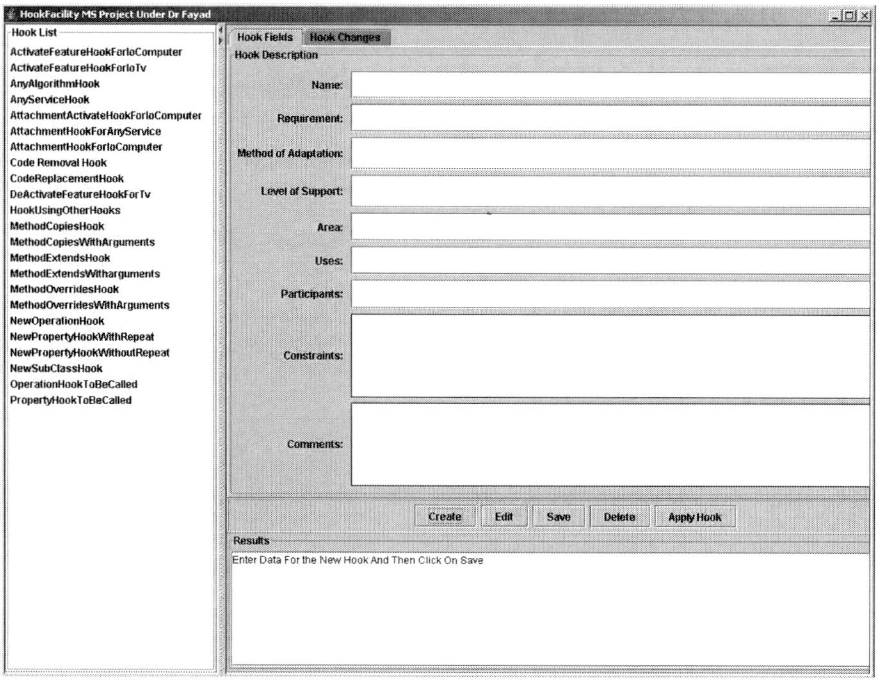

FIGURE 1.7 Creating new hook templates.

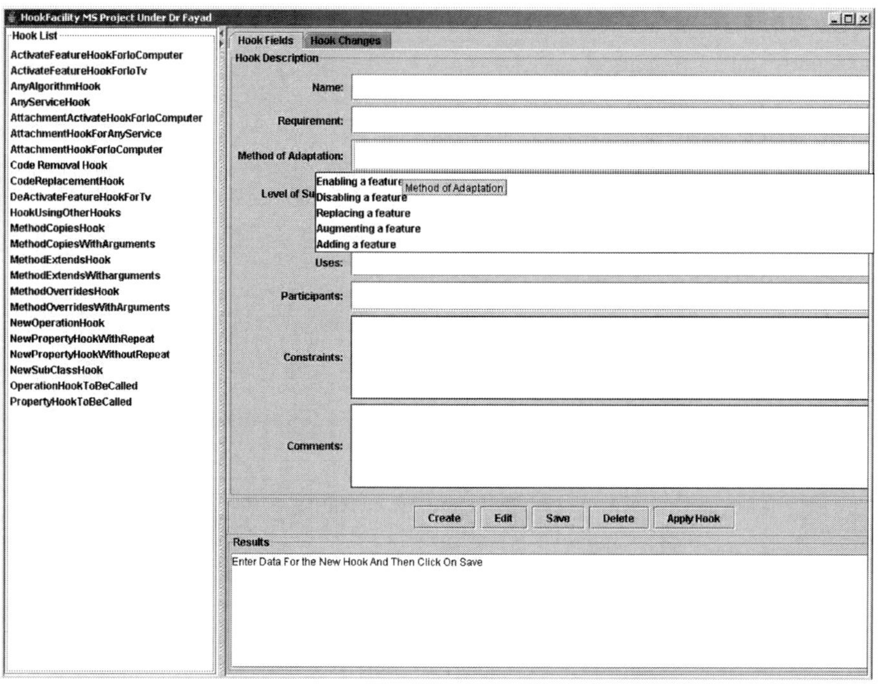

FIGURE 1.8 Editing existing templates.

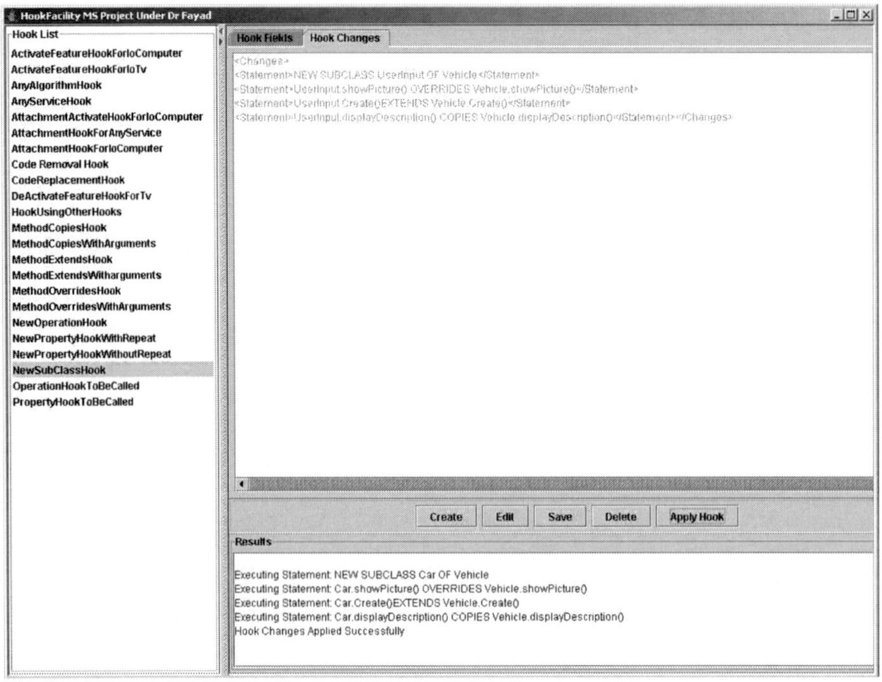

FIGURE 1.9 Creating new hook templates—execution sequence.

Deleting existing hooks. One can delete unused templates from the repository by selecting it in the list and clicking the Delete button.

Applying changes in hook template. The changes or modifications specified in the change section of the hook template may be applied by selecting or creating the template satisfying the requirement and hitting the Apply Hook button. The text area component shows the result of execution.

The engine or tool executes each change statement individually and the text area depicts the results (Figure 1.9).

WHAT ARE THE FEATURES OF THE HOOK?

The hook engine or the hook facility tool has the following useful features:

- The tool supports the creation of hook templates and it allows editing and deletion of templates.
- A registry maintains the repository of hook templates.
- The changes or transformations in the hook template are written by using hook grammar. The tool parses the changes or modifications according to the hook grammar rules.
- The engine or tool generates code corresponding to the changes specified in the hook template.
- The engine or tool guides the users through the process of applying the changes.
- The engine or tool asks the users for inputs required to apply the change statements.
- The engine or tool keeps the users well informed through appropriate messages, while applying these changes.

- The engine or tool also allows the users to search the hook.
- The engine or tool has a number of special purpose hooks, such as adaptability, customization, traceability, accessibility, personalization, extensibility, integration, and configuration abilities.

REFERENCES

Niklaus, W. *Programming in Modula-2*. Berlin, Heidelberg: Springer, 1982.

Peter, N. ed. "Revised Report on the Algorithmic Language ALGOL 60." *Communications of the ACM* 3, no. 5 (1960): 299–314.

Scowen, R. S. Extended BNF—A generic base standard. In Proceedings of the Software Engineering Standards Symposium, 30 August, 1993.

2 Abstraction
Knowledge Maps, Stability, and Patterns

Abstraction is real, probably more real than nature.

Josef Albers, 2008
New York Times

2.1 INTRODUCTION

Chapter 1 briefly discussed the knowledge, or skills required, to design, format, and create highly innovative knowledge maps. Important building blocks were described in that chapter, along with an approach that systematically integrates them to develop better and meaningful software products. This chapter provides additional information, details, and tidbits of useful advice to support and buttress the process of creation of knowledge maps and thus ease the process of creating diverse software products. In other words, the chapter's rationale and scope is to provide a profound exploration of the significant elements that compose our methodology in the creation of knowledge maps.

The main purpose of this exploration is to uncover the basic rules of the game, those golden rules software creators must follow and obey, in order to ease and simplify their software development activities. This set of golden rules is represented by knowledge core sets or patterns. These core sets are allocated into a knowledge map on the basis of their purpose or tasks regarding software development, for example, analysis. There are five areas of interest in knowledge maps: analysis concerns, design concerns, knowledge concerns, development concerns (application concerns), and deployment concerns. Each one of these concerns isolates the core sets that are significant to the domain in question and suppresses ones that are unimportant or too application specific. For instance, analysis concerns deal with the elicitation and understanding of requirements (the problem space), whereas design concerns concentrate on recipes for handling defined requirements (the solution space). Knowledge concerns represent the experience achieved by creating synergies between analysis concerns and design concerns; they also convey knowledge of architectural or styles. Development concerns deal with the formation of the contexts to be used in the implementation of knowledge maps. Deployment concerns deal mainly with the quality factors that must be fulfilled when deploying the knowledge maps, such as performance, scalability, and adaptability.

The intricate process of extracting the core sets pertinent to the domain under discourse and omitting the unimportant ones is nothing different from the traditional abstraction process. *Abstraction* refers to the process of extracting the significant details of an entity, or group of entities, and suppressing the unimportant ones (Berard n.d.). In knowledge maps, this abstraction process, along with its enclosed levels and participants (core sets), is restrained by certain rules or boundaries; this restriction is imposed to help software practitioners better understand and use knowledge maps and avoid any misconception or wrong examination of the core sets pertinent to the domain addressed.

Although creating knowledge maps may seem to involve complex and tedious processes, as well as an extensive learning curve, actually, it is not so. To systematically describe the process of creating knowledge maps, along with the set of rules that guides their creation, we need to describe first their elements, or building blocks, and then proceed with the actual process of knowledge maps creation.

A knowledge map consists of a series of goals, capabilities, and transient aspects that are insightfully woven to specify the groundwork for any domain of interest. Goals are the essential concerns that determine the rationale of a problem of interest. Capabilities are twofold here: they are defined as the capabilities to achieve the goals, and they encapsulate abstraction processes that are internally enduring and externally adaptable, via hooks (see Sidebars 1.2 and 1.3; Fayad, Hamza, and Sanchez 2005). Transient aspects are those requirement-centric classes that create the context for goals and capabilities as a whole unit.

Before examining the elements that are pertinent to knowledge maps, we need to specifically describe the abstraction process that is used to represent and define the elements of knowledge maps. After that, we will be able to better understand why these elements are so important and critical for knowledge maps. To help better understanding of the process of creating and understanding knowledge maps, we will also provide an intuitive visual representation of knowledge maps in subsequent chapters. This visual representation will allow us to understand the underlying language that helps the integration of elements in knowledge maps.

2.2 LEVELS OF ABSTRACTION IN KNOWLEDGE MAPS

Abstraction is a complex and tedious concept that leads users to increased confusion and antagonism, because it can be used interchangeably, as either a process or an entity. As a process, abstraction concentrates on the isolation of the significant aspects from a domain, all the while overlooking the less important ones. As an entity, abstraction takes a different meaning altogether: the representation of a significant part of the aspect of interest via a view, a model, or some sort of focused representation (Berard n.d.). On the one hand, a model is a simplified construct of a complex entity, which is used to enable understanding of the relevant elements that form this entity of interest. On the other hand, a view is merely a simplified representation of a model. In other words, both model and view represent the outcomes of an abstraction process, as well as representing the different levels of abstraction a problem can take.

Representation of knowledge maps steadfastly relies on the application of certain levels or degrees of abstraction, which drives the conception of its building blocks. These levels of abstraction vary from higher levels, where practitioners are involved and aware of one or more aspects that represent the purpose of the domain of interest, to lower levels, where users take abstraction in knowledge maps one step further, by investigating the core set itself as a domain of interest and its representation as a view or model. The application of these levels of abstraction can go on forever, which has the potential to result in a set of details that does not have anything to do with the context in which the knowledge map was created. To prevent the possible occurrence of these irrelevant aspects, we propose that each of the elements pertinent to the knowledge maps be exploited by using two levels of abstraction: atomic aspects and nonatomic aspects.

Atomic aspects are the business-centric classes that define the constructs of the problem core. They are the same classes used in any of the traditional object-oriented class diagrams (Hamza and Fayad 2003), but with a significant quantum of business focus embedded in them. These classes are tagged by using two stereotypes: enduring business theme (EBT) and business object (BO). Nonatomic aspects are business-centric patterns that present a second level of abstraction. This second level of abstraction consists of other classes and, in some rare cases, other patterns as well. The tag names used to represent nonatomic aspects are Pattern-EBT and Pattern-BO. The next section will provide details about the elements that these tags represent.

2.3 MAPPING ELEMENTS IN KNOWLEDGE MAPS TO SOFTWARE STABILITY CONCEPTS AND PATTERNS

Before delving deeper into the underlying thesis of knowledge maps, we need to describe the tight relationship between the following elements: goals, capabilities, development scenarios, deployment, and software stability concepts and patterns. Table 2.1 represents an overview of this relationship.

In the world of knowledge maps, everything is classified in terms of goals, capabilities, and transient aspects. These aspects, however, are directly mapped into other fields of study, as in the case of software stability concepts and patterns. In Table 2.1, the goals of knowledge maps are directly mapped to software stability concepts such as EBTs, because they represent a domain-independent knowledge that contains enduring contracts or rules under which the concept is applied. Because of the enduring and reusable quality of goals and their conceptual nature, goals can also be directly mapped into the domain of patterns as stable analysis patterns. The same direct mapping process occurs with capabilities, which are mapped to software stability concepts such as BOs, because they also are enduring and reusable and their rationale is the achievement of goals. Due to their embedded properties, they also form the basis for pattern representation.

TABLE 2.1
Mapping of Elements in Knowledge Maps

Knowledge Maps	Stability	Patterns
Goals	EBTs	Stable analysis patterns
Capabilities to achieve each goal	BOs	Stable design patterns
Synergy of goals and capabilities	EBTs + BOs	Knowledge maps and many stable architectural patterns
Development scenario	IOs	Process patterns
Deployment	EBTs + BOs	Stable analysis patterns, stable design patterns, and stable architectural patterns
Dynamic analysis/the business language	Stability model/ one-shot software development	Building systems of patterns

BOs, business objects; EBTs, enduring business themes; IOs, industrial objects.

Therefore, in the world of patterns, these BOs are known as stable design patterns. Goals and capabilities depend on each other: a goal must have one or more capabilities associated with it and a capability must have a well-defined goal to fulfill. When we have two or more goals, along with their combined capabilities, a knowledge map essentially takes shape. Knowledge maps are directly mapped in software stability concepts as the synergy between EBTs and BOs. Because knowledge maps consist of goals and capabilities and their nature is enduring and reusable, the overall outcome of their association in the world of patterns is stable architectural patterns. Knowledge maps convey architectural styles that adapt or acclimatize to new requirements or contexts via extension points. These extension points tell us not only how knowledge maps will be used here but also what the context of deployment really is (which is possible by means of hooking a set of transient classes to them). Due to the volatile and changing nature of transient classes, they are mapped as industrial objects in software stability concepts. In the world of patterns, they are also known as *process patterns.*

One important point is that regardless of the different names attributed to these concepts, their characteristics, values, purposes, and behavior remain almost the same throughout their use. Therefore, these terms will be interchangeably used in the book. The rationale of this nomenclature is to bridge the existing communication gap between technical and business people by use of a shared language. This means that a nontechnical manager, for example, can understand or exercise control over the processes in place relating to a particular software product the same way a developer can, because both speak the same language.

In the world of software stability concepts, the enduring quality and reusability of EBTs and BOs are determined mainly by examining the underlying knowledge, sometimes overlooked or assumed by practitioners, mainly found in business issues and rules. Thus, EBTs and BOs represent a set of norms and rules on how to understand and solve a set of recurrent problems that require immediate attention from practitioners. From a knowledge map's perspective, goals and capabilities share almost the same vision as EBTs and BOs. They are all business-centric and within-context aspects that provide a retrospective of a domain's rationale. The process of identifying the EBTs and the BOs of a problem can be explored in detail in Fayad (2002a, 2002b), Fayad and Altman (2001), and Hamza and Fayad (2003).

The following sections describe one important element of knowledge maps: goals. One's understanding of the basic definition of goals is critical for the construction of the knowledge map of the selected domain, because goals specify the fundamental themes that drive the understanding of the selected domain. Therefore, practitioners and developers must pay great attention to them when building their own environments or knowledge maps.

2.4 THE SOFTWARE STABILITY MODEL

The *stability model* (Fayad 2002b) represents an innovative method of designing, creating, and modeling any system, including software systems. It is an extension of object-oriented software design methodology, but it has its own set of suggested rules, format guidelines, procedures, and several heuristics to arrive at a more advanced and complex object-oriented software system. Designing and building high-quality software systems has been the focus of immense interest among the proponents and designers of software systems. One of the most desirable quality attributes, yet the most difficult to achieve, is stability. A stable basis in software design technology provides us a solid foundation for building high-quality software systems.

The overall goal is to achieve innovative criteria, such as the following:

- *Stability criteria.* Objects meeting this criterion will be stable and robust over time and will not need incorporation of any changes.
- *Reusability criteria.* The majority of the objects meeting this criterion can be reused for a huge number of applications.
- *Maintainability criteria.* Maintainability is an object-oriented valid tool in stability model applications, because the objects will rarely need of maintenance and updates.
- *Wide-applicability criteria.* Patterns meeting this criterion have wide recurrence and represent a base block for modeling in any context with an appropriate level of flexibility, so that the developer can apply the pattern to the desired application. This also includes generality criteria, where the objects become domain-independent and can be applicable to any context regardless of the domain, according to which the context of any pattern should be general enough to form a base for developing any context in any application.

2.4.1 Goals

From the standpoint of knowledge maps, goals are business-centric themes that provide an enduring aspects of a domain's rationale. They also represent the essential themes of any domain, themes that are free of irrelevant or insignificant details not pertinent to the domain under consideration. For example, the concept of *friendship* is a universal theme applicable across different cultures and beliefs. Basically, it is not bound to any one of the possible contexts of applicability, because its meaning is pervasive and universal.

Generally speaking, goals are difficult to discover, because they are basically implicit themes that are hidden within the complexity and lack of understanding of the problem of interest. For example, imagine being involved in a project that requires the development of a scalable biometric system for a top-security company. As a developer, you will automatically suggest certain tangible objects that you think will represent the correct solution for the desired system. These tangible objects may be the following: Suspect, SystemOperator, FacialScanning, FaceImage, Fingerprints, RetinalPattern, IrisPattern, DNASequence, SuspectDetection, CameraSensor, FacialShape, and Database. These tangible objects are specific to a particular context; therefore, they cannot be the main focus of your solution, since they will change in the short or long term, because of the introduction of new technologies and subsequent changes in needs and requirements. Thus, even if you include them, any investment in software product containing these tangible objects will be lost.

In order to overcome these unique problems, we need to start thinking in terms of goals, because it is their pervasive nature that will help software products survive constant changes or modifications in requirements or needs. Following the above example, we will try to tackle the same problem, but focus on the main goals of the problem. Let us first define the area of interest, that is, biometrics. Biometrics is a science that measures the physical or behavioral features of an entity. This definition in conjunction with the question *what is biometrics for* will assist us in the process of defining one of biometrics main goals. This first goal will be *branding* (Sanchez 2005). Branding is a construct that readily creates a close association between an entity and its brands, by forming a unique identity that differentiates this entity from its peers (Sanchez 2005). This branding's strange behavioral characteristics strongly complement the use of biometrics. Therefore, if you want to solve the existing project successfully, you must try to address the *branding* goal in your solution. Branding is a conceptual model or solution and consists of the following elements: Branding type; AnyEntity, as the handler for any type of entities; AnyBrand, to deal with any type of brand; AnyMechanism, to control all possible mechanisms involved in branding; AnyIdentity, to represent the sum of all essential qualities that will be used by a brand; and AnyParty, which can be the branding practitioner and the spectator.

In summary, a goal mainly answers the question, What is the concept used for? However, a goal does not indicate how the concept can perform this operation; this task is handled by the capabilities of the concept. Goals and capabilities depend on each other and provide the foundation for an unimaginable set of architectures. Therefore, before getting into the subjects of how goals are associated with capabilities, we need to define exactly what capabilities are in a real sense.

2.4.2 Capabilities

A goal without a capability would not be useful or worthwhile. We will use these rules during the creation of knowledge maps. Capabilities are the business-centric workhorses that support the realization or fulfillment of a goal. Like goals, capabilities are enduring artifacts, but with a minor difference: they are externally adaptable, via hooks (Fayad, Hamza, and Sanchez 2005). Their adaptable nature can be determined only by examining the relationships between the underlying business and direct application and by applying the right *hooking code*. First, these relationships can be inheritance, aggregation, or associations. Second, the hooking code is solely responsible for weaving business and industry together, rather than focusing on the generalization–specialization principle. An important point here is that BOs are not directly adapted by the industry (transient aspects); in fact, they are not. Rather, hooks create an environment, where capabilities are able to attach any transient aspect without changing the internal structure of the capabilities and without a bit of chance of a collapse.

Capabilities are less difficult to find compared to the task of determining the underlying goals of a domain, because capabilities tend to represent any knowledge skill, process, or ability required for the execution of a specified course of action or work flow. Answering a few simple questions helps in identifying capabilities. For instance, how can we approach the underlying goal? What do we need to fulfill it? Who is it that is going to use it? As an example, imagine that the goal of interest is sampling (Sanchez, Lai, and Fayad 2003). By asking the above important questions, we will arrive at the following results: How can we approach the underlying goal? SamplingEntity, SamplingType, Applicability. What do we need to fulfill it? AnyMechanism, AnyCriteria, AnyMedia. Who is it that is going to use it? AnyParty.

In the methodology of knowledge maps, capabilities are adaptive concerns that ensure a reduced cycle time for coping with a vast number of transient requirements and handling of other goals and capabilities. This feature enables both on-demand adaptations and flexibility to transient aspects and on-demand scalability of the environment to expand the abilities needed to achieve a goal. The above behavior of capabilities is at the end introduced as faster return of investment, while still maintaining a high-quality solution.

In essence, then, the domain capabilities are important aspects that attempt to encapsulate the business processes and categories of a business-centric theme or goal. When these capabilities are directly associated and linked with their goals, they form a synergetic force that would represent the groundwork for the understanding of any domain.

2.4.3 Knowledge Maps: Formation and Stable Architectural Patterns

In Chapter 1, we provided a visual representation of knowledge maps to give an idea of how a knowledge map is structured. From the point of view of a positioning level, that representation would be just enough; however, from the point of view of a research level—the main objective of this work—such a representation is not simply enough. Therefore, we will provide a more technical representation of knowledge maps. Figure 2.1 shows such a representation.

As we can observe here, knowledge maps reach far beyond an organized visual representation. Underneath their images and symbols lie a set of well-elaborated pieces of code (i.e., Java code) that determines how goals, capabilities, development scenarios, and so on are developed in terms of Java classes. In Chapter 10, we will go through in a detailed manner how these building blocks are implemented in Java. However, for now, an important thing to remember is that for academic purposes, we have used Java as the target programming language, because it is a fully object-oriented programming language. Software practitioners, however, can also use other programming languages, such as C#, VB.NET, C++, and AspectJ, if they so desire.

The creation of knowledge maps requires considerable advanced skills and knowledge and steps beyond the identification of tangible aspects that are bound to specific contexts of applicability. It also requires a systematic capture and full understanding of the domain in which it resides. This includes describing the domain not regarding its tangible side but regarding its conceptual side as well as describing its underlying affairs or essentials and the elements required to fulfill these underlying affairs.

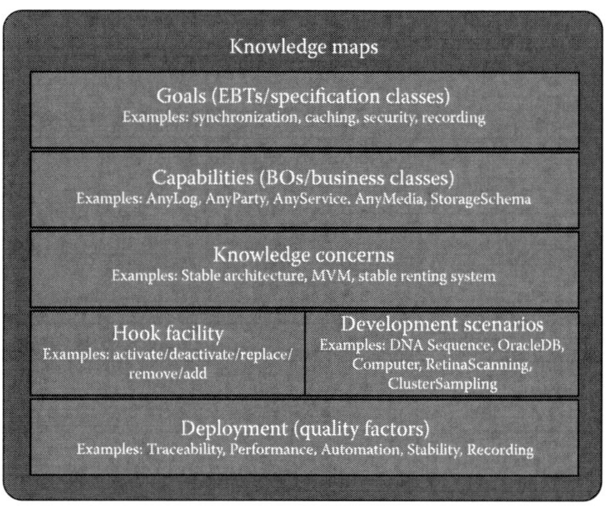

FIGURE 2.1 The knowledge maps representation.

The process consists of seven important steps:

1. *Domain knowledge partitioning.* This is an interactive process, where we will decompose the domain or domain into distinct levels of abstraction. By doing this, we can deal with domain knowledge with ease.
2. *Single domain focus.* This step concentrates on selecting a subdomain (a product of the domain partitioning) of interest at a certain time (this subdomain will be represented as a knowledge map). Then, we will extract the goals that drive the rationale of this subdomain. This step will be executed with all the subdomains generated after the partitioning of the main domain.
3. *Place goals.* This step focuses especially on placing the extracted goals into the goals section of the knowledge map (i.e., three to five goals per knowledge map).
4. *Identify capabilities associated with goals.* The main focus of this step is to identify, filter, and evaluate the potential capabilities that will fulfill the goals that were identified in our knowledge map.
5. *Connect capabilities to their goals.* The sole purpose of this step is to specify how the domain's rationale will be fulfilled once and for all.
6. *Branch out to other knowledge maps.* This step is twofold:
   a. Because the knowledge maps of partitioned subdomains may have overlapping capabilities, we may connect two knowledge maps that were once part of the same domain, before they had been partitioned.
   b. One domain's partitioned subdomain can be associated with another domain's subdomain (remote knowledge). This remote knowledge serves as both a usage indicator of our current domain knowledge and a position indicator of the subdomain that we are dealing with.
7. *Formation of knowledge maps.* In this step, we establish a set of knowledge maps that will realize the rationale of particular domains. The number of knowledge maps will depend upon how deeply we can explore or partition our domain of interest.

The results of these steps will be a set of interrelated goals and capabilities that serve a particular purpose. Figure 2.2 shows this process.

The next section will describe the development scenarios of knowledge maps and how they are attached to the core formed by goals and capabilities.

2.4.4 Development Scenarios

Development scenarios are determined by examining or inspecting how capabilities cope with determined contexts, full of transient or industry details, by using extension points or some sort of *hooking code*. Contexts tend to be very volatile, unstable, and fickle, because they are driven by current business and cultural responses, not future ones. This reality makes them unstable and replaceable.

Reference the example where we needed to develop an efficient biometric system. Developers generally tend to overlook or ignore the sometimes *off-content* essence of the problem and proceed with details of the problem with which they are quite familiar. This prompt response ended up with certain objects that are internally and externally unstable: FaceScanning, FaceImage, FacialShape, Database, Suspect, Operator, and so on. From the standpoint of knowledge maps, these unstable aspects are known as *industrial objects*.

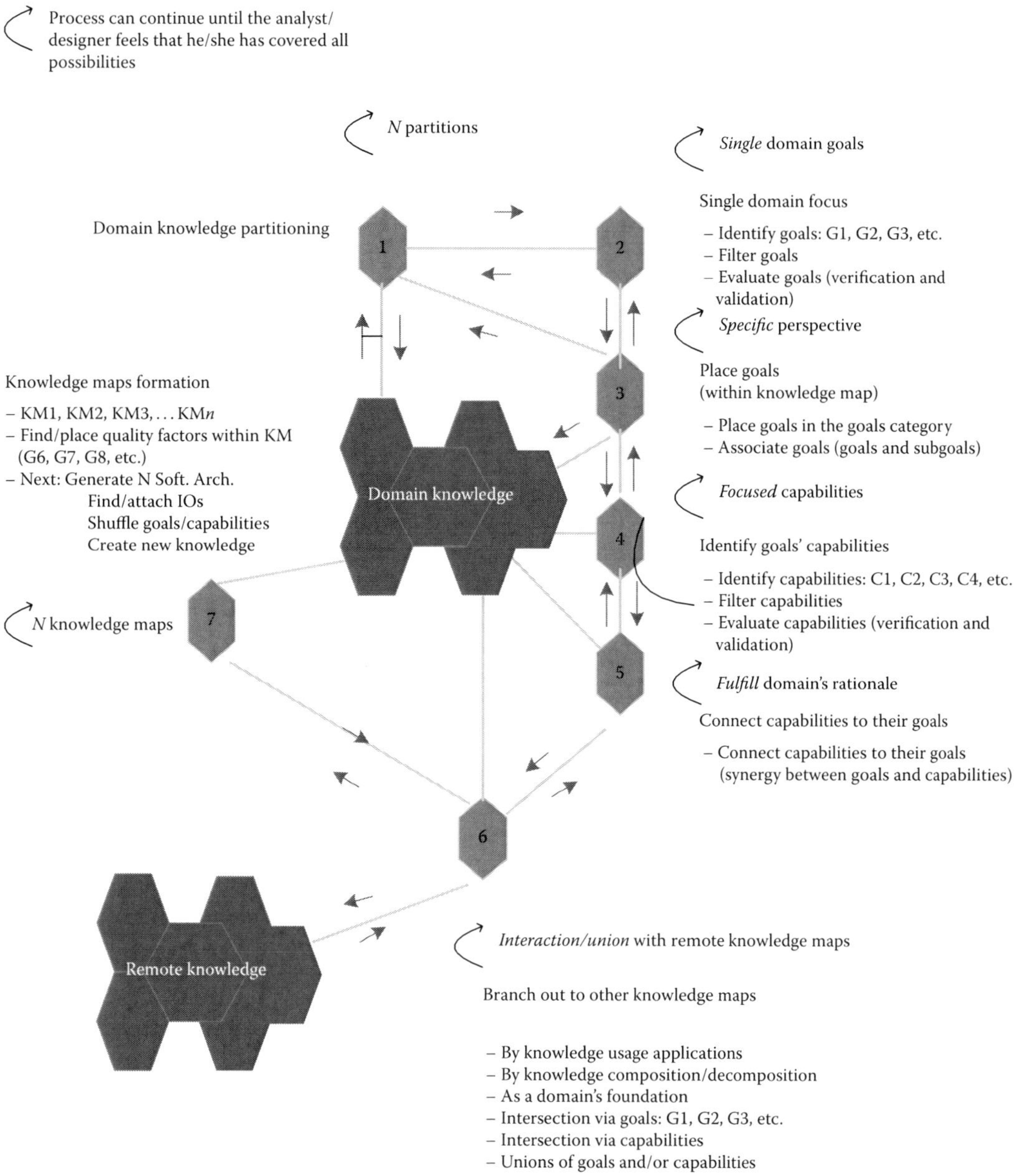

FIGURE 2.2 The formation of knowledge maps.

The intricate process of finding these classes is common and straightforward. One special way is to follow traditional methods such as Abbot's approach (Abbot 1983, pp. 882–894) or other methods, where the target is the finding of verbs as candidate classes. Another preferable way is to first examine the capabilities and directly map them to the physical world (i.e., develop ontologies), then apply the just-mentioned software engineering methods, and find the objects that complement the mapped one from the capabilities.

This process may not sound earth shattering, but it is quite effective, especially when dealing with a vast number of problems, each different in nature and composition, and a reduced notion of the context under discourse exists. The effectiveness of this process lies in the utilization of these transient aspects, via hooks, to complement the performance of

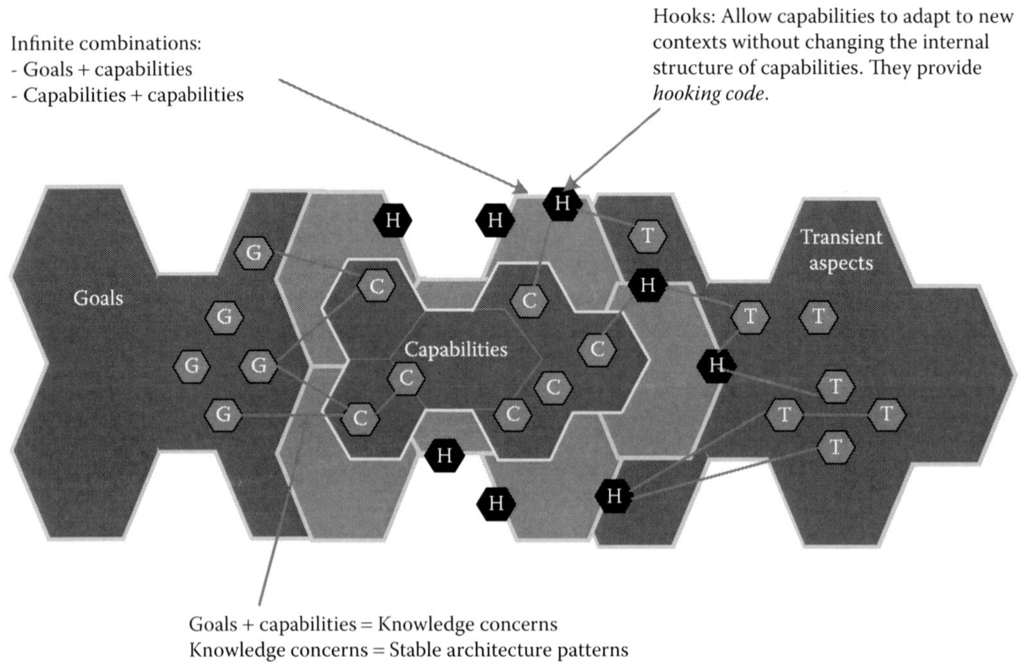

FIGURE 2.3 The adaptation of capabilities via hooks.

certain capabilities without changing the internal structure of these capabilities. Figure 2.3 illustrates the aforementioned process.

Figure 2.3 details the transitioning between goals and capabilities to the physical world. Remember that adaptation to new contexts is done at the hook level and not at the capability level. This allows us to change the transient aspects at will, without posing threats to the capabilities integrity and enduring work flow.

The next section describes the last element in knowledge maps: deployment and verification and validation concerns.

2.4.5 Deployment and Verification and Validation

Deployment and verification and validation focus mainly on the definition of robust knowledge maps, by using a set of quality factors pertinent to the domain of interest. Providing a definition of these robust knowledge maps is a special and unique challenge, especially when dealing with domains with unique characteristics and behavior and a lack of complete and systematic processes to support their creation.

The deployment and utilization of knowledge maps is intended to serve as a mirror of the collaborative experience gained over the definition of goals and capabilities. This collaborative experience is used to define the quality factors governing the deployment of knowledge maps.

These quality factors are identified and detected by following the same process that is used to identify various goals. In fact, quality factors are also the goals of a domain, with a central focus on how to use this domain (and how it should not be used) in larger or more specific domains. These quality factors usually represent nonfunctional requirements, such as reliability, performance, scalability, traceability, and usability. However, the definition of these quality factors is highly dependent upon the domain or subject of interest. Due to the reusable and domain-independent nature of goals and quality factors, some cases may already exist wherein a few of the quality factors of one domain appear in another. But this is not always the case.

Clearly, to make knowledge easily accessible, complete, and accurate, we must explain or note how the context is set and what the purpose of its use is. One unique and special way to represent this usage notion is to identify quality factors and indicate which capabilities are required to complete them (this also includes the attachment of a set of industrial objects). Along with the synergy between goals and capabilities, quality factors and capabilities also provide the basis for the generation of an unimaginable number of software architectures. This particular behavior is noted in Figure 2.3. In the end, we would recommend applying the same heuristics and assessment indicators used for goals identification and verification and validation. As a result, we emerge here with a set of enduring software architectures that comply with identified sets of quality factors of a domain of interest, along with the evaluation and assessment methods that will guide its examination and validation process.

In general, the rationale of providing a deployment concern in the knowledge maps guarantees and ascertains the proper forming and usage of core sets and the enclosed patterns pertinent to the domain of interest. In Chapters 9 through 13, we will expand and elaborate upon the idea of knowledge maps, along with their concerns.

SUMMARY

This chapter described in detail the structure of knowledge maps and the distinct concerns it develops. These concerns are analysis, design, knowledge, development or application, and deployment. We implemented these objectives by outlining the essence of a knowledge map in proper order.

OPEN RESEARCH ISSUES

The following are some of the open research issues that need to be examined and require future work and experimentation:

1. Using or employing knowledge maps to develop suitable knowledge representation schemes for storing nonmonotonic knowledge or skills that allow computationally efficient manipulation (see Sidebar 2.1).
2. Using knowledge maps to generate core knowledge to be utilized in aspect-oriented architectures and programming. Here, aspect-oriented programming is a programming paradigm, and aspect-oriented software development is used to aid and assist programmers in the separation of concerns, specifically cross-cutting concerns, as an advance in modularization.
3. Identifying a broader base for software reuse through knowledge maps and avoiding having to reinvent the wheel all the time. We believe that knowledge maps provide high levels of reuse in software development: analysis, design, documentation templates of architectural patterns, code, test procedures, test cases, manual reports, and so on. We also believe that the success of any type of software system or application largely depends on whether its capability may be reused in different collaborative scenarios in broad application areas, without requiring significant software redevelopment efforts or any overhead involvement.

4. Using the stability model as a way for knowledge elicitation is a process of obtaining knowledge from any source, for example, human and literature sources. This can involve the use of reading, researching, interviews, observation, and protocol analysis.
5. Utilize the concurrent software development model or knowledge map methodology as a way for developing ontologies of any application or domain, where "ontology is a specification of a conceptualization" (Gruber 1992, 1993; see Sidebar 2.2).

REVIEW QUESTIONS

1. List the different concerns of interest in knowledge maps.
2. What do the following topics deal with?
   a. Analysis concern
   b. Design concern
   c. Knowledge concern
   d. Development concern
   e. Deployment concern
3. Is creating knowledge map an abstraction process? Explain.
4. What aspects make up a knowledge map?
5. Is the following statement true or false? Goals determine the rationale of the problem domain.
6. What are the two important aspects of BOs (capabilities)?
7. What is meant by transient aspects of a knowledge map?
8. Explain how abstraction can be used, as either a process or an entity depending on the context under consideration.
9. What strategy is used in the knowledge map methodology to prevent occurrences of irrelevant aspects?
10. What is meant by atomic aspect of an element pertinent to a knowledge map?
11. Define the nonatomic aspect in the context of knowledge maps.
12. What stereotypes are used to tag atomic classes?
13. Nonatomic aspects can be represented by using ________ tag names.
14. Is the following statement true or false? Capability might not have a goal to fulfill.
15. What aspects of knowledge maps are mapped into the software stability concept?
16. Provide a mapping of knowledge maps with the software patterns.
17. Explain the term *goal* in the context of knowledge maps.
18. Is loyalty a goal? Explain: why or why not?
19. Is the following statement true or false? Goal provides answer to the question, What is the concept for?
20. What is role of the capabilities of a concept?
21. Goals and capabilities have a symbiotic relation. Explain how?
22. Explain the term *capability* in the context of knowledge maps.
23. Finding capabilities for a concept is easier than finding the goals. Is it true? Justify.
24. How can capabilities be identified? Explain by giving an example.
25. Why do capabilities result in higher returns of investment?

26. What are the goals and capabilities of the following?
    a. Marriage
    b. Doing a project
    c. A stamp collection process
    d. A banking system
27. Sketch the knowledge map representation and describe it in brief.
28. List the steps that must be carried out for creating a knowledge map. Explain each step.
29. How can development scenarios be identified?
30. List the ways of finding industrial objects.
31. Can the transient aspect of knowledge maps be changed at will? If so, give reasons to support your answer.
32. What do you mean by quality factors? How are they identified?
33. Is the following statement true or false? Quality factors are also goals of a domain, but with a central focus on how to use this domain in large or more specific domains.
34. What is nonmonotonic knowledge?
35. What is aspect-oriented programming? Give an example.
36. What is ontology?
37. What are the differences and similarities between knowledge maps and aspect-oriented programming?
38. What are the differences and similarities between knowledge maps and ontology?

EXERCISES

1. Name three ultimate goals (EBTs) of sample requirements D1, which is titled "Ocean Resources Management System."
2. Name three ultimate goals (EBTs) of sample requirements D2, which is titled "Dengue Fever Prevention and Outbreak Management System."
3. Name all the capabilities (BOs) of sample requirements D1, which is titled "Ocean Resources Management System."
4. Name all the capabilities (BOs) of sample requirements D2, which is titled "Dengue Fever Prevention and Outbreak Management System."

PROJECTS

1. Use the abstracted EBTs and BOs from your responses to problem statement E1, which is titled "Ocean Resources Management System" to form a knowledge map.
2. Use the abstracted EBTs and BOs from your responses to problem statement E2, which is titled "Dengue Fever Prevention and Outbreak Management System" to form a knowledge map.

SIDEBAR 2.1 Formal Methods and Formal Languages (Supratik Mukhopadhyay)

In the past, formal abstraction-refinement techniques (Morgan 1994) have traditionally been used to support model/specification-driven development of software. Stepwise refinement techniques (Morgan 1994) map an abstract model of the system to concrete software through a set of small refinement steps. In the model-driven architecture framework, a platform-independent model of the system is effectively mapped to a platform-specific model through model transformations.

In knowledge-based software engineering, knowledge representation techniques are used for representing domain knowledge. Popular knowledge representation techniques include various monotonic and nonmonotonic logics (Barwise 2006), such as description logics (Baader et al. 2003) and default logics (Besnard 1989). Goals may be specified as intentional knowledge within a knowledge base. Capabilities can be added to provide extensional definitions. Goals may be reified by automatically *connecting* the intentional goals to the extensional knowledge by using the deductive reasoning capabilities of the underlying logical framework (Boddu et al. 2004). Reifying goals results in ontology. Consider the following (simplified) requirement from the Bay Area Rapid Transit project (van Lamsweerde 2001): "if a train is on a track, then its speed should be less than the specified speed on the track." A first-order discourse corresponding to this requirement (Boddu et al. 2004) is given below (the discourse can be created from the natural language requirement automatically).

```
EX X1
EX X2
end referent
isa (X1, train)
isa (X2, track)
ison (X1, X2)
end discourse
= >
EX X3
EX X4
EX X5
end referent
isa (X3, speed)
of (X1, X3)
isa (X4, speed)
of (X4, X2)
isa(X4, specified)
shouldbelessthan (X3, X4)
end discourse
end discourse
```

This is a discourse of the form D1 ⇒ D2, where D1 and D2 are discourse representation structures. Notice that the atomic formula `shouldbelessthan (X3, X4)` is undefined in the requirement, as well as the discourse representation structure. Such undefined atomic formulas need to be interpreted in the current context. A knowledge map system should first search the knowledge base for a definition of the atomic formula. If a definition is found, it should consult the user about whether to interpret the atomic formula with a definition from the knowledge base. (It presents the user with all the definitions of the atomic formula found in the knowledge base if there are more than one.) If either no definition is found in the knowledge base or the user does not agree with definitions in the knowledge base, the user will be asked to specify what it means by *should be less than* in the sentence—"If a train is on a track, then its speed should be less than the specified speed of the track." The user might specify "X should be less than Y if X < Y." This input will be used to interpret the atomic formula _ `shouldbelessthan (X3, X4)` as `shouldbelessthan (X3, X4) ← (X3 < X4)` _ . (The user might be less precise in specifying the meaning of *should be less than*; in this case, more refinement is needed and the user will be prompted to refine his or her specification.) Thus, for the sentence "Managers can access the database," the user will be asked the meaning of the word *managers*. If the user specifies "Tom and Jim are managers," the interpretation for isa(X, Managers)_ will be isa(X, Managers) ← (X = Tom or X = Jim).

We can use a *closed world assumption* to interpret the atomic formulas. Hence, lack of extra information would mean that in a closed world isa(X, Managers) = (X = Tom or X = Jim) _ and. The definitions of _shouldbelessthan (X3, X4)_ and _ isa (X, Managers) are then stored in the knowledge base along with their English interpretations "X should be less than Y if _X < Y" and "Tom and Jim are managers," respectively, for use in future sessions. Thus, a refinement of the requirements goals results in a model-theoretic interpretation of the atomic formulas.

REFERENCES

Baader, F., D. Calvanese, D. L. McGuinness, D. Nardi, and P. F. Patel-Schneider. *The Description Logic Handbook: Theory, Implementation, and Applications.* Cambridge, UK: Cambridge University Press, 2003.

Barwise, J. ed. *Handbook of Mathematical Logic.* North Holland, the Netherlands: Elsevier, 2006.

Besnard, P. *An Introduction to Default Logic.* Berlin; Heidelberg, Germany: Springer, 1989.

Boddu, R., L. Guo, S. Mukhopadhyay, and B. Cukic. "RETNA: From Requirements to Testing in a Natural Way." Paper presented at the IEEE International Conference on Requirements Engineering, Kyoto, Japan, 2004.

Morgan, C. *Programming from Specifications.* 2nd edn. New Jersey, NJ: Prentice Hall, 1994.

van Lamsweerde, A. "Goal Oriented Requirements Engineering: A Guided Tour." Paper presented at the IEEE International Conference on Requirements Engineering, Toronto, ON, Canada, 2001.

SIDEBAR 2.2 The Definition of Ontology

Ontology is a specification of a conceptualization (Gruber 1993, 2009). The word *ontology* seems to generate a lot of controversies in discussions about artificial intelligence. It has a long history and tradition in philosophy, in which it refers to the subject of existence. It is also often confused with epistemology, which refers to knowledge and knowing.

Ontology (Gruber 2009) is defined as an "explicit specification of a conceptualization," which is, in turn, "the objects, concepts, and other entities that are presumed to exist in some area of interest and the relationships that hold among them." While the terms specification and conceptualization have caused much debate, the essential points of this definition of ontology are

- An ontology defines (specifies) the concepts, relationships, and other distinctions that are relevant for modeling a domain.
- The specification takes the form of the definitions of representational vocabulary (classes, relations, etc.), which provide meanings for the vocabulary and formal constraints on its coherent use.

This definition does not distinguish between tangible and conceptual objects and all the common components of the any giving ontology are domain dependent.

In both computer science and information science, ontology is a formal representation of a set of concepts within a domain and the relationships between those concepts. It is used to reason about the properties of that domain and may be used to define the domain.

Ontologies are used in artificial intelligence, the semantic web, software engineering, biomedical informatics, library science, and information architecture, as a form of knowledge representation about the world or some part of it. Common components of ontologies as shown in (Gruber 1995) include the following:

- *Individuals.* Instances or objects (the basic or *ground-level* objects)
- *Classes.* Sets, collections, concepts, or types of objects
- *Attributes.* Properties, features, characteristics, or parameters that objects (and classes) can have

- *Relations.* Ways that classes and objects can be related to one another
- *Function terms.* Complex structures formed from certain relations that can be used in place of an individual term in a statement
- *Restrictions.* Formally stated descriptions of what must be true in order for some assertion to be accepted as input
- *Rules.* Statements in the form of an if-then (antecedent-consequent) sentence that describe the logical inferences that can be drawn from an assertion in a particular form
- *Axioms.* Assertions (including rules) in a logical form that together comprise the overall theory that the ontology describes in its domain of application. This definition differs from that of *axioms* in generative grammar and formal logic. In these domains, axioms include only statements asserted as a priori knowledge. As used here, *axioms* also include the theory derived from axiomatic statements.
- *Events.* The changing of attributes or relations

Ontologies are commonly encoded using ontology languages. Ontologies resemble or look like faceted taxonomies, but use richer semantic relationships among terms and attributes, as well as very strict rules about how to specify terms and relationships. Because ontology does more than just control vocabulary, they are thought of as knowledge representation. An oft-quoted definition of ontology follows: "the specification of one's conceptualization of a knowledge domain" (see http://www.ksl.stanford.edu/people/dlm/papers/ontologies-come-of-age-mit-press-(with-citation).htm). An example of concepts and relationships in ontology is shown in Figure 2.4.

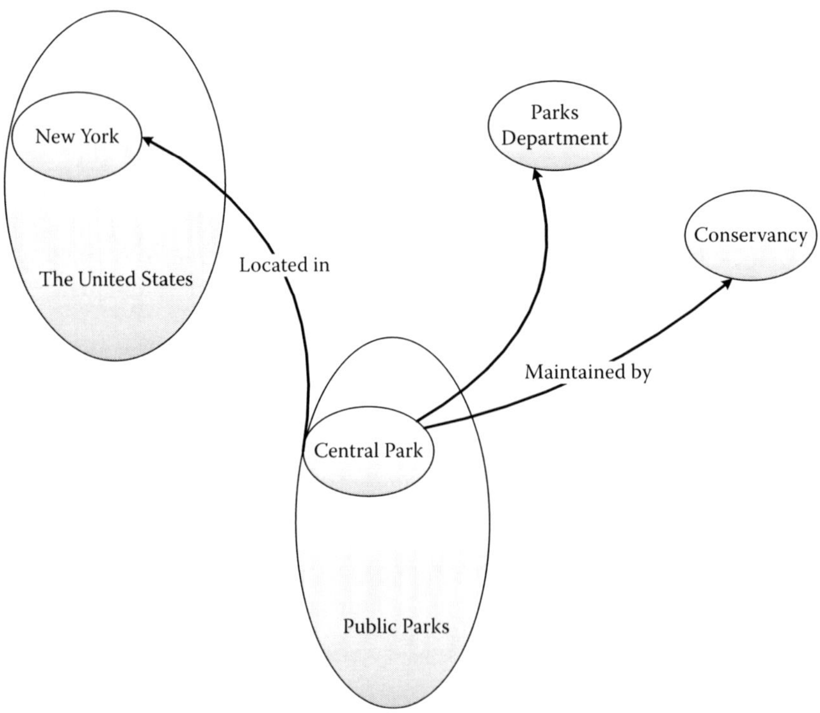

FIGURE 2.4 An example of concepts and relationships in ontology.

Ontologies, because they are machine readable, allow applications to be standardized, while domain-specific information can be customized over time. The goal of Ontologies is to move the complexity of the system into how the information is organized, rather than in the application that processes that information.

REFERENCES

Gruber, T. R. "A Translation Approach to Portable Ontologies." *Knowledge Acquisition* 5, no. 2 (1993): 199–220.

Gruber, T. R. Toward principles for the design of ontologies used for knowledge sharing. *International Journal of Human-Computer Studies*, Vol. 43, Issues 4–5, November 1995, pp. 907–928.

Gruber, T. R. "Ontology." in the *Encyclopedia of Database Systems*, L. Liu and M. T. Özsu, eds., Springer-Verlag, 2009.

Section II

Goals of the Knowledge Maps

Goals always represent what a concept is for, but not how a concept actually does it (capabilities). They also convey the enduring business rules under which the capabilities of a concept must function and act. Goals are important and critical pieces of formation of the knowledge maps.

Section II discusses goals or/and enduring business themes, the origin of goals, several perspectives of goals that are related to people, business, and projects; describes guidelines for ultimate goals, the impact of goals on problem understanding; and documents two major goals of the knowledge maps as stable analysis patterns: discovery and knowledge. Section II contains three chapters and six sidebars.

Chapter 3 is titled "The Goals: Significance and Identification," and it defines goals and their origin, discusses the goal significance, shows how to deal with goals: extraction and assessment, briefly lists goals of the knowledge maps, and shows short pattern documentation templates for a few goals. This chapter concludes with a brief summary and numerous open research issues. This chapter also provides review questions, exercises, and some projects.

Chapter 4 is titled "Discovery Stable Analysis Patterns," and it discusses, models, and documents this pattern by using Fayad's stable pattern documentation template as shown in Appendix A. This chapter concludes with a summary and many open research issues. This chapter provides numerous review questions, exercises, and some projects.

Chapter 5 is titled "Knowledge Stable Analysis Pattern," and it repeats with the same headers as shown in Chapter 4.

Sidebar 3.1 is titled "Goal-Oriented Requirements Engineering," and it views any system as a collection of active components (*agents*). Agents may restrict their behavior to ensure the constraints that they are assigned (Lapouchnian 2005). In GORE, agents are assigned responsibility for achieving goals (Lapouchnian 2005). A *requirement* is a goal whose achievement is the responsibility of a single software agent (Lapouchnian 2005).

Sidebar 3.2 is titled "Goal Programming," which is a fanciful or exotic nomenclature for a very simple and straightforward concept: the thin fine line between stated objectives and listed constraints is never completely crystallized.

Sidebar 3.3 is titled "Goal-Oriented Development," which involves traversing through a goal tree, an And–Or tree, whose root is associated with the system-wide goals.

Sidebar 4.1 is titled "Knowledge Discovery," and it derives special knowledge from the available set of input data.

Sidebar 4.2 is titled "Business Rules," and it discusses the business rules as well-defined rules through a set or collection of well-calibrated processes to achieve certain goal(s).

Sidebar 5.1 is titled "Knowledge Definition," and it defines in simple words what exactly knowledge is?

3 The Goals
Significance and Identification

Setting goals is the first step in turning the invisible into the visible.

Anthony Robbins

Goals always represent what a concept is for but not how the concept is applied to achieve those goals (capabilities). They also convey the enduring business rules under which the capabilities of a concept must function and act. Goals are important and critical pieces for the formation of knowledge maps. They are important and mandatory, because they encapsulate a discipline's rationale and retrospective. This rationale and retrospective analysis will embed the appropriate axioms or rules under which a knowledge map will be exploited. By discovering these goals, we can have a more precise idea of what problem domains and their nature really is and the elements that are necessary to solve it.

3.1 INTRODUCTION

The main use of goals to wrap up the outcomes of the analysis phase is not entirely new or fresh. Goals were originally suggested and recommended in Anton (1996), Anton and Potts (1998), and van Lamsweerde (2001). However, in Anton (1996), Anton and Potts (1998), and van Lamsweerde (2001), goals were simply defined as the functions (activities) and constraints bound to an organizational process. In other words, these goals will change proportionally to any change or modification in the processes of an organization. Usually, the number of functions or activities can be quite large in an organization, especially if the documentation of the organizational processes is analyzed or explored in a cautious manner (Anton 1996; Anton and Potts 1998). Consequently, this number can be overwhelming for the stakeholders and the technical staff of an organization to handle and manage.

In this book, we will share a different view and perspective of what goals really are. In this view, goals are simply the enduring themes that justify why a software solution, area of study, and so forth is needed in a determined environment (e.g., organization and software project). They are neither functions nor constraints bound to organizational processes, as suggested in Anton (1996) and Anton and Potts (1998).

Indications or suggestions of goals usually become obvious in meetings and situations, where there is some sort of negotiation between one or two parties, for example, requirements elicitation meeting. In this instance, you have a group of software analysts, designers, architects, and maybe some developers who are trying to learn and understand what the needs, scopes, and trade-offs of a subject or project really are. For example, an individual (project manager) tries to convince another individual (customer) to accept a particular software product. If the project manager can satisfy or meet the needs and requirements of a customer, there is a certainty that the customer will accept the software product. Otherwise, the customer will just walk away, even when the product is cheaper. This is because there is no need for using that product if it is not going to satisfy the customer.

Another example of when goals become obvious and noticeable is a situation when developers deal with businesspeople and/or stakeholders. Entrepreneurs always tend to think about a problem in terms of principles, things that describe what the problem is. Developers, however, tend to think in terms of concrete elements that can be adapted and used to satisfy a determined process. Principles, by nature, are more enduring and stable than processes. For instance, let us use branding as the subject of discourse. Branding is an enduring principle that has been part of our daily lives throughout time. The earliest known shepherds effectively branded their cattle to identify, detect, and differentiate them from other shepherds cattle, by using a set of machinery, such as iron brands and paint tars. Today, branding is specially used in other fields such as marketing, human computer interaction (HCI), and biometrics. The process to achieve branding is totally different and separate in each field, but the principle of *what branding is for* remains exactly the same; for example, it allows us to identify an entity and to differentiate among the entities peers.

It is important to remember that any type of business or activity is driven forward mainly by enduring principles that determine its function and success. Unfortunately, many companies overlook these principles and depend solely on available machinery or industry, to temporarily extend their businesses lives or overhaul these activities (patching your business or activity). Software development process is a good practical example of such behavior. This unique behavior can be turned into an expensive viscous cycle that will never end. In fact, it will constantly incur additional cost, time, and effort every time and whenever new patches come along. These issues show and demonstrate how important it is to focus on those aspects that are likely to endure in any business than just focusing on patching existing deficiencies with new machinery proportionally to the appearance of new requirements.

The terms *principles*, *essential themes*, or *enduring themes of the subject* will be interchangeably used to represent goals. In the rest of the chapter, we will illustrate why goals are important and critical elements of a problem space, and what the processes of identifying them are. Later in the chapter, we will also introduce the essential themes driving the realization and understanding of knowledge maps. These essential themes will be described using a short-pattern documentation template. Two complete pattern documentation templates are provided in Appendix A.

3.2 SIGNIFICANCE OF GOALS

Goals have always been the essential part of object-oriented analysis or problem space; however, their use and deployment was always implicit and, in several cases, totally ignored by software practitioners (van Lamsweerde 2001). For instance, software practitioners put more effort and energy into trying to solve a problem than trying to learn and understand it first. Therefore, requirements and needs were never synchronized with the machinery or elements used to fulfill them. Consequently, businesses all over the world experienced great losses due to software systems that was not completed on time and software developed wrongly. The use of goals to overcome the aforementioned problems is starting to be recognized (Anton 1996; Anton and Potts 1998) now, especially during the requirements acquisition phase. Software practitioners are beginning to think in terms of *why do we need this subject (system)* or *what are the objectives that we are planning to achieve with this subject.*

The most important and critical problem with the above approach is that practitioners have a peculiar tendency to think in terms of the application's elements that are most likely to change and transform over time, due to the emergence of new problems or new business requirements (Fayad, Hamza, and Sanchez 2005). The approach being introduced in this chapter concentrates mainly on those goals that will remain the same over time. These goals will overcome application changes, new business requirements, and so on, because they are always created at a knowledge level and not at an application level.

Alex van Lamsweerde (2001) provides a set of reasons and causes as to why the definition of goals is so important and critical to the requirements elicitation process. Some of these reasons are also applicable to the approach that is being introduced here. The special ones that apply to our approach are products of certain properties of goals that have determined the value of a subject matter's rationale. (We have included some of our reasons too.) Following are these important properties (Hamza and Fayad 2003) (Some properties were omitted, because they merely apply to a model that consists of goal and capabilities.):

- *Stable.* A goal must represent a stable and conceptual structure that determines a subject matter's rationale.
- *Natural.* It is important to present, in a cohesive and natural manner, no less important, language to assure its reusability elsewhere.
- *Domain-independent.* A goal must represent a conceptual structure that appears in multiple domains of applications.
- *Single enduring business theme (EBT).* A goal must represent a single EBT. This means that we are focusing on one problem at a time.

These properties were described to help practitioners understand in detail numerous reasons as to why goals are so critically important. These reasons are as follows:

1. The definition of a goal implies adequate requirements, specification, and completeness. A requirement specification was said to be complete if a determined goal was achieved.
2. Goals are the subject's rationale retrospective that provides the high- and low-level essentials or principles that the management and technical personnel can easily understand and apply.
3. Goals facilitate a natural mechanism to allow management and technical groups to be on the same page with respect to the design objective of the subject (system).
4. The process of definition of a goal is a focused process—irrelevant details are always avoided.
5. Goals are enduring themes that are not bound to volatile information. Their enduring nature is determined by specially focusing on aspects that will remain stable over time (knowledge) and not on aspects that are application specific.
6. A proper identification of goals will drive the discovery of their capabilities. In other words, once we have successfully found (and evaluated) a determined and set goal, we will be able to determine its capabilities.
7. A goals identification process facilitates great team dynamics and vibrancy. Because goals are described in a simple, straightforward, and natural language, they are clearly understood by managers and technical staff. Therefore, managers can actively participate in identification of goals, which is the area with which they are most familiar.

3.2.1 An Example: A Simple E-Commerce Application

To illustrate the importance of using goals to understand the purpose of a subject's existence, we will discuss a simple e-commerce example. This example will convey why the goals are so much needed.

Let us now imagine that a business firm is requesting JustACompany to develop the firm's new e-commerce application. Without any hesitation or doubt, JustACompany accepts the project and it gets ready for its development. Now, JustACompany's software development team proceeds with the requirements elicitation process. During this process, the team reviews the problem statement. Then, it looks out for the candidate objects of the problem by using traditional software engineering approaches (Abbot 1983, pp. 882–894; Fayad, Hamza, and Sanchez 2005). A typical result of this process will consist of objects such as Customer, ShoppingCart, CreditCard, Database, Catalog, Order, and Product. These objects are usually extracted from a problem statement that was given to JustACompany. The model is illustrated in Figure 3.1.

Consider the above illustrated e-commerce example as a specific business case. We have a simple model that has the tendency to be redone or recast every time new requirements appear on the loop. Proportional to the occurrence of new business requirements,

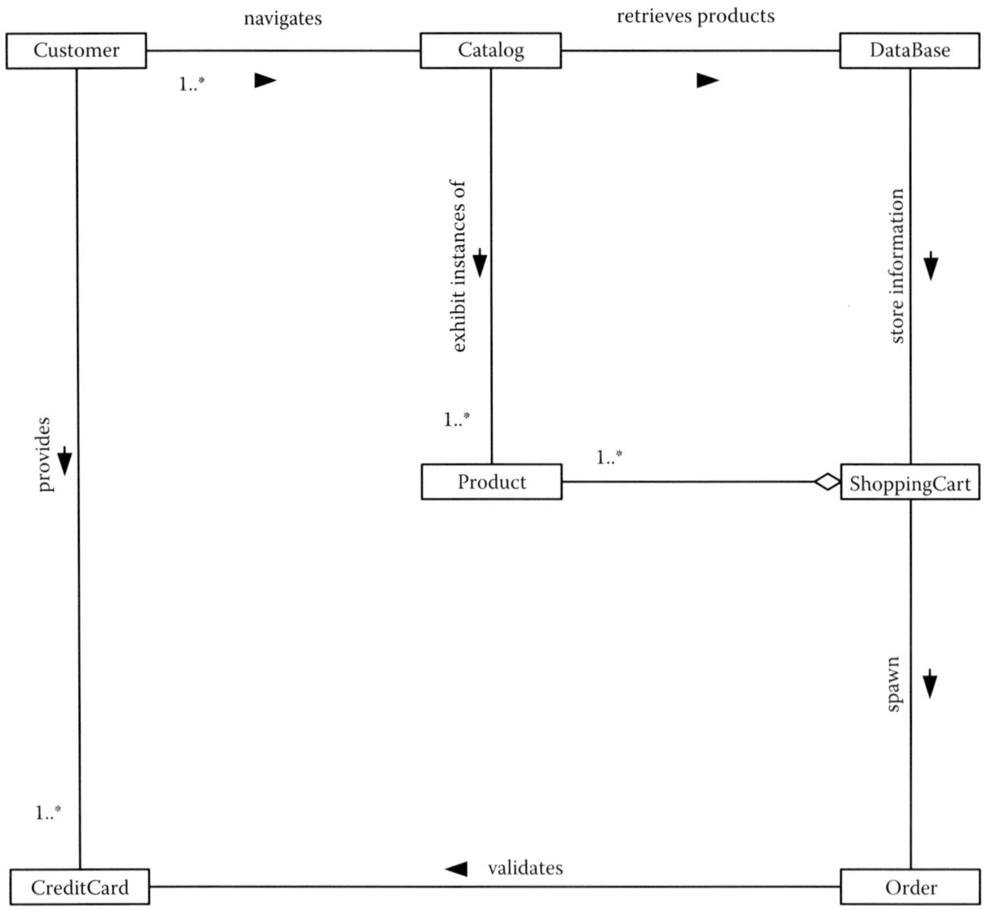

FIGURE 3.1 A simple e-commerce application.

this e-commerce solution will be susceptible to drastic and sudden changes and adaptations of the elements that form it (Fayad, Hamza, and Sanchez 2005). These drastic changes or transformations may also jeopardize the life of the e-commerce application from a business perspective, because businesses are usually reluctant to keep financing a project that produces more recurring costs than benefits. Therefore, our e-commerce application has two possible ends: abandonment or reengineering. Neither one of those states is positive and beneficial for the business. In fact, no pleasing effects will arise unless if we focus on the company's business themes likely to endure throughout its life. So, JustACompany will ask these questions: What are those enduring themes of e-commerce? Is there a way to capitalize on the declared e-commerce's enduring principles and trace them over the success of the business? How and in what way, can we find these enduring themes? To answer such questions, we need to describe the enduring business's identification process.

These critical and important questions will facilitate a complete understanding of the e-commerce subject's problem space. For instance, we will be able to characterize and acknowledge the e-commerce subject's scope, nature, and its core elements. Those core elements are essential outcomes, purely conceptual, for which we are actually looking. They are the single themes of interest; they are the *true what* of the e-commerce subject. The following section will describe the process for identifying the goals of a subject matter or discipline. Our case argued here would be the identification of the goals of an e-commerce application.

3.3 DEALING WITH GOALS: EXTRACTION AND ASSESSMENT

Focusing mainly on goals during the analysis phase is very important and vital for understanding the problem space of any subject matter. Instead of focusing on aspects that come and go proportionally with the appearance of new requirements and technology, we focus here on elements that we know will remain enduring or stable over time. Within the realm or domain of knowledge maps, these goals are classified into three categories: personal goals, business goals, and project goals. The determination of a goal's category is finalized by examining the nature and target context of the subject matter. This classification will set the boundaries or perimeters for the identification and assessment of future goals.

Goals are the special domains where businesses, projects, and persons (hosts) meet when trying to understand a particular subject of interest. Goal achievement resides within the harmony or equilibrium between these goal's hosts. This harmony is recognized by the relationships and levels of organization among each one of the concerns that form these host's rationale. These concerns may include businesses, projects, and persons own values, desires, and constraints, such as mission, vision, needs, meaning, return on investment (ROI), trade-offs, current state, and future destination. Each one of these concerns can be addressed or discussed in isolation; however, the relationship among them defines the overall flow or guidance for how these hosts are handled, and how they complement their high- and/or low-level neighbors (e.g., business and persons).

Business goals are, in fact, the mission constructs of any business entities. They define the path that any business has to follow if it wants to succeed in a corporate sense. This path consists of high-level abstractions that will drive a business performance and focus on a determined society (e.g., a market-driven society). However, because we are dealing with high-level abstractions, it is common that these concerns are built from other fine-grained (level of detail) concerns. For example, the fulfillment of business goals depends on realization of project goals, the project goals depend on achievement of personal goals, and so on.

Project goals are need-driven concerns formulated to reach a future deliverable for a set of customers (internal or external). They always reside in the satisfaction of determined needs, such as project completion, availability, solution scope, and time to market, which are transformed into products (e.g., software applications and services). Project goals are just "the compass guiding the direction of your project" (National Leadership Grants [NLG] Project Planning: A Tutorial; Woodley 2008). The fulfillment of project goals will lead us closer to achieving overall business goals. For example, let us imagine that two of your business's goals are to ensure your products *acceptance* by possible customers and to achieve a greater ROI. In order to fulfill these particular goals or needs, we have to ensure customer *satisfaction*, customer *support* in case they have concerns or questions regarding your product, and *efficiency* and *efficacy* for how to *sell* and *deliver* your products to these customers. After acknowledging these project goals, we would have a better idea or notion of type of infrastructure needed, prerequisites required to implement it, and information about the solution scope. This may ultimately result in the definition of a set of products that will eventually achieve these goals. A customer relationship management system may be required, for example. Here is where the responsibility of technical and business staff comes into place to formulate and implement the infrastructure. *Infrastructure* is a set of products or services required to assure the completion of business goals.

Dealing and acquainting with personal goals is a little bit different from dealing with business and project goals. This raises a sensitive question—*why*? Here, we are directly dealing with the goals of each one of the individuals participating in the company or project of interest. In fact, they are the most delicate or sensitive goals that we have to address and imply during a project's lifecycle, because they implicitly determine the success of accomplishing the project goals. Here is where you, as team leader or manager, try to bring or import each individual on the same boat or journey, while still ensuring each individual's goals are achieved in an efficient manner. These goals include incentives or rewards that will motivate, cajole, and urge individuals in multicultural environments to possess a proud feeling of being a key element in the team and will promote efficient collaboration and support among all team members. The management staff and team leaders must discern the required steps to set the right type of environment where these goals will be addressed in detail and then give them enough attention and care to engage these individuals (e.g., technical staff).

Within a business environment, each one of the goals categories can be addressed in isolation. However, the harmonized relationship between them will determine the success or failure of any business. Throughout the course of this chapter, we will concentrate completely and specifically on project goals, especially the ones related to project development (e.g., products and services). However, we will certainly acknowledge and note the importance of personal and business goals in the achievement of project goals. Failing to acknowledge the critical importance of personal and business goals during exploration of project goals will certainly affect their accuracy and achievement. Hence, one should never ignore these goals during the process of definition of project goals.

In a nutshell, extracting all underlying goals from a discipline or subject matter is a challenging and strenuous task for both novice and experienced software/business/project practitioners. The software stability concept approach (Fayad 2002a, 2002b; Fayad and Altman 2001) shows greater and immense promise for the extraction of goals of a discipline. In spite of a detailed process highlighted here, mastering its basic principles requires a lot of practice and skills, because of a common tendency to think in terms of application's objects rather than knowledge's objects. The next section describes the process of extracting goals in a cohesive and a seamless way, so that practitioners can learn, define, comprehend, and understand it quite easily.

3.4 EXTRACTING THE GOALS OF A DISCIPLINE: THE PROCESS

Through a set of straightforward questions and enquiries, we will also illustrate and demonstrate how you can extract, assess, and filter main goals of a discipline. The high-level process for extracting these goals is illustrated in Figure 3.2. Please note that you can perform this process in parallel with the capabilities identification process.

Using the e-commerce example cited before in the chapter, we will now illustrate how the e-commerce project's goals and ambitions are extracted, filtered, and evaluated. One of the important questions that JustACompany asked when they were facing a potential collapse

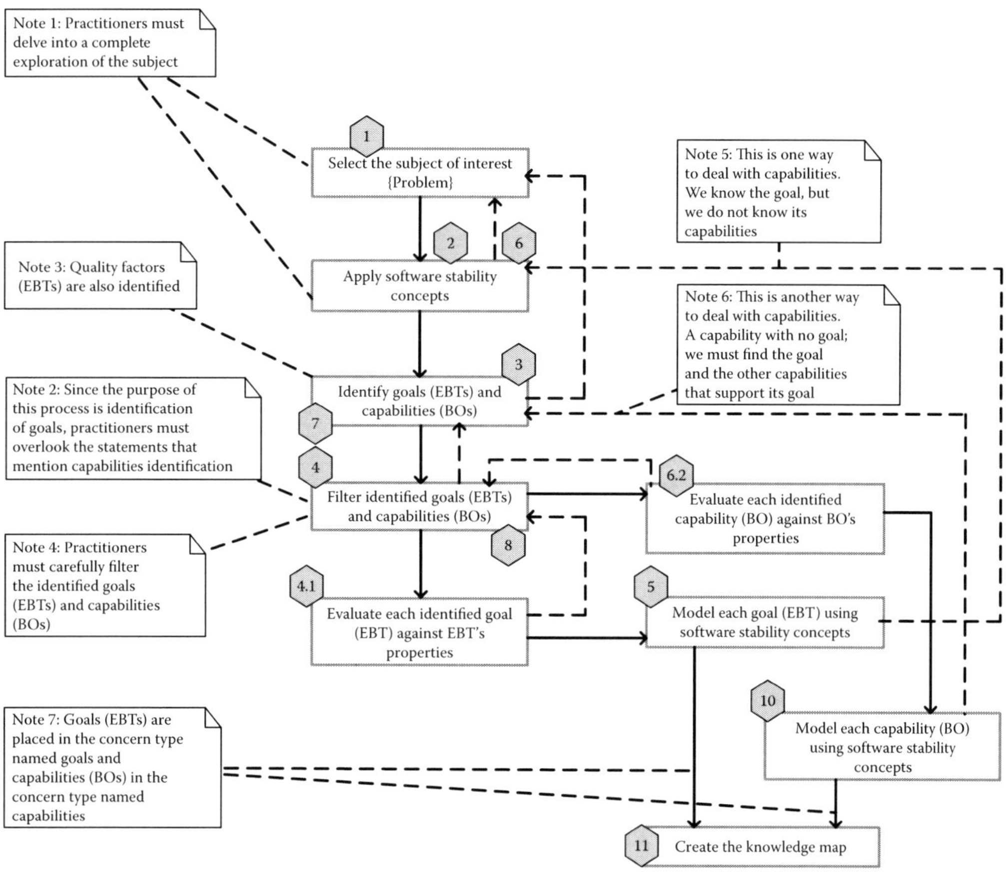

FIGURE 3.2 The high-level process for identifying goals.

or abandonment of their product was—*what are those enduring themes of e-commerce?* The answer to this question is found by following this simple process:

1. *Selecting subject of interest.* This step is designated to help guide software practitioners to examine and probe the subject of interest. Irrelevant details, such as application details, must be ignored totally during this subject's exploration. Practitioners must focus only on the aspects that define the core knowledge of e-commerce. We would support this step by asking the following questions:
   a. What is the subject or problem? The subject is e-commerce.
   b. Are you aware of scenarios or situations, where the subject (e-commerce) appears? This will help us narrow down the scope of the subject being explored. In other words, we have a focused problem.
2. *Applying the software stability concepts approach.* This step follows the heuristics provided in Fayad (2002a, 2002b), Fayad and Altman (2001), and Hamza and Fayad (2002) to identify and detect the initial list of goals of a discipline. This step also uses the following questions to get this initial list.
   a. What is the subject/problem (e-commerce)? What are the reasons for this subject matter/concept to exist?
   b. What does the subject (e-commerce) do? (See Chapter 4.)
3. *Identifying goals and capabilities.* This step's outcome will be a conglomeration of potential goals, such as ROI, product navigation, product selection, trading, order handling, customer service, convenience, and security.
   a. These goals are potentially useful, because they surfaced or appeared by solely examining what we know or what we have found about the subject in current literature.
   b. In addition, some of them may be still referencing the subject matter from an *industrial objects* perspective.
4. *Filtering identified goals and capabilities.* This step concentrates mainly on filtering the list of potential goals we just identified in the previous step. By filtering down these goals, we are getting and inching closer to the stage, where we may say that we have the right set of goals of the discipline. This step is driven by the following questions:
   a. Does the goal depend on adjacent goals to exist? If yes, the goals must be removed. Is the goal part of a wish list of a stakeholder(s)? If yes, the goals must be eradicated. By a *wish list*, we mean the list of goals that you may like to achieve or have at certain point of time.
   b. The resulting list of goals will be smaller than the initial list. The new list will be trading, order handling, convenience, and security.

4.1. *Evaluating each identified goal (EBT) by using the EBTs' properties.* This step is driven by the following questions and heuristics from Cline and Girou (2000), Fayad (2002a, 2002b), Fayad and Altman (2001), and Hamza and Fayad (2002):
   a. Can we replace or exchange any of the remaining goals with another goal? If yes, the affected goal must be removed.
   b. Is the goal internally and externally enduring and stable? In other words, does the goal reflect the essence of the subject matter's existence? If no, the goal must be removed.

   c. Does the goal have a direct physical representation? If yes, the goal must be removed.
   d. The final list will consist of three goals: trading, convenience, and security.
   e. These goals will determine *what e-commerce is*. The resulting list will also answer the middle question of JustACompany: is there a way to capitalize on the declared e-commerce's enduring principles and trace them over the success of the business?

4.2. *Evaluating each identified capability (business object [BO]) against the properties of BOs.* See Chapter 4 for more details.

The rest of the steps 5 through 11 shown in Figure 3.2 are addressed in Chapters 4 through 6, respectively. Regarding the question "Is there an efficient way to capitalize on the declared e-commerce's enduring principles and trace them all over the success of a business?" the answer is yes, there is a way! This way is twofold: the definition of the goals capabilities, and establishing a synergy between the goals and their capabilities without losing generality. This means that this unique synergy will provide the foundations for the development of a set of applications on an on-demand basis. This will also answer the aforementioned question. The critical process for identifying these capabilities will be explained in Chapter 4.

3.4.1 Dealing with Subgoals

Similar to goals, subgoals are those enduring principles that determine the very rationale and nature of a discipline or domain. They are also represented as EBTs. However, their existence and satisfaction are not as required and mandatory as the ones of the main goals are. Subgoals can be considered as the *extras* you may wish to have or satisfy within a given determined event or situation. Therefore, bringing subgoals or the extras as part of the essential elements of a discipline may bring you more problems than actual benefits. First, you are extending and redefining the context or boundary of your study. This extended context will certainly contain additional axioms or constraints and events that may, in some cases, contaminate the direction of your subject study and the overall rationale of that particular subject. Second, the inclusion of subgoals implies the consideration of certain trade-offs that the designing team (or any software practitioner) must consider and accept, when a particular need or requirement appears. For example, do you think your goals were achieved without satisfying your subgoals? Please think about this question in a critical manner.

Let us use the concept of *marriage* as a goal that contains or implies a set of extras or subgoals, such as love, friendship, companionship, and harmony. If we ask the question "can marriage exist with the satisfaction of any of its subgoals?" honestly, we know that the answer is *yes*, marriage can exist per se without the satisfaction of any of its subgoals. This does not mean, however, that it is wrong to satisfy marriage's extras or subgoals. What we are trying to say here is that unless you really need the existence of a subgoal to clarify or focus the purpose of your study, you should not deal with subgoals or include them in your subject's rationale study.

The next section covers the main goals of the knowledge maps. Because we have covered the subject of what goals are and how they can be identified, readers will be able to understand the goals of the knowledge maps and get an accurate picture of what they are and how they were identified.

3.5 GOALS OF KNOWLEDGE MAPS

This section covers in detail the goals of knowledge maps. They are illustrated as the stable analysis patterns (see Chapter 2). Seven goals drive the function and rationale of knowledge maps. Table 3.1 summarizes these goals.

3.5.1 Goal 1: Learning

- *Name.* Learning stable analysis pattern (Fayad and Telu 2005).
- *Context.* Learning stable patterns can be applied to various patterns in various day-to-day fields or applications. For example, in a formal learning experience, a student has a set of syllabus from which he or she is taught and then tested later on. In the workplace, people often learn through previous experiences. Generally, a person with a graduate degree and five years of experience in a particular field is more knowledgeable about that field than a student who has just graduated in the same field because, while working in the field, the experienced individual has learned enough through the process of encountering and conquering challenges presented in the workplace.

  We can easily see a sense of learning occurring here as a routine activity in nonhumans too. Birds always learn by observation and experience. A baby bird starts learning by watching its mother or father fly and trying to fly small distances by hopping from one tree to another. As it grows, the bird begins fly better until it masters the talent.
- *Problem.* How to create a conceptual model for the learning concept that is general enough to be applicable to any domain, which incidentally includes users' requirements that will apply to all possible users of learning applications.

TABLE 3.1
Goals of Knowledge Maps

Goal	Description	Provided?
Learning	In this domain, learning is the cognitive process of acquiring skills or knowledge about a specific discipline.	Yes
Discovery	It is defined as the process or act of discovering something or somebody unexpectedly or after research.	Yes
Knowledge	In this domain, it portrays the distinct aspects (goals and capabilities) of the knowledge maps.	Yes
Abstraction	A view of a problem that extracts the essential information relevant to a particular purpose and it ignores the remainder of the information.	No
Visualization	In this domain, it visualizes the existence of certain goals, capabilities, and transient aspects in the knowledge maps, as well as their relationship with each other.	No
Synthesis	It is the ability to create new knowledge out of pertinent aspects of a particular domain. This includes the association of direct knowledge and remote knowledge.	No
Leveraging	This refers to the reuse of source and/or target patterns from one set of core sets in remote knowledge maps and/or domain-specific applications.	No

- *Solution and participants.*
  - *Solution.* See Figure 3.3.
  - *Participants*
    - *Classes*
      - *Learning.* It represents the learning of any party. This class consists of behaviors and attributes that control the learning process.

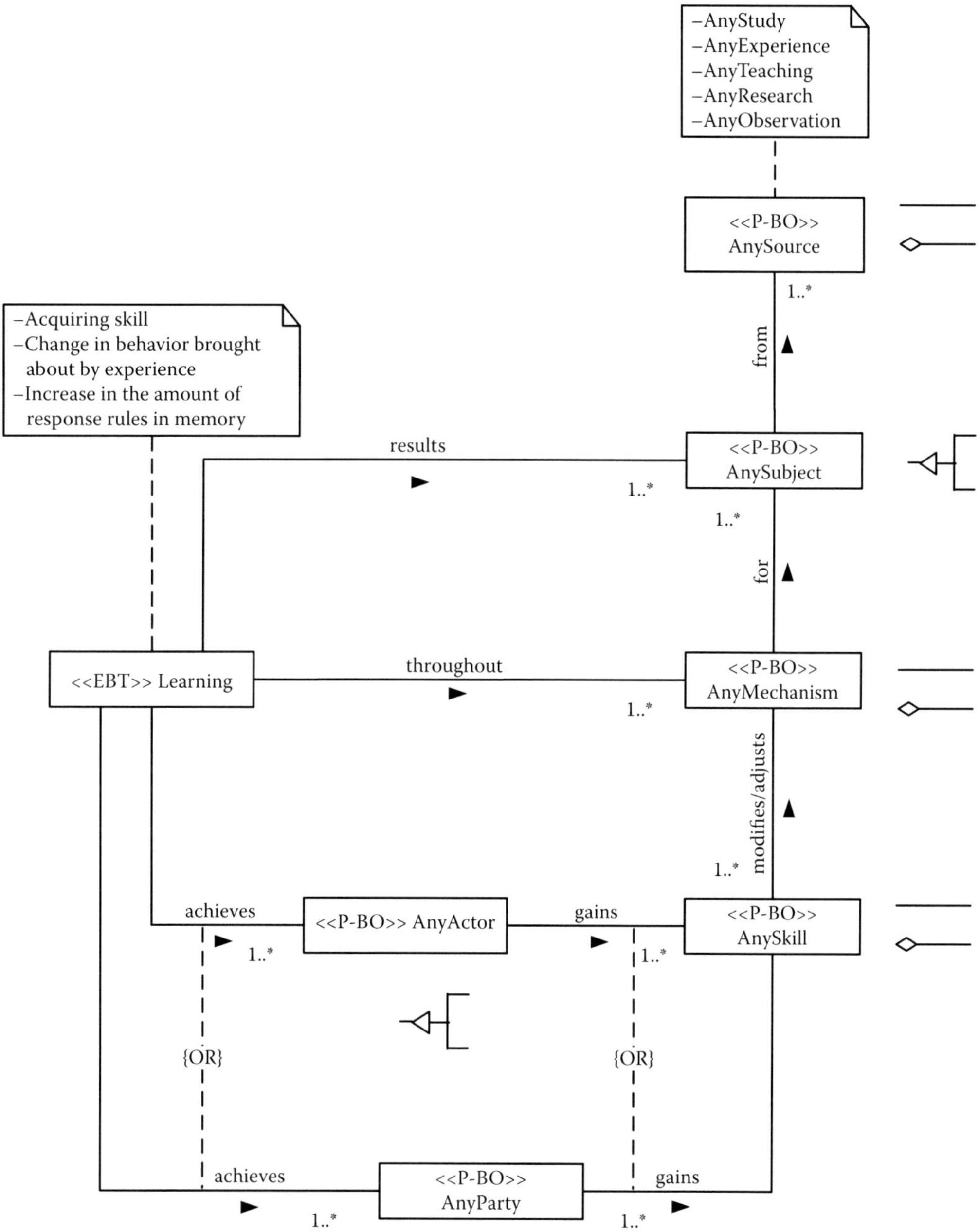

FIGURE 3.3 The learning stable analysis pattern (culled from Fayad's pattern archive).

- *Patterns*
  - *AnyActor.* It represents any actor, who gains any kind of skill or achieves learning through various means.
  - *AnySkill.* It represents the skill that is gained by the actor, by adopting the learning process.
  - *AnyLearningProcess.* This is the methodology an actor adopts to gather a special skill. This can be experience, trial and error, formal education, or just watching.
  - *AnySubject.* It represents the subject or the topic of learning.
  - *AnySource.* It represents the source from which this learning process takes place. This can be a book, the Internet, a paper, or a journal. This is optional. In some cases, such as group discussion, this might not actually exist, unless the group discussion is about a book or a journal.

3.5.2 Goal 2: Discovery

- *Name.* Discovery stable analysis pattern (Khadpe 2005).
- *Context.* This pattern can be used in several applications and scenarios, where a discovery concept is used in the system. It can be used for the discovery of facts, patterns of any artifacts or discovery of anything in a multitude of domains.
- *Problem.* Discovery is an enduring concept, whose application can range from the discovery of the universe to the discovery of a mathematical formula, to the discovery of patterns in data. It asks how to make the discovery analysis pattern general enough to be applicable to any domain?
- *Solution and participants*
  - *Solution.* See Figure 3.4.
  - *Participants*
    - *Classes*
      - *Discovery.* It describes the discovery process.
    - *Patterns*
      - *AnyDiscoveryType.* It represents the different types of discoveries in different application areas or domains.
      - *AnyDiscoveryMechanism.* It represents the BO, which deals with different kinds of discovery mechanisms.
      - *AnyDiscovery.* It represents the BO, which represents the desired discovery.
      - *AnyActor.* It represents a person or a group of people who interact or a scientific group responsible for the discovery process.
      - *AnyEvidence.* It represents the proof of the discovery.

3.5.3 Goal 3: Knowledge

- *Name.* Knowledge stable analysis pattern (Fayad and Telu 2005).
- *Context.* Knowledge can be gained through experience or conducting studies. It represents a collection of facts, rules, tips, or lessons learned with respect to anything that must be synthesized to create knowledge. This pattern will be used to represent knowledge synthesis and acquisition.

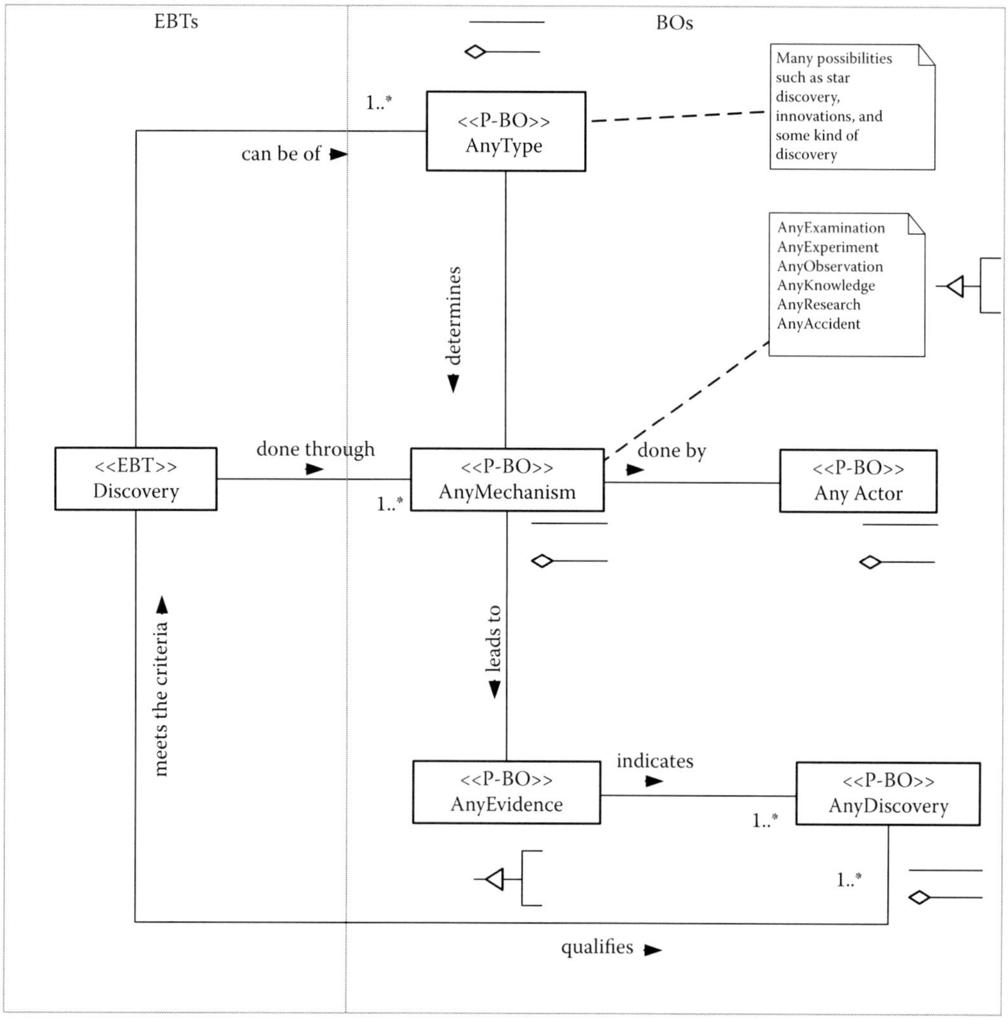

FIGURE 3.4 The discovery stable analysis pattern (culled from Fayad's patterns archive).

- *Problem.* How do we build an effective model that encloses the common core knowledge of knowledge?
- *Solution and participants*
  - *Solution.* See Figure 3.5.
  - *Participants*
    - *Classes*
      - *Knowledge.* It describes the synthesized knowledge.
    - *Patterns*
      - *AnySkill.* It represents different types of abilities that an actor can gain via any mechanism.
      - *AnyMechanism.* It represents the BO, which deals with different kinds of discovery mechanisms.
      - *AnySubject.* It represents the class, where all facts and other information pertinent to a domain are located for future reference.

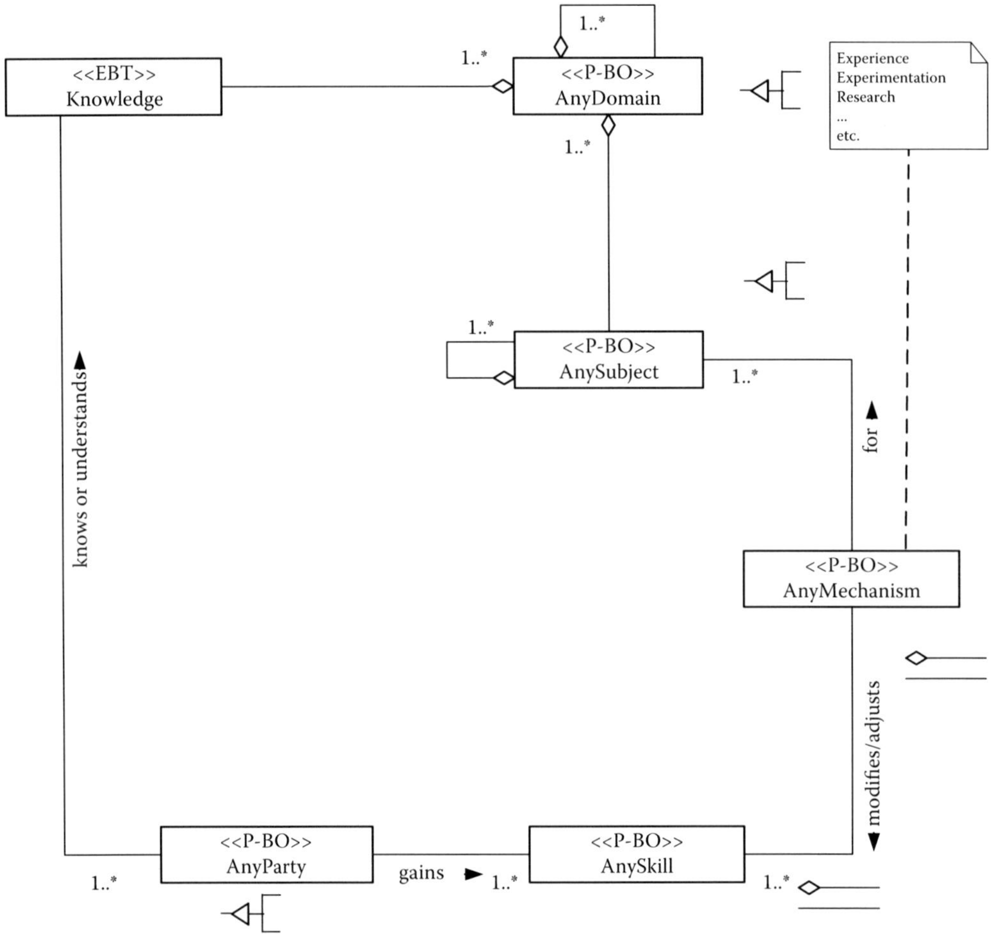

FIGURE 3.5 The knowledge stable analysis pattern (culled from Fayad's patterns archive).

- *AnyDomain*. It represents the environment that will be explored in search of rules, constraints, and knowledge pertinently related to one or more subjects.
- *AnyActor*. It represents a person or a group of people who is/are or will be gaining knowledge.

3.6 SUBGOALS

3.6.1 Use an Analogy: Marriage or Friendship

A good and feasible way to describe subgoals is through forwarding a number of meaningful analogies. Having written that, we will use the concept of friendship throughout this section to explain what subgoals are in the context of knowledge maps. Subgoals can be seen as the wish list elements that come along with the utilization of a determined goal. In other words, imagine the goal *friendship*. You will always expect from the word *friendship* several other aspects to be achieved along with it, such as trust, companionship, assistance, forgiveness, and love. As we discussed before, subgoals are the wish list elements of a determined goal. It does not mean that by achieving *friendship* we are guaranteeing the achievement of other aspects associated with it. We can have *friendship* without any or all of the aspects involved in the process. Because of this, you must be very careful when including subgoals as part of any solution formed by specific goals, because they accompany different rules, axioms, and/or problems that would undermine your original solution.

The inclusion of subgoals to your original solutions has its advantages and disadvantages. Some of the advantages are as follows:

1. It will allow the representation of alternative ways of satisfying a particular goal.
2. It will also represent successful connections or trajectories toward the achievement of goals.
3. It will increment the criteria that feed available verification and validation methods.

Some of the disadvantages are as follows:

1. It will increase the level of complexity and difficulty of a solution, because there are no reasonable limits to the number of alternatives to satisfy a particular goal.
2. It will also increase the number of rules and conditions, which influence the accuracy of solution derived that come along with the inclusion of more subgoals.

As you can read, the use of subgoals can bring a lot benefits to your proposed set of solutions. However, if their use is not properly addressed during the design of your solution, they might result in additional problems than presumed benefits. Having written that, we always favor and prefer nonuse of subgoals as part of your solution.

SUMMARY

The main objective and goal of this chapter was to provide a set of heuristics to ease the process of identifying and assessing the goals of a discipline. The second objective was to state and define the basic understanding of what a goal really is and why we need goals in the analysis phase of any problem. The third objective was to provide some of the goals that drive the formation of knowledge maps. These goals were provided by using a short-pattern documentation template. We have implemented these objectives by using the software stability concept as the main approach for identifying goals. By using this unique approach, we also provided the behind-the-scenes knowledge about goals usage, and in what manner they make a fundamental part of a complete analysis of the problem or subject of interest.

We have included a very short template for documenting stable analysis patterns for the sake of simplicity. By doing so, the reader will cover the provided goals description in a short time.

OPEN RESEARCH ISSUES

The following are some of the open research issues that need to be examined and require future work and experimentation:

1. *Using stability model as a method for goal elicitation.* It is a process of obtaining goals from any source, such as human and literature sources. This can involve the use of reading, researching, interviews, observation, and protocol analysis.
2. Utilize the concurrent software development model or knowledge map methodology as an effective way for requirements engineering of any application or domain, instead of using goal-oriented requirements engineering approaches, such as NFR Frameworks, KAOS, and GBRAM (see Sidebar 3.1).

REVIEW QUESTIONS

1. What is a goal?
2. What are the types of goals?
3. Which type is concerning software development?
4. Which ones of the following concepts are goals and why?
   a. Project
   b. Account
   c. Ownership
   d. Agreement
   e. Range
   f. Measurement
   g. Evaluation
   h. Performance
   i. Trust
   j. Love
   k. Anger
   l. Revenge
   m. Pleasure
   n. Acknowledgment
   o. Acceptance
   p. Warning
   q. System
   r. Entity
   s. Log
5. Describe what goals are from two different perspectives—traditional and stability.
6. What other terms are used interchangeably with the term *goal* in the book?
7. List the properties of goals that can be applied to the model.
8. State why goals are so important.
9. Goals are classified into ______________, ______________, and ______________.
10. Describe what business goals are.
11. Define project goals.
12. "Personal goals are the most delicate of all the goals that one needs to address during the projects lifecycle." Justify this statement.
13. Is the following statement true or false? Goal extraction can be carried out in parallel with the capabilities identification process.
14. Enlist the steps carried out for identifying goals.
15. What are subgoals? How are they different from goals?
16. "Adding subgoals makes the problem more complex." Do you agree with this statement? Support your stance with appropriate arguments.
17. Summarize the goals that drive the knowledge map's rationale.
18. ____________________ is the wish list elements of a determined goal.
19. What are the pros and cons of using subgoals to a solution?
20. What the following terms stands for:
    a. NFR
    b. GORE

   c. KAOS
   d. GBRAM
21. Define GORE.
22. Name the GORE approaches.
23. What are the benefits of using GORE?
24. What are the differences between GORE and software stability model as a way for de-engineering requirements?

EXERCISES

1. Model or create a class diagram of each of the goals of the following concepts (not all goals) by using software stability model, resulting stable analysis pattern for EBT, and stable design pattern for BO.
   a. No documentation
   b. No meta models
   c. No implementations
      For each goal do the following:
      i. Create three scenarios (usage) for each of the following concepts.
      ii. Extract common properties.
      iii. Create the pattern based on software stability/concurrent-oriented software development model in Chapter 1.
         A. Activity
         B. Diagram
         C. Decision
         D. Action
         E. Friendship
         F. Health
         G. Condition
         H. Employment
         I. Swimlane
         J. Object
         K. Acting
         L. Negotiation
         M. Trade
         N. Event
         O. Transition
         P. Trust
         Q. Workflow
         R. Splitting
         S. Merging
         T. StartingPoint
         U. EndingPoint
         V. Concurrency
         W. Constraint
         X. Synchronization

Y. Recording
Z. Traceability

2. Each movie has one or more specific and ultimate themes (goals). Name the ultimate goal(s) of the following classic movies:
   a. *Lagaan*
   b. *The Lord of the Rings* trilogy (2001–2003)
   c. *Titanic* (1997)
   d. *Toy Story* (1995)
   e. *The Silence of the Lambs* (1991)
   f. *Crumb* (1995)
   g. *The Lion King* (1994)
   h. *Shrek* (2001)
   i. *The Breakfast Club* (1985)
   j. *Speed* (1994)
   k. *Scarface* (1983)
   l. *Fatal Attraction* (1987)
   m. *Ghostbusters* (1984)
   n. *Dirty Dancing* (1987)
   o. *Back to the Future* (1985)
3. Each book has one or more specific and ultimate themes (goals). Name the ultimate goal(s) of the following classic books:
   a. *The Road*, Cormack McCarthy (2006)
   b. *Harry Potter and the Goblet of Fire*, J.K. Rowling (2000)
   c. *Beloved*, Toni Morrison (1987)
   d. *The Liars' Club*, Mary Karr (1995)
   e. *American Pastoral*, Philip Roth (1997)
   f. *Mystic River*, Dennis Lehane (2001)
   g. *Cold Mountain*, Charles Frazier (1997)
   h. *Watchmen*, Alan Moore and Dave Gibbons (1986–1987)
   i. *Black Water*, Joyce Carol Oates (1992)
4. Each TV show has one or more specific and ultimate themes (goals). Name the ultimate goal(s) of the following classic TV shows:
   a. *The Simpsons*, Fox (1989–present)
   b. *The Sopranos*, HBO (1999–2007)
   c. *Seinfeld*, NBC (1989–98)
   e. *The X-Files*, Fox (1993–2002)
   f. *Sex and the City*, HBO (1998–2004)
   g. *Survivor*, CBS (2000–present)
   h. *The Cosby Show*, NBC (1984–92)
   i. *Friends*, NBC (1994–2004)
   j. *The Oprah Winfrey Show*, Syndicated (1986–present)
   k. *American Idol*, Fox (2002–present)
   l. *Beverly Hills, 90210*, Fox (1990–2000)
   m. *Star Trek: The Next Generation*, Syndicated (1987–94)
   n. *Miami Vice*, NBC (1984–89)
   o. *L.A. Law*, NBC (1986–94)
   p. *Moonlighting*, ABC (1985–89)

q. *Planet Earth*, Discovery Channel (2007)
r. *The Golden Girls*, NBC (1985–92)
s. *Prime Suspect*, ITV (1991–2006)

PROJECTS

1. Identify and model two to four ultimate goals and connect them together for the following domains:
   a. Manufacturing
   b. Modeling
   c. Requirement analysis
   d. Customer relationship management
   e. Database
   f. Project
   g. Kitchen
2. Identify and model the class diagrams using two to three ultimate goals for each of the sample requirements in Appendix D.

SIDEBAR 3.1 Goal-Oriented Requirements Engineering

Goal-oriented requirements engineering (GORE) regards any system as a collection of active components (*agents*). Agents may restrict their behavior to ensure the constraints that they are assigned (Lapouchnian 2005). In GORE, agents are assigned responsibilities for achieving goals (Lapouchnian 2005). A *requirement* is a goal, whose achievement is the responsibility of a single software agent (Lapouchnian 2005). Agent-based reasoning is central to requirements engineering, because the assignment of responsibilities for goals and constraints among agents in the software-to-be and in the environment is the main outcome of the RE process (van Lamsweerde 2000). There are a number of important benefits associated with explicit modeling, refinement, and analysis of goals (mostly adapted from van Lamsweerde, 2001) such as the following:

- Goals provide a precise criterion for *sufficient completeness* of a requirement specification.
- Goals provide a precise criterion for requirements *pertinence*. A requirement is pertinent with respect to a set of goals in the domain, if its specification is used in the proof of at least one goal at Yue (1987).
- A goal refinement tree provides *traceability* links from high-level strategic objectives to low-level technical requirements (Lapouchnian 2005).
- Goal modeling provides a natural mechanism for structuring complex requirements documents (van Lamsweerde 2001).
- Goals can be used to provide the basis for the detection and management of conflicts among requirements (Robinson 1989; van Lamsweerde 1996).
- A single goal model can capture variability in the problem domain using alternative goal refinements and alternative assignment of responsibilities.
- Quantitative and qualitative analysis of these alternatives is possible (Lapouchnian 2005).
- Goal models provide an excellent way to communicate requirements to customers.
- Goal refinements offer the right level of abstraction to involve decision makers for validating choices being made among alternatives and for suggesting other alternatives (Lapouchnian 2005).

The main approaches of GORE are as follows:

- *The NFR framework (nonfunctional requirements).* NFR framework provides a *process-oriented* approach for dealing with nonfunctional requirements (Chung et al. 2000; Mylopoulos, Chung, and Nixon 1992).
- *i*/Tropos. i** (Yu 1997) is an agent-oriented modeling framework that can be used for requirements engineering, business process reengineering, organizational impact analysis, and software process modeling. The *i** modeling framework is the basis for *Tropos*, a requirements-driven agent-oriented development methodology (Castro, Kolp, and Mylopoulos 2000). The Tropos methodology guides the development of agent-based systems from the early requirements analysis through architectural design and detailed design to the implementation. Tropos uses the *i** modeling framework to represent and reason about requirements and system configuration choices. Tropos has an associated formal specification language called *Formal Tropos* (Fuxman et al. 2001) for adding constraints, invariants, pre- and post-conditions capturing more of the subject domain's semantics to the graphical models in the *i** notation. These models can be validated by model checking.
- *KAOS (Knowledge Acquisition in autOmated Specification)* (Dardenne, van Lamsweerde, and Fickas 1993) *or keep all objects satisfied* (van Lamsweerde and Letier 2003). A KAOS specification is a collection of the following core models:
  - *Goal model* is where goals are represented and assigned to agents.
  - *Object model* is a UML model that can be derived from formal specifications of goals because it refers to objects or their properties.
  - *Operation model* defines various services to be provided by software agents.
- *GBRAM (goal-based requirements analysis method).* The emphasis of GBRAM (Anton 1996, 1997) is on the initial identification and abstraction of goals from various sources of information. It assumes that no goals have been documented or elicited from stakeholders and thus can use existing diagrams, textual statements, interview transcripts, and so on. GBRAM involves the following activities: *goal analysis* and *goal refinement.*

REFERENCES

Anton, A. "Goal-Based Requirements Analysis." Paper presented at the Proceedings of Second IEEE International Conference on Requirements Engineering, Colorado Springs, CO, April 1996.

Anton, A. "Goal Identification and Refinement in the Specification of Software-Based Information Systems." PhD Thesis, Georgia Institute of Technology, Atlanta, GA, 1997.

Castro, J., M. Kolp, and J. Mylopoulos. "Towards Requirements-Driven Information Systems Engineering: The Tropos Project." *Information Systems* 27, no. 6 (2000): 365–89.

Chung, L., B. Nixon, E. Yu, and J. Mylopoulos. *Non-Functional Requirements in Software Engineering*. Kluwer Academic Publishing, 2000.

Dardenne, A., A. van Lamsweerde, and S. Fickas. "Goal-Directed Requirements Acquisition." *Science of Computer Programming* 20, nos. 1/2 (1993): 3–50.

Fuxman, A., M. Pistore, J. Mylopoulos, and P. Traverso. "Model Checking Early Requirements Specifications in Tropos." Paper presented at the Proceedings of 5th International Symposium on Requirements Engineering, Toronto, Canada, August 2001.

Lapouchnian, A. "Goal-Oriented Requirements Engineering: An Overview of the Current Research Department of Computer Science." University of Toronto, White Paper, June 28, 2005.

Mylopoulos, J., L. Chung, and B. Nixon. "Representing and Using Non-Functional Requirements: A Process-Oriented Approach." *IEEE Transactions on Software Engineering*, Special Issue on Knowledge Representation and Reasoning in Software Development 18, no. 6 (1992): 483–97.

Robinson, W. "Integrating Multiple Specifications Using Domain Goals." Paper presented at the Proceedings of 5th International Workshop on Software Specification and Design, Pittsburgh, PA, May 1989.

van Lamsweerde A. "Divergent Views in Goal-Driven Requirements Engineering." Paper presented at the Proceedings of Workshop on Viewpoints in Software Development, San Francisco, CA, October 1996.

van Lamsweerde, A. "Requirements Engineering in the Year 00: A Research Perspective." Paper presented at the 22nd International Conference on Software Engineering (ICSE'2000), Limerick, Ireland, June 2000.

van Lamsweerde, A. "Goal-Oriented Requirements Engineering: A Guided Tour." Paper presented at the Proceedings of 5th IEEE International Symposium on Requirements Engineering, Toronto, Canada, August 2001.

van Lamsweerde, A., and E. Letier. "From Object Orientation to Goal Orientation: A Paradigm Shift for Requirements Engineering." Paper presented at the Proceeding of Radical Innovations of Software and Systems Engineering, Post-Workshop Proceedings of the Monterey'02 Workshop, Venice, Italy, Springer-Verlag, LNCS 2003.

Yu, E. "Towards Modeling and Reasoning Support for Early-Phase Requirements Engineering." Paper presented at the Proceedings of 3rd International Symposium on Requirements Engineering, Washington, DC, January 1997.

Yue, K. "What Does It Mean to Say that a Specification is Complete?" Paper presented at the Proceedings of Fourth International Workshop on Software Specification and Design, Monterey, CA, 1987.

SIDEBAR 3.2 Goal Programming

According to Johnson and Trick (1996), *goal programming* is a fanciful or exotic nomenclature for a very simple and straightforward concept: the thin fine line between stated objectives and listed constraints is never completely crystallized. One needs to synthesize this slender difference to arrive at a proper goal programming. Specially, when a number of objectives surround the problem, it is usually a feasible idea to consider some or all of them as real constraints instead of stated objectives.

Goal programming is thus very simple, basic, and flexible: change, alter, modify, or bring some objectives into feasible constraints, by adding or introducing slack, extra, and/or surplus number of variables to represent a departure from a goal. Charnes, Cooper, and Ferguson (1955) first used goal programming in 1955, although the actual name first appeared in a 1961 treatise by Charnes and Cooper (1961). Seminal works by Lee (1972), Ignizio (1976), Ignizio and Cavalier (1994), and Romero (1991) are also followed by research personnel world over. The first real engineering application of goal programming, due to Ignizio in 1962, was the design and placement of the antennas employed on the second stage of the *Saturn V*. This was employed to launch the Apollo space capsule that landed the first men on the moon.

REFERENCES

Charnes, A., and W. W. Cooper. *Management Models and Industrial Applications of Linear Programming*. New York, NY: Wiley, 1961.

Charnes, A., W. W. Cooper, and R. Ferguson. "Optimal Estimation of Executive Compensation by Linear Programming." *Management Science*, 1 (1955): 138–51.

Ignizio, J. P. *Goal Programming and Extensions*. Lexington, MA: Lexington Books, 1976.

Ignizio, J. P., and Cavalier, T. M. *Linear Programming*. Upper Saddle River, NJ: Prentice Hall International Series in Industrial and Systems Engineering, 1994.

Johnson, D. S., and Trick, M. A. Cliques, coloring, and satisfiability: Second DIMACS implementation challenge, *DIMACS Series in Discrete Mathematics and Theoretical Computer Science*, American Mathematical Society 26, 1996.

Lee, S. M. *Goal Programming for Decision Analysis*. Philadelphia, PA: Auerbach, 1972.

Romero, C. *Handbook of Critical Issues in Goal Programming*. Oxford: Pergamon Press, 1991.

SIDEBAR 3.3 Goal-Oriented Development (*Supratik Mukhopadhyay*)

The notion or idea of goals was originally derived from the artificial intelligence literature (Russell and Norvig 2004). In a goal-oriented development method, goals are stated as intentional specifications that later, during the software development process, are reified by getting associated with extensional specification (c.f. remote knowledge) (van Lamsweerde 2001). In a multistakeholder software development environment, each agent states its goals that might be competing or contradicting. It is the duty of the requirements engineering process to weed out these contradictions. A goal-oriented methodology can be used to reduce the complexity of the software development process. In software stability model, goals are simply the enduring themes (Fayad 2002a, 2002b) that justify why a software solution, area of study, and so on is needed in a determined environment (e.g., organization and software project).

Goals can be used to identify a number of aspects in a software project. (The goal of consistency is a correctness aspect of a bank transaction system.) They can be stated (in natural language or formally) or can be discovered and abstracted out, during the requirements analysis phase, manually or through a semiautomated specification mining process. Formalisms for stating goals in a formal manner are usually some versions of temporal and modal logics (Blackburn, de Rijke, and Venema 2002). Goals can refer to either system-wide objectives (e.g., high throughput) or low-level aims describing the requirements of a particular module. They can arise from both functional and nonfunctional requirements of a system (e.g., real-time constraints). A goal can be refined to one or more subgoals.

A goal-oriented software project involves traversing through a goal tree, an And-Or tree, whose root is associated with the system-wide goals. The goal corresponding to each node is refined by simpler subgoals corresponding to the children of the node. Refining a goal should take into account constraints imposed on it by other goals. The leaves of the tree correspond to atomic goals that can be implemented as *services* in such a way that the top-level goals are *satisfied*. Each level of the tree corresponds to complex *services* and meeting goals associated with the nodes of that level those are built by combining *services* corresponding to the nodes at the next level. Goals can aid and assist in verification and validation. They can either serve as specifications in a formal verification process or test cases can be derived out of them (von Mayrhauser, Scheetz, and Dahlman 1999).

A goal-oriented software development methodology easily integrates with other existing methodologies and software architectures. For example, the system model can describe the goals in a model-driven development environment. A service-oriented architecture (Singh and Huhns 2005) can be viewed as a goal-oriented framework, where the existing services are the leaves of a goal tree and services that are more complex are successively built by composing simpler services. A goal-oriented methodology fits well into an agent-based software development project (Cheong and Winikoff 2005). An agent-oriented architecture is essentially a goal-oriented one where each agent implements a system goal. Interactions between agents can also be formulated as goals and can be described either through interaction diagrams or through temporal logic constraints. Goal-oriented development can be integrated into object-oriented development methodology through the process of refinement of subgoals for nonfunctional requirements, goals for functional requirements, and conflict analysis (Mylopoulos, Chung, and Yu 1999).

Classical goal-oriented software development assumes that requirements are available a priori and are frozen at the initial stage of the project. Hence, intentional specification about goals can be expressed through monotonic predicates. This creates problems in real-life software projects where requirements, and hence goals, continually change during the lifecycle of the project. For projects with dynamically changing requirements, goals can be described and refined formally using nonmonotonic knowledge representation schemes (Makinson 2005).

REFERENCES

Blackburn, P., M. de Rijke, and Y. Venema. *A Course in Modal Logic*. Cambridge, UK: Cambridge University Press, 2002.

Cheong, C., and M. Winikoff. "Hermes: A Methodology for Goal-Oriented Agent Interactions." Paper presented at the Proceedings of Fourth International Joint Conference on AAMAS, Utrecht, the Netherlands, 2005.

Fayad, M. E. "Accomplishing Software Stability." *Communications of the ACM* 45, no. 1 (2002a): 111–15.

Fayad, M. E. "How to Deal with Software Stability." *Communications of the ACM* 45, no. 4 (2002b): 109–12.

Makinson, D. *Bridges from Classical to Nonmonotonic Logic*. Vol. 5. Texts in Computing Series, London: College Publications, 2005.

Mylopoulos, J., L. Chung, and E. Yu. "From Object-Oriented to Goal Oriented Requirements Analysis." *Communications of the ACM* 42, no. 1 (1999): 31–37.

Russell, S., and P. Norvig. *Artificial Intelligence: A Modern Approach*. Upper Saddle River, NJ: Prentice Hall, 2004.

Singh, M. P., and Huhns, M. N. *Service-Oriented Computing: Semantics, Processes, Agents*. New York, NY: Wiley, 2005.

van Lamsweerde, A. "Goal-Oriented Requirements Engineering: A Guided Tour." Paper presented at the Proceedings of 5th IEEE International Symposium on Requirements Engineering, Toronto, Ontario, Canada, August 2001.

von Mayrhauser, A., M. Scheetz, and E. Dahlman. "Generating Goal-Oriented Test Cases." Paper presented at the Proceedings of the 23rd Annual International Computer Software and Applications Conference, October 27–29, Phoenix, AZ, 1999.

4 Discovery Stable Analysis Pattern

All truths are easy to understand once they are discovered; the point is to discover them.

David Whitehouse
Renaissance Genius: Galileo Galilei & His Legacy to Modern Science, 2009

Discovery is defined as the act of finding or discovering something. It could be a disease, a drug or hidden patterns, and so on. It can be applied in areas as diverse as data mining, space exploration, forensics investigation, and humans' growth and development. For example, in data mining, we are dealing with the discovery of hidden patterns and knowledge from the widely available set of data. In space exploration, we are dealing with the discovery or anything that leads humans to have a better understanding of what is beyond our planet. In forensics investigation, the goal is to present a set of findings to a judge, jury, or opposition to help defend or blame a suspect. In human's growth and development, discovery is seen, when children start discovering and understanding the environment that surrounds them. Every single piece of the environment, such as events, noises, and other humans, is a new discovery for them. As you could see, there is nothing different in the way discovery is handled in any of those areas of application. In fact, discovery is the same in all of them. Therefore, the reason for the analysis of this concept, with the sole purpose of extracting its core knowledge, is a worthwhile reason, especially if you are planning to reuse it in numerous applications, while still maintaining a cost-effective nature.

4.1 INTRODUCTION

Discovery is the elaborate process of finding information or inventing something. This information can range from associations, trends, hidden patterns, or any meaningful knowledge. Discovery in the data mining scenario, for example, is called as *knowledge discovery*, which is the process of finding meaningful patterns in a set of data that explain past events in such a way that one can use the patterns to help predict future events (see Herbert 1970).

Discovery is a stable analysis pattern as well. Analysis patterns are conceptual models that model the core knowledge of the problem (Fayad and Wu 2002; Hamza and Fayad 2002, 2006). Stable analysis pattern is one of the building blocks of constructing a stable pattern language, which we call a *knowledge map*. A nonformal language is composed of stable analysis and stable design patterns. The knowledge map and the software stability model (SSM) go hand in hand, and in tandem, to develop stable architectures and stable frameworks. Stable architectures are simply a collection of two or more stable analysis patterns and numerous stable design patterns (usually between 4 and 5). Stable frameworks can be seen as the skeleton upon which stable architectures, as well as the application-specific aspects (industrial objects [IOs]), are integrated for a given software solution. SSM is a uniquely layered approach for developing software systems (Fayad and Wu 2002; Hamza and Fayad 2002, 2006). The layers of SSM comprise of enduring business themes (EBTs) which are the classes that present the enduring and core concepts of the underlying industry or business. In this case, discovery is an EBT, as it deals with the core knowledge of the system. The second layer, business object (BO), is similar to the design patterns of stable pattern language. BOs are the workhorses that map the EBT to concrete objects. The notion about EBTs and BOs will be more comprehensible in the next few sections.

Currently, there are a number of existing discovery patterns (see Mobasher 1996; Sain and Tamrakar 2012); however, the solutions they portray/portrayed are quite different when compared to our solutions. The existing patterns usually concentrate on one particular algorithm or technique at a time, such as data mining, psychology, information theory, or algorithms used for finding maximal forward references and large reference sequences. Focusing on one technique at a time, as well as on the constraints that stem from this technique, makes these patterns rigid and stiff, and hard to reuse, especially when each solution handles discovery in a unique way. Our solution, however, originates from ultimate stability in mind. We intend it to represent the core knowledge of discovery, regardless of its context of applicability. In other words, we will build our patterns in terms of conceptual aspects rather than detail-specific ones, which are highly coupled to volatile requirements. The rest of the chapter introduces the novel discovery analysis pattern.

4.2 DISCOVERY STABLE ANALYSIS PATTERN

4.2.1 Pattern Name: Discovery Stable Analysis Pattern

The term *discovery* represents an act of finding or discovering something. It may be a productive insight causing a breakthrough in some domains or a compulsory revelation of facts. Discovery is an interesting, elementary, and complex concept. It is an interesting concept that has always been part of human activities or events like life evolution, science creation and management (i.e., math and physics), and human growth. The basic fact that we are constantly exposed to its intricacies makes it elementary to our sensorial system. However, as elementary as it appears to our senses, it is a very complex concept, especially when we want to analyze and extract its core knowledge. The proof of this is the fact that each discovery solution (see Mobasher 1996; Sain and Tamrakar 2012) is unique and has few elements that overlap each other. Discovery is an enduring aspect and is always conceptual; discovery stable patterns always encapsulate the discovery process.

4.2.2 Known As

Discovery, also called *innovation* or *invention*, has different meanings in different scenarios (see Herbert 1970); discovery in the data mining scenario is knowledge discovery, but discovery of the number *0* by Aryabhatta is an invention. In other words, the connotation of this concept will depend on the area where it is applied and whether the evidence or factual event has already existed by its own means (i.e., nature, life evolution). Regardless of the suggested meaning, the structure of this concept will remain the same.

Sometimes, *breakthrough* is used interchangeably with *discovery* depending on the context of usage. A discovery is called a breakthrough only when the finding/learning so achieved paves way for future success by removing the barrier to progress. However, not all discoveries result in significant advancements. Thus, there is a subtle difference between discovery and breakthrough, and hence, you cannot use the core knowledge of discovery as you do it from a breakthrough pattern.

Discovery is the act of observing, reasoning, and analyzing aspects that are unknown to humans. Discovery may use preexisting knowledge or fact, but it still has certain amount of creativity associated with it. Creativity is the ability to present new or existing ideas and one's imagination in a very concise way, so that everybody can assimilate or absorb it very easily. However, discovery requires much more than just creativity; it needs knowledge and power to think beyond the obvious and to discover things that are yet unknown. This means creativity is an essential ingredient for discovery, and discovery may be incomplete without creativity. Nevertheless, you cannot replace discovery with creativity to represent the same idea.

Sometimes, the usage of word *discovery* is mixed up with the usage of *revelation*. Revelation is the act of uncovering information about a certain entity. Discovery definitely involves disclosing the findings from a particular discovery. However, this act of disclosing is not as important during a discovery as is the actual discovery. Revelation is just a small part during discovery, so that everyone is aware of the finding. Thus, it does not make any sense to reuse the core knowledge of discovery for representing revelation.

4.2.3 Context

One can apply the principles of discovery in areas as diverse as data mining, space exploration, forensics investigation, and humans' growth and development. For example, in data mining, we are dealing with the discovery of hidden patterns and knowledge from the data that are available widely. In space exploration, we are dealing with the discovery or anything that leads humans to a better understanding of what is beyond our planet. In forensics investigation, the goal is to present a set of findings to a judge, a jury, or an opposition to help defend or blame a suspect. In human growth and development, discovery is seen when children start discovering and understanding the environment that surrounds them. Every single piece of the environment, such as events, noises, and other humans, is a new discovery for them. As you could see, there is nothing different in the way discovery is handled in any of those areas of application. In fact, the concept of discovery is the same in all of them. Therefore, the reason for the analysis of this concept is extracting its core knowledge; it is a worthwhile reason, especially if you are planning to reuse it in numerous applications, while still maintaining a cost-effective nature.

4.2.4 Problem

Discovery is an enduring concept, whose application can range from the discovery of the universe to the discovery of a mathematical formula, to the discovery of patterns in data. However, current solutions strive in providing a stable solution that is applicable at any time, when there is a necessity for discovery, such as the discovery of an event or thing. The struggle, the effort, cost, and time spent to handle forthcoming adaptations (product of changes in requirements) will increase at exorbitant and astronomical levels, which are not accepted or handled by any company. Hence, how can we model a stable pattern that is easily adaptable to any kind of discovery application?

The aforementioned problem becomes more acute and serious when the problem's abstractions or the core knowledge of existing solutions depend on application-specific or low-level details, rather than the opposite, where low-level details depend on abstractions. So, how can we guarantee a solution that represents the core knowledge of discovery without being specific or without following one particular discovery process style? For example, some discoveries are made successful by experimentations, while others happen by sheer accident; some discoveries are made by observations. Discovery requirements include the following.

4.2.4.1 Functional Requirements

Involvement and ownership. The person or the investigator, who indulges in the act of discovery, has to be dedicated and devoted to the assigned task. Discovery is performed by observing and identifying things that a common person cannot. Thus, discovery is a process that demands countless hours of active involvement, and years of perseverance and belief in what one is investigating, where AnyParty has exclusive rights and control over their discovery. AnyParty (owner) can gain, transfer, and lose their ownership of their discovery in a number of ways, such as selling it for money or giving it away for AnyReason. Ownership of discovery is self-propagating in that when AnyParty owns a discovery, any other additional goods produced using those discoveries will also be owned by the parties. Ownership implies responsibility for actions regarding the property of the discovery. A *legal shield* is said to exist if the discovery's properties legal liabilities are not redistributed among the discovery's owners. Ownership allows for parties sharing gains and use of their discovery. Ownership also includes the intellectual property (IP), which refers to a legal entitlement, which sometimes pegs to the expressed form of a discovery. This legal entitlement generally enables its holder (AnyParty) to exercise exclusive rights of use in relation to the discovery of the IP. IP laws are designed to protect different forms of discovery, they include copyright, patent, trademark, trade secret, and so on.

Ownership/party types. The ownership or AnyParty type classifies all the legal parties: individuals, organizations, countries, political parties, and/or a combination of some or all of them.

Source/root. It names the initiators of the idea or provides an accurate reference of who contributed to the idea over the years.

Discovery mechanisms. It describes all the techniques and approaches used to lead to discovery.

Characteristics/properties. Characteristics are feature-prominent attributes or aspect of the discovery. List and utilize all the distinctive properties of the discovery, such as aspects, attributes, services, test cases, and outstanding features.

Domains or fields of knowledge. The discovery belongs to a domain knowledge, which has its own context, terms, laws, rules, and vocabulary, but utilized in many different domains.

Proof. We live in an era that boasts a number of discoveries and inventions. However, each one of these discoveries needs a sufficient amount of *proof* or *evidence* to prove that a particular discovery is indeed a new discovery. Because there is an immense need for *proof*, most of us think that it is a simple affair. Obviously, the path or road to proof is neither easy and simple nor always straightforward. Under normal circumstances, one will need to take a long journey before reaching a valid or right conclusion.

Assessments lead to indicators or evidences. A discovery is never proved or ascertained, unless one can measure or assess it, and there are sufficient quantities of indicators or evidences to prove that the discoverer has actually made the act of discovery. Indicators or evidences could be those buttressing points or tools that can vouch for the act of discovery. Verified and validated metrics/results will need to be made available as formidable evidence of discovery. Failure to do so may result in unrecognized discovery. Thus, providing evidence as a proof for discovery is the most important requirement for any discovery.

Classification. Once an act of discovery has been made, we need to associate each set of proof to one class or a set of predefined classes based on the values of some definite attributes. A discovery that is of no use at all is useless because it does not catch the attention of people, nor does any institution recognize it. Thus, the overall goal is to discover knowledge that is not only useful and beneficial but also interesting to the users.

Impacts. Notable discoveries of the past have had a deep impact on all aspects of human lives, be they personal or professional. For example, the discovery of penicillin served as a cure for various infective diseases and benefited thousands of people. Similarly, accidental discoveries of fire and wheel permanently changed the way people lived. However, because man always craves for more and more knowledge discoveries—accidental or planned—discoveries keep happening every now and then. Thus, discovery is a continuous and ongoing challenge for all of us; its innumerable impacts affect our daily lives with a pronounced effect (from extremely positive to truly negative). Some effects can be far reaching and deep, whereas others may exert a very small impact. However, it is a certainty that those who master the art of discovery will eventually win the race. For the winner, the impact could be monumental, like winning a Nobel prize for the invention or the intangible satisfaction of helping millions of people with the discovery.

Identification. Recently, there has been a large quantum of information and details obtained, indexed, and deposited in various databases and data banks. Identifying, detecting, and interpreting interesting and significant patterns from this rich repository of information has become an essential part in directing advanced software and documentation research. Identifying discovery tools and techniques seems to be the biggest problem and a perennial bottleneck.

Recording. Discovery is worthless or useless without proper and meaningful recording, but people just record what they see, by using their perception. What they usually see in front of them is always decided by what they consider or think to be significant or prominent. Bias is a sure certainty, but one can reduce or eliminate it by researching and probing all concepts and ideas in a proper manner. Thus, a proper recording that results in the essence of the discovery is an important requirement for successful discovery.

Releasing. Announcing or promoting a discovery is a thing of uncertainty and insecurity. It is often very difficult and tedious to advertise a particular incident of discovery, as without proper investigation and assertion, the validity of discovery may be a big question.

4.2.4.2 Nonfunctional Requirements

Accuracy. It is the state of being accurate and the discovery conforms to known rules and facts or recognized standards. In the domains of applied and pure science, engineering, industry, and statistics, the term *accuracy* is defined as the degree and extent of closeness of a measured, estimated, or calculated quantity to its actual, real (true) value. Accuracy is closely related to the degree of precision, also called the *factors of reproducibility* or *repeatability*, the extent to which further or advanced measurements or calculations show or repeat the same or similar results. The term *accuracy* is also defined as the degree to which a given quantity is correct and free from error (see Taylor 1999; Broderson et al. 2010).

Advantageous. It is the quality of the discovery being encouraging or promising of a successful outcome. The quality of a discovery is said to be advantageous when people find the discovery very beneficial and useful in their daily life. A discovery is also highly advantageous when it provides a number of benefits and advantages that could eventually help humankind. The word *advantageous* means furnishing convenience or opportunity, favorable, profitable, useful, beneficial, *an advantageous position*; *an advantageous discovery.*

Advancement. The discovery represents an advancement of an entity and brings improvement or enhancement of an existing entity. A particular discovery will also help the advancement of further development of a given domain of expertise. Advancement: an act of moving forward, development; progress: *the advancement of knowledge.* The act of advancing, or the state of being advanced; progression; improvement; furtherance; promotion to a higher place or dignity; as, the advancement of learning.

Reliability. It is the quality of being dependable or reliable. It is closely connected with the quality of discovery. In its general explanation, reliability is the factor of *consistency* or *repeatability* of discovery parameters. *Reliability* is also the consistency of a set of quality factors used to check the veracity of the announcement of a particular discovery.

Usability. Discovery is easy to use and/or to be utilized. A discovery made or announced should be easily available for daily use. A discovery is said to be usable when its use becomes flexible and practical. *Usability* is also used to denote the ease with which people can employ a particular discovery in order to achieve a particular goal or objective (Norman 2002).

4.2.5 Challenges and Constraints

Knowledge discovery systems (Sidebar 4.1) always face a number of difficult challenges and constraints. In fact, constraints can challenge good pattern designs. Current stable analysis discovery systems are armed with many techniques and processes that can be potentially applied to find solutions to a new problem. However, this system may face a challenge or constraint of selecting or choosing the most appropriate and practically feasible technique(s) for a problem that is pending at hand. This is due to the real domain area that is difficult to perform a deep comparison of all practically applicable techniques.

Because of the generic nature of discovery, one may need to exercise enough caution and care to handle stability patterns. Another important thing that a software developer must focus on is the number of types of discoveries that tend to pose innumerable difficulties and obstacles, especially in devising a practically feasible mechanism.

4.2.5.1 Challenges

Current knowledge discovery systems (Sidebar 4.1) are equipped with a number of techniques that have the ability to relate to a new problem. Nevertheless, an existing system may face a unique task of selecting the most appropriate technique or method, because it is so difficult to make a comparative study of all applicable techniques.

- The discovery pattern plans to model discovery performed in numerous domains or fields that need some sort of discovery process. However, the very definition of discovery needs to be constrained for this pattern, that is, informational perspective.
- Discovery is a generic concept—it can mean any breakthrough by serendipity or a product of persistent research and innovation. These flavors of discovery need to be handled carefully by the stable pattern.
- Discovery has various forms in different domains; for example, discovery of water on Mars is conducted by observation and research and not by experimentations. All these different aspects need to be taken care of while building the model.
- Different parties like a scientist, a child, or a lawyer may handle discovery. Therefore, this pattern must be able to capture all the different roles that a party can take.
- There are different mechanisms used to perform a discovery, such as experimentation, observation, and research. Therefore, this pattern must be able to encapsulate the canonical knowledge and semantics shared among all the different mechanisms for discovery.
- Regardless of the mechanism selected to perform discovery, there must be any type of evidence available that leads to the ultimate proof of the discovery—the feeder of information. The pattern, then, must be able to cope with different types of evidence that will induce a proof for the discovery by any type of mechanism.

4.2.5.2 Constraints

- Discovery has one type or more than one types. Some discovery types go hand in hand with other discovery types to discover a number of artifacts.
- Depending on the type of discovery, there could be one or more discovery mechanisms to handle this aspect. For example, discovery of a medicine requires observation, experimentation, research, and so on.
- An individual or a group of individuals conducts a discovery process. For example, scientists collaborate with doctors, analysts, and researchers to arrive at a solution.
- An individual or a group of individuals may define a special set of criteria for their aim of discovery that influence the discovery mechanisms.
- One or more than one discovery mechanisms may require zero or more than zero things, events throughout zero, or more than zero media to produce a specific discovery.

- The discovery mechanism leads to one piece of evidence or more. Therefore, in order to use this pattern, we have to acknowledge any type of evidence that will lead us to the actual discovery.
- Any evidence supports the actual existence of a thing and event that is to be discovered.
- Evidences qualify the actual discovery or discoveries made during the evaluation of them.
- Because discovery is a crosscutting concern, there must be one sole instance of this concept shared across the entire patterns reminding classes.
- There must be a selection of a particular process before the usage of the discovery aspect.

4.2.6 Solution

The solution to the above problem is demonstrated in the form of a model, followed by the participants. See Figure 4.1.

- *Pattern structure*
  - *Classes*
    - *Discovery*. It describes the discovery process. It is the EBT. It must follow certain rules and regulations, so that no laws are violated while making a discovery. For example, patents must be taken into consideration when claiming discovery.

Discovery has many flavors and it comes in many forms and domains. In the domain of legal studies, the act discovery is an essential process of pretrial litigation proceedings, when both sides demand correct and fitting information, details, legal papers, and documents from each other, in a well-calibrated attempt to discover and seek out legally correct and admissible facts. In a general sense, a number of legal devises and tools make the part of judicial proceedings to discover something, such as depositions, witnesses, arguments, hearings, requests for admissions and document production (http://www.lectlaw.com/def/d058.htm).

Within the realm of the software development process, the act of discovery refers to seeking out bugs and errors, during the developmental or maintenance phase. For discovering errors, the developer needs to search and find out through the heaps of source code by running simultaneous and multiple tests. Discovery can mean a geological discovery, medical discovery, or data mining discovery. Depending on the type of discovery, there could be one or many discovery mechanisms.

These mechanisms could be examinations, experimentations, observations, analysis, research, and so forth. An individual or a group of individuals, or a company, conducts the discovery process. Every discovery requires the evidence to prove itself. The discovery mechanism implemented later leads to the body of evidence that finally asserts or proves the discovery or discoveries. This discovery should meet the criteria for which it was discovered and needs to be qualified to be a valid discovery. Irrespective of the domain in which discovery is carried out, it is always difficult to carry out the process and is a time-consuming and laborious process. The immense difficulty associated with discovery is because the information/knowledge sought by the discoverer is not available in an easily understandable form.

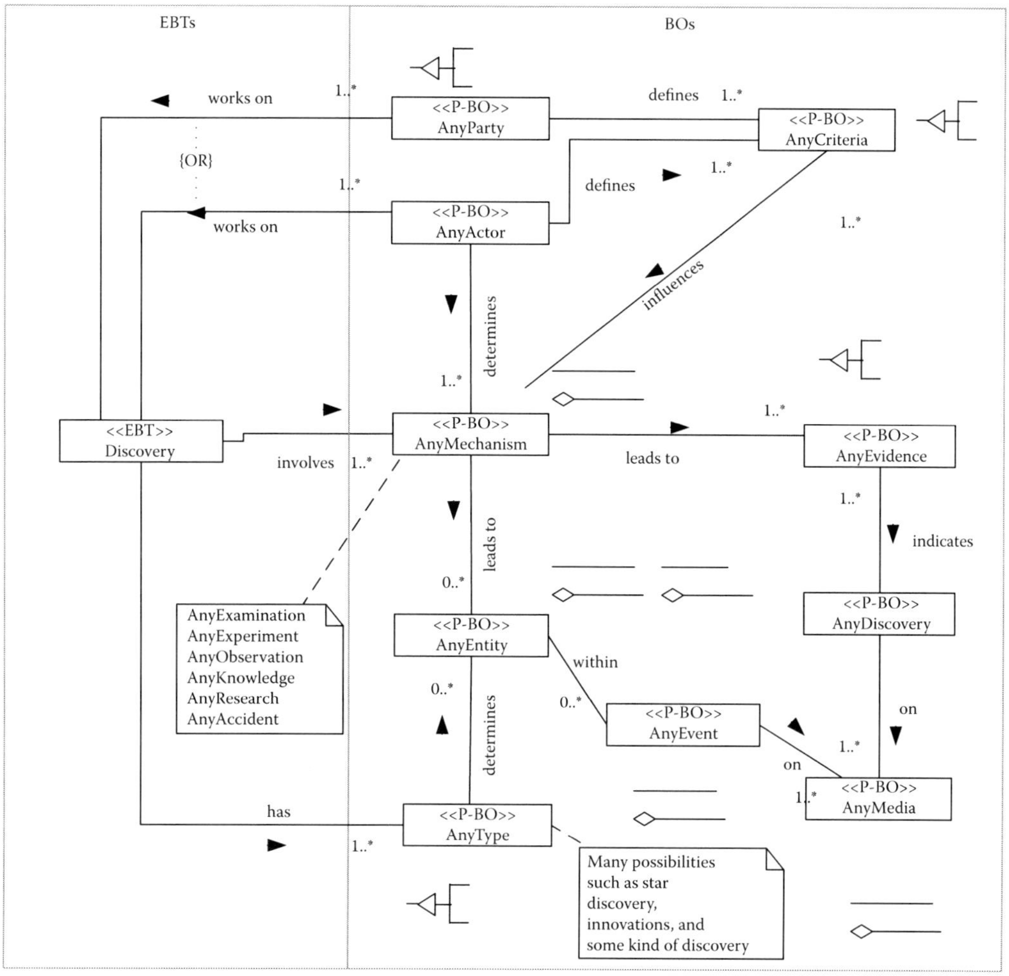

FIGURE 4.1 Discovery stable analysis pattern.

The participants of discovery pattern are classified into two categories as follows:

- *Patterns*
  - *AnyParty.* It represents a person or a group of people or an organization or a group of scientists responsible for the discovery process. AnyParty is involved in the whole process of discovery from the very beginning, and thus AnyParty is solely responsible for any impact the discovery may have. AnyParty stable design pattern is a very common pattern and it is provided in [x, y].
  - *AnyActor.* This class represents any individual, hardware, software, or creatures that utilize or discover the different forms of the discovery. The AnyActor stable design pattern is a very common pattern and it is provided in [x, y].
  - *AnyCriteria.* It represents the criteria specified by AnyActor or AnyParty. AnyCriteria influences AnyMechanism followed to verify the trustworthiness of AnyActor or AnyParty. AnyCriteria stable design pattern is a very common pattern and it is provided in [x, y].

- *AnyMechanism*. It represents the BO, which deals with different kinds of discovery mechanisms. Some discovery mechanisms used to make discovery can be examination, experimentation, observation, knowledge, research, accident, and investigation. The AnyMechanism stable design pattern is a very common pattern and it is provided in [x, y].
- *AnyType*. It represents the different types of discoveries in different application areas or domains. For example, discovery of life on Mars or discovery of medicines for curing cancer could be different discoveries. The AnyType stable design pattern is a very common pattern and it is provided in [x, y].
- *AnyEntity*. It represents an entity used by AnyMechanism for generating AnyDiscovery. Every entity has certain properties and characteristics that can be quantified and verified against any criteria. The AnyEntity stable design pattern is a very common pattern and it is provided in [x, y].
- *AnyEvent*. It is something that takes place or an occurrence. The AnyEvent stable design pattern is a very common pattern and it is provided in [x, y].
- *AnyMedia*. It represents the media through which the discovery will take place. For instance, one can discover certain patterns on the Internet. Others may discover patterns in email messages, by using text mining. The AnyMedia stable design pattern is a very common pattern and it is provided in [x, y].
- *AnyEvidence*. It represents the proof of the discovery. AnyParty may provide evidence in terms of compiled research results or actual test results or the evidence may be photographs. Evidence must be concrete, tangible, and verifiable in order for the discovery to be meaningful. The AnyEvidence stable design pattern is a very common pattern and it is provided in [x, y].
- *AnyDiscovery*. It represents the BO, which represents the desired discovery. It is the actual discovery that is being made or carried out. The AnyDiscovery stable design pattern is a very common pattern and it is provided in [x, y].

- *Class Diagram Description*. The class diagram provides visual illustration of all the classes in the model, along with their relationships with other classes. Description of the class diagram is as follows:
  - AnyParty and/or AnyActor use the discovery process for achieving AnyDiscovery (BO) by defining a zero or more than zero set of AnyCriteria (BO) that influence the selected AnyMechanism (BO).
  - Discovery is the EBT of this model. Discovery (EBT) must have one or more than one AnyType (BO).
  - AnyType (BO) represents all the domains in which one can carry out the discovery (EBT). AnyType (BO) can be represented as discovery of a star or some other innovations like finding another planet that has life.
  - AnyType (BO) plays an important role in determining a zero or more than zero set of AnyEntity (BO).
  - Devising AnyMechanism (BO) specific to the discovery in context involves discovery (EBT) too.
  - AnyCriteria influences this AnyMechanism (BO), which is defined by AnyParty (BO) or AnyActor after thorough analysis. Depending on the domain of discovery, AnyMechanism (BO) varies and can range from simple examination to full-fledged experimentation.

- AnyMechanism (BO) must lead to creation of AnyEvidence (BO), because without demonstrable and confirmable evidence discovery cannot be complete.
- The presence of AnyEvidence (BO) indicates that AnyDiscovery (BO) qualifies for being called discovery (EBT), as it now meets the criteria.
- AnyDiscovery may utilize zero or more of AnyEntity, within zero or more of AnyEvent on AnyMedia.

4.2.6.1 CRC Cards

Discovery (Discovery) (EBT)

Responsibility	Collaboration	
	Client	**Server**
Discovering	AnyType	implement (), qualifyAnyDiscovery()
	AnyMechanism	discover(), selectType(), satisfyConditions()
	AnyDiscovery	provideAdvancement(), innovate(), limits(), requires(), applyTo(), qualify()

Attributes: properties, conditions, qualityGuides, importance, fieldOfDiscovery, applications, requirements, states, limitations

AnyActor(AnyActor) (Pattern-BO)

Responsibility	Collaboration	
	Client	**Server**
To perform	Discovery	agree(), disagree(), participate(), group(),
	AnyCriteria	interact(), associate(), organize(), join(), discover(), monitor(), switchRole(), request(), explore(), playRole(), conduct(), carryout(), analyze(), find out(), initiate(),

Attributes: id, name, type, role, member, affair, activity, category

AnyParty(AnyParty) (Pattern-BO)

Responsibility	Collaboration	
	Client	**Server**
To perform	Discovery, AnyCriteria	participate(), playRole(), interact(), leave(), group(), associate(), organize(), request(), setCriteria(), switchRole(), partake(), join(), monitor(), explore(), receive(), collectData(), integrate(), agree(), disagree()

Attributes: id, partyName, type, role, member, affair, activity, partiesInvolved, id, activity, category (or orientation), purpose

AnyCriteria (AnyCriteria) (Pattern-BO)

Responsibility	Collaboration	
	Client	**Server**
Present a set of requirements and constraints	AnyParty AnyActor AnyMechanism	define(), verify(), apply(), priority(), parse(), exhibit()

Attributes: id, name, condition, property, priority

AnyMechanism (AnyMechanism) (Pattern-BO)

Responsibility	Collaboration	
	Client	**Server**
To implement	Discovery AnyCriteria AnyEvidence	execute(), provideEvidence(), status(), performActions(), activate(), deactivate(), attach(), detached(), pause(), return()

Attributes: context, id, name, status, application, components, description

AnyType (AnyType) (Pattern-BO)

Responsibility	Collaboration	
	Client	**Server**
To classify the types of discoveries	Discovery AnyEntity	determine(), change(), operateOn(), pass(), resume(), label(), classify(), attached(), nameAttributes(), specify()

Attributes: id, name, properties, interfaceList, methodList, clientList

AnyEntity (AnyEntity) (Pattern-BO)

Responsibility	Collaboration	
	Client	**Server**
To be utilized in or as an evidence	AnyType AnyEntity AnyMechanism AnyEvent	performfunction(), status(), type(), update(), new(), relatedTo()

Attributes: id, entityName, entityType, status, position, states, type

AnyEvent (AnyEvent) (Pattern-BO)

Responsibility	Collaboration	
	Client	**Server**
To facilitate	AnyEntity AnyMedia	takeplace(), to be reported(), startDate(), duration(), setActivity(), organize(), facilitate()

Attributes: id, eventName, eventType, status, states, type, size, duration, startingDate, startingTime, endingDate, endingTime, location, commonality, activities

AnyMedia (AnyMedia) (Pattern-BO)

Responsibility	Collaboration	
	Client	**Server**
Place to store, perform, and reside	AnyEvent AnyDiscovery	connect(), broadcast(), capture(), store(), display(), access(), select(), remove(), navigate(), secure(), defineProperties(), identify()

Attributes: id, mediaName, mediaType, capability, entry, securityLevel, status, sector, security

AnyEvidence (AnyEvidence) (Pattern-BO)

Responsibility	Collaboration	
	Client	**Server**
To provide proof	AnyDiscovery	provide(), lookFor(), indicate(), present(),
	AnyMechanism	pointsTo(), consistsOf(), prioritize(), interpret(), validate()

Attributes: id, name, number, description, type, status, conditions, limitations, upperLimit, lowerLimit, importanceFlag, attributes, relationships, impacts, context, components

AnyDiscovery (AnyDiscovery) (Pattern-BO)

Responsibility	Collaboration	
	Client	**Server**
Store information about itself	Discovery	type(), new(), generate(), gain(), loses()
	AnyEvidence	performActions(), impact(), value(), meetCriteria()

Attributes: id, name, type, status, value, states, impacts, conditions, numberOfApplications, ownedBy, ingredients, properties

4.2.7 Consequences

The pattern supports the motivation behind its modeling.

- The pattern supports different types of discoveries in different application areas. This means that one can use the described discovery patterns as an essential framework for building any type of discovery. For example, one can use this pattern to represent discovery of computers, as well as discovery of new planet, by just attaching the discovery-specific IOs. Thus, diverse discoveries in varied domains can be modeled using the depicted discovery pattern.
- This pattern is a proper level of abstraction, which tries to cover only the shareable characteristics and behavior of all the elements it portrays. This is because the pattern is designed by extracting common characteristics of the process of discovery. The pattern is extremely stable because of the existence of core components—EBT and BO form the basis of patterns.
- Because the pattern is abstract when used in a specific application, its scope must be constrained in accordance with a category of employment, that is, child development and growth. However, the pattern itself offers a way to constraint its scope with AnyDiscoveryMechanism pattern. By restricting the number of mechanisms used in the pattern, one can easily control and manage its scope.
- This pattern always supports the motive behind its modeling. It depicts a generic pattern, which can be utilized in applications across diverse domains. This generalized usage is possible, because the pattern has a stable core consisting of BOs and EBT, which results in reusability.
- To use this pattern in accordance with a specific discovery, its scope must be significantly narrowed down. For instance, is the discovery process involved in data mining, or is the discovery involved in child development and growth? The discovery pattern will be constrained to the selected category where it will exist. However, this is not a limitation, because for the new discovery its type and mechanism can be easily added.

Given above points, one can conclude that the discovery design pattern is scalable and can be used in a number of scenarios, without requiring any change in the base framework. Only the scenario-specific objects need to be wired with the base discovery patterns framework, in order to fit this design pattern in the scenario. This is possible because the discovery pattern is derived by keeping stability in mind and this is the good thing about discovery design pattern. However, the discovery pattern will require integration of verification systems to verify the correctness of the discovery. Many people will think this as a negative feature about the discovery pattern, as the pattern looks incomplete. However, this is not true, because patterns can never exist alone and they should be used in conjunction with other components.

4.2.8 Applicability

4.2.8.1 Case Study 1: Discovery of a New Vitamin K

This case study demonstrates the applicability of the discovery pattern in the discovery of vitamin K. Vitamin K discovery is a part of AnyDiscovery type. The different discovery mechanisms determined are experiments, investigation, and research.

The scientists involved in this discovery process are responsible for initiating the discovery process and conducting the research, investigation, and experiments (see Figure 4.2).

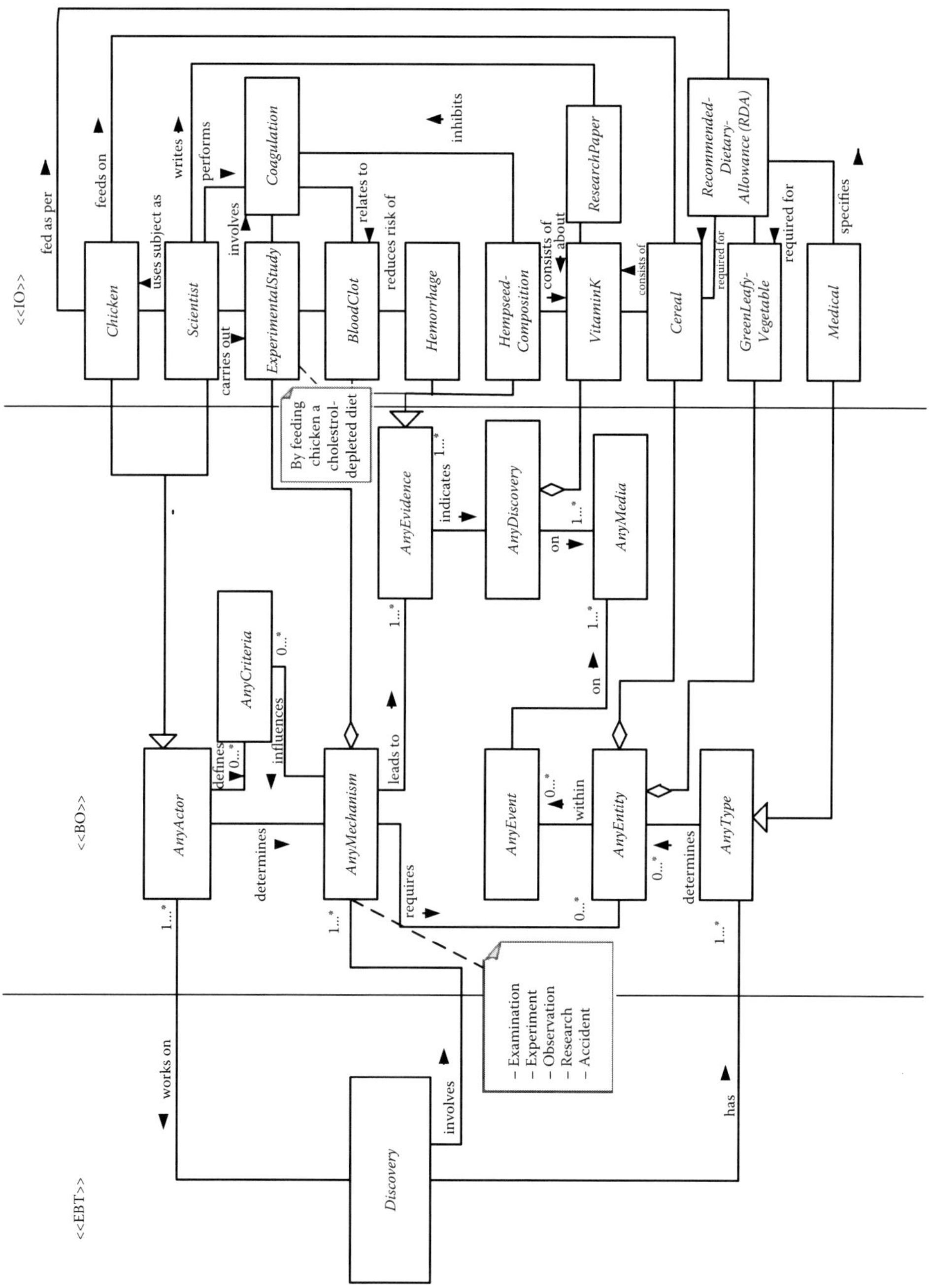

FIGURE 4.2 Class diagram for case study 1.

A scientist inherits information from the super class AnyActor. The results of these mechanisms act as evidence and vitamin K is a part of AnyDiscovery.

Use Case Title: Discovery of Vitamin K

Actors	Roles
AnyActor	1. Scientist 2. Chicken

Class Name	Type	Attributes	Operations
Discovery	EBT	1. discoveryType 2. discoveryMechanism 3. discoveryCriteria 4. discoveryMedia	1. followsMechanism() 2. finds()
AnyCriteria	BO	1. criteriaName 2. levelOfStandard	1. decidesMechanism()
AnyActor	BO	1. actorType 2. actorCategory	1. performsTask()
AnyMechanism	BO	1. mechanismName 2. mechanismDescription 3. mechanismProcedure	1. leadsToEvidence() 2. dependsOnCriteria()
AnyType	BO	1. typeName 2. typeDescription 3. properties	1. determines() 2. categorizes()
AnyEntity	BO	1. entityName 2. entityCategory 3. entityDescription	1. relatesToEvent() 2. takesPart()
AnyEvent	BO	1. frequencyOfOccurence 2. eventLocation 3. eventDescription	1. executes() 2. involvesEntity()
AnyConsequence	BO	1. consequenceFactor 2. consequenceImpact 3. consequenceDescription	1. occurs()
AnyMedia	BO	1. mediaType 2. mediaDescription 3. mediaName 4. mediaPurpose	1. servesAsMedium()
AnyDiscovery	BO	1. discoveryType 2. evidence 3. discoveryDescription 4. discoveryConsequence	1. supportedByEvidence() 2. happens()
Chicken	IO	1. breed 2. birthDate	1. lives() 2. servesAsSubject()
Scientist	IO	1. name 2. workHrs 3. skills 4. knowledge	1. discovers() 2. writes() 3. performs()

(Continued)

Class Name	Type	Attributes	Operations
ExperimentalStudy	IO	1. type	1. involves()
		2. domain	2. proves()
		3. subject	
		4. performer	
Coagulation	IO	1. type	1. relatesToBloodClot()
		2. reason	
Hemorrhage	IO	1. type	1. servesAsEvidence()
		2. affectedArea	
		3. cause	
BoodClot	IO	1. intensity	1. reducesRisk()
		2. affectedOrgan	2. prevents()
Hempseed Compostion	IO	1. type	1. providesNutrition()
		2. breed	2. consistsVitaminK()
		3. ingredients	
VitaminK	IO	1. type	1. maintainsHealth()
		2. composition	2. preventsHemorrhage()
		3. typeOfBonds	
ResearchPaper	IO	1. domain	1. tellsAboutDiscovery()
		2. author	
		3. topic	
		4. title	
Cereals	IO	1. type	1. servesAsFood()
		2. nutrientContent	2. providesVitaminK()
Green Leafy Vegetable	IO	1. type	1. providesVitaminK()
		2. breed	
		3. vitaminType	
Medical	IO	1. type	1. specifiesRDA()
Recommended Dietary Allowance	IO	1. nutritionLevel	1. aidsInDiscovery()
		2. quantity	2. definesNutritionNeed()

Use Case Description

1. *Discovery* involves *AnyMechanism*, and *ExperimentalStudy* provides a mechanism for discovery.
   What kind of mechanism is needed for discovery? How is appropriateness of mechanism assured? How experimental study provides a mechanism? What kind of experimental study?
2. *ExperimentalStudy* involves *Coagulation* that relates to *BloodClot*, and BloodClot reduces risk of *Hemorrhage* that forms a part of *AnyEvidence*.
   How blood clot reduces risk of hemorrhage? What forms evidence? What kind of evidence?
3. *HempseedComposition* consists of *VitaminK*, inhibits coagulation, and forms *AnyEvidence*.
   What is Vitamin K? How coagulation is inhibited?
4. *AnyEvidence* indicates *AnyDiscovery* and discovery is done on *AnyMedia*.
   What kind of media? How evidence indicates discovery?
5. *AnyEvent* occurs on *AnyMedia* that comprises of *AnyEntity*.

6. *AnyType* determines *AnyEntity*.
   What are types of entity?
7. *Medical* specifies *RecommendedDietaryAllowance (RDA)* that is required for *GreenLeafyVegetables and Cereals*.
   How Medical specifies RDA? What is the purpose of RDA? Who defines medical?
8. *Cereals* and *GreenLeafyVegetables* consist of *VitaminK*, which is *AnyDiscovery*.
   What is the proof that cereals and green vegetables consists of Vitamin K?
9. *Scientist* uses *Chicken* as subject for *ExperimentalStudy*.
   Why scientist uses chicken? What kind of experimental study?
10. *Chicken* feeds as per *RDA*.
11. *Chicken* and *Scientist* forms *AnyActor*, who determines *AnyMechanism*.
    What kind of mechanism? Why actor determines mechanism? What is the purpose?
12. *AnyActor* defines *AnyCriteria*, which influences *AnyMechanism*.
    How criteria influence mechanism? What criteria actor defines? On what basis?
13. *AnyMechanism* requires *AnyEntity* and leads to *AnyEvidence*.
    How mechanism leads to evidence?
14. *AnyActor* works on *Discovery* that has *AnyType*.
    What are the types of discovery? How the actor starts for discovery?

4.2.9 Related Patterns and Measurability

4.2.9.1 Related Pattern

The discovery process exhibits different forms under different applications and under different domains. Although these different types of discovery mechanisms are implemented in tools, discovery as a pattern does not exist per se. The discovery process exists as a part of different tools. Some tools implement pattern discovery process, some tools implement association rules, while others implement other mechanism (see http://maya.cs.depaul.edu/~mobasher/webminer/survey/node7.html). These tools are domain specific and deal with one particular algorithm or discovery mechanism, but the solution we have provided is generic in nature, which can be adaptable in any domain and any application. Thus, the existing tools in the market can be used only for a specific purpose and use. The tool will fail when you try to use it in another context. However, the tools created on top of the stable discovery pattern depicted here can be used for as many domains as needed. In short, the single tool can serve as a key to all the available tools in the market for discovery. Based on this definition, the following metamodel can be modeled for the discovery problem.

In law, *discovery* is the pretrial phase in a lawsuit, in which each party through the law of civil procedure can request documents and other evidence from other parties or can compel the production of evidence by using a subpoena or through other discovery devices, such as requests for production of documents and depositions. In other words, discovery includes interrogatories, motions or requests for production of documents, requests for admissions, and depositions.

Traditional Model (Business as Usual) versus Stable Model (Pattern)

- The basis of the traditional model is entirely IOs, which are physical objects and are unstable. However, the stable model is based on three different concepts—EBT, BO, and IO. The EBTs represent elements those remain stable internally and

externally overtime. The BOs are objects that are internally stable but externally adaptable; IOs are replaceable, are unstable, and are application classes/objects.
- The traditional model is hard to reuse, when requirements change. Any changes in the requirements might cause a complete reengineering of the project. The stable model is highly flexible and is reusable in wide domains and applications.
- The traditional model requires high maintenance cost in terms of time, labor, and money. The system built by using the traditional model cannot be extended or adapted. The stable model is easily maintainable and extendable.

To summarize, the features of stability model like stability, scalability, understandability, reusability, maintainability, and simplicity make it far better as compared to the traditional model.

4.2.9.2 Measurability

- *Quantitative Measure*
  - Quantitative metrics refer to the quantity aspect of EBTs, BOs, and IOs. The more the number of classes, the more it will result in lines of code while developing the system. In addition, as lines of code increase, error propagation rate will also increase and it will be difficult to maintain accuracy in the pattern development.
  - The quantitative aspect shows that EBTs, BOs, and IOs should be selected in such a way that it should cover all the necessary patterns required in modeling, and yet it should be developed in a manageable number of lines of code, which will result in lesser error propagation. The second aspect of quantitative metrics is when compared to a traditional model the stability model has less number of classes with the focus on explicit as well as implicit factors.
  - Traditional models are specific to an application as they are based on classes specific to one application only. A stable model is generic, and it can be extended to develop any application by just hooking the application-specific IOs to the stable pattern. This makes the stable model highly flexible. Also in the stable model, as the base pattern is known well in advance, determining and developing estimations or measurement metrics is far easier and less time consuming, as compared to that observed in the traditional model.
- *Qualitative Measure*
  - A stable model being very generic can be reused to apply to any application, whereas the traditional model is built on application-specific tangible objects and thus cannot be reused. Reusing traditional model requires a lot of reengineering, effort, time, and cost. This makes the stable model more scalable and flexible. Moreover, it is easy to maintain the stable model as compared to traditional model, because it is flexible and easily adaptable.
  - For software requirement specificity, we can formulate one formula; to create it, we just need to define a few terms. We can use Q1 for specificity of requirements. By specificity of requirements, we mean lack of ambiguity. The second value is completeness. By completeness, we mean how well they cover all the functions of classes to be implemented. We refer to it as Q2. So, now to determine specificity for requirements, we will use following formula:

$$Q1 = \frac{Nui}{nr}$$

where:

Ql is specificity of requirements
Nui is number of common requirements identified
nr is total number of requirements

$$nr = nf + nnf$$

where:

nf is number of functional requirements
nnf is number of nonfunctional requirements

Thus, the lower the value of requirement specificity, the greater will be the ambiguity. Therefore, the value of requirement specificity should always be optimal.

4.2.10 Modeling Issues, Criteria, and Constraints

4.2.10.1 Modeling Heuristics

4.2.10.1.1 General Enough to Be Reused in Different Applications

The stable design pattern so developed can be applied to a wide range of applications. The pattern has been developed keeping generality in mind. Discovery has different meanings under different contexts. BOs defined for the pattern are general enough, such that they can be hooked to IOs of any application, and the pattern is capable enough to derive the specific functionality of the application. This part has been well explained in our discussion of applicability, where discovery is used to define legal discovery. Similarly, we can use the same pattern to develop a model for drug discovery.

4.2.11 Design and Implementation Issues

For a design pattern to be useful and applicable across many different problem domains, it must represent an EBT, which defines the core value of the pattern and which can withstand changes over time. For the discovery pattern, the EBT is identified to be that of discovery. Discovery is the enduring concept of observing, finding, and noting things that are not known to anyone before. This enduring concept must maintain its characteristics in different applications. While analyzing the mapping concept, we could identify a set of BOs that form the basis of the discovery pattern. These BOs are also very stable and extendable into different IOs depending on the type of application. Therefore, discovery is the process of making AnyDiscovery by AnyParty. How any discovery is made is encapsulated by AnyDiscovery and AnyMechanism and can be influenced by AnyEvidence. All these BOs are stable and generic enough, because they will not change when the discovery is applied in various contexts or over time.

To apply the discovery pattern to a particular application, we also look for IOs that are extensible from the BOs and are tangible objects that reflect the true problem domain. IOs generally are not stable and you may need to modify them over time. Because the BO layer remains stable and the EBT lasts over time, any possible changes are restricted to the IO layer only.

Stability model is based on EBTs, BOs, and IOs. The EBTs used are general so that they can be applied in various domains. But there are a few implementation issues that we must tackle and manage. They are as follows.

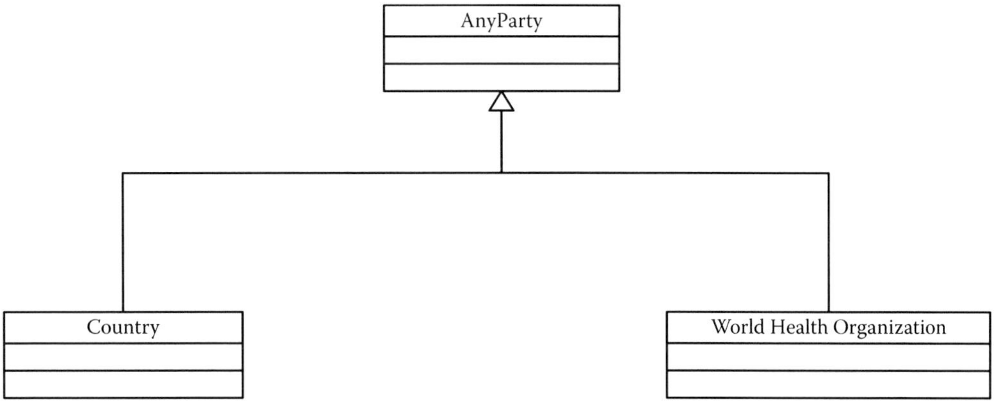

FIGURE 4.3 Inheritance.

4.2.11.1 Delegation versus Inheritance

Figure 4.3 shows the model implemented with inheritance.

The model shown in Figure 4.3 is static and fixed. Country and WHO are the subclasses, and they inherit attribute, operations, and methods from AnyParty. If any change occurs in AnyParty, then it will reflect in all the subclasses, even when that change is not needed for all subclasses concerned here. In other words, superclass will not hide any methods from its subclasses.

4.2.11.2 Model Implemented with Delegation

The model in Figure 4.4 shows the use of delegation instead of inheritance. How it affects the modeling pattern is an interesting feature. Delegation provides dynamism, that is, run-time flexibility, which is one of its distinct features. The rest of the characteristics are similar to inheritance as it also provides a reuse technique. Dynamic coupling between superclass and subclass is the key feature.

In this case, the same submodel is implemented by using delegation instead of inheritance. Now, even if superclass adds some changes, it will not reflect in all subclasses because of delegation, as it provides dynamic run-time linking by invoking a call from one object in superclass to the object in the subclass concerned. Now, if some additional rules need to

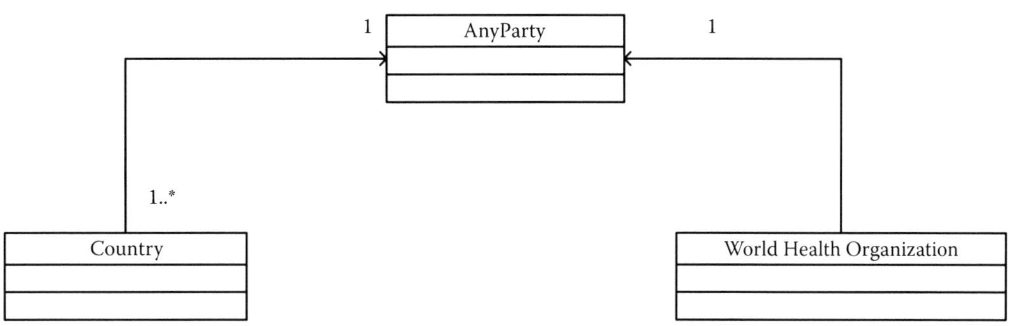

FIGURE 4.4 Delegation.

be implemented for WHO guidelines, then all we need to do is create a separate method for WHO rules and then pass the object to the WHO subclass. So, in this case that particular change will not be seen in the Country subclass. In other words, a superclass can hide its methods from subclasses.

The code for the delegation example taken is as below

```
public class anyparty{
      public void publichealth(system.out.println('public health
             issue'));
      public void population(system.out.println('number of people
             in the country'));
}
public class who
{
      anyparty a = new anyparty();
      public void publichealth(a.publichealth());
}
public class country
{
      anyparty b = new anyparty();
      public void population(b.population());
}
```

The above code shows how the class WHO creates an object and delegates the class AnyParty by using that object to invoke the method in class AnyParty. Thus, it will use the relevant methods from class AnyParty for its own class. In this way, class AnyParty can hide its methods from other classes that do not require those methods.

4.2.12 Testability

If the discovery analysis pattern can be used as it is in its original form, without changing the core design and by only plugging IOs for infinite number of applications, then the discovery pattern can be said to be testable. In Section 4.2.8, two widely different applications are illustrated, and they do not require changing the core design of the pattern. Using the scenarios listed in this paper, many such scenarios can be deduced and proved to say that the discovery pattern is indeed testable. Another alternative to test the discovery pattern is to come up with such cases where discovery is being used/can be used but the given discovery pattern solution does not model the given problem.

In general, patterns designed by using the stability model are more easily testable when compared to the traditional model. This is because the EBTs and BOs rarely change and they can be applied to other applications without any major changes. In this project, the discovery stability pattern is tested by applying the pattern to many different applications without introducing any changes to the core pattern. This is achieved by plugging in the necessary IOs to the core discovery pattern. In the same way, other application's IOs from any context can be plugged to BOs. The above pattern so developed will be considered testable only when it can be applied to any scenario/application.

4.2.13 Formalization Using Object Constraint Language, Z++ or Object Z, and/or Extended Backus–Naur Form

We can describe the discovery pattern in a many-sorted first-order language (Yang, Chin, and Chung 1992). A many-sorted first-order language consists of a set of sorts or types. Each sort is associated with a universe and a set of relation and function symbols whose definition would come from the domain. In addition, each sort can have a one or more subsorts. The universe of a sort is the union of the universes of its subsorts. We use an XML-based schema for describing a discovery pattern. The XML description can be used for generating code in a language such as C++ or C#. We can briefly describe the syntax for discovery patterns below. Because the full description of the schema is space consuming and reduces readability aspects, we are providing only a part of the schema. The remaining portions of the schema can be developed along similar lines.

```
<pattern>
      <title>
      "discovery"
      </title>
      < sort>
            <title>
      "Discovery"
            </title>
            <sort>
                  <title>
                  "discoveryType"
                  </title>
                  <sort>
                              <title>
                              "name"
                        </title>
                              <type>
                              String
                              </type>
                              <universe>
                              {life on Mars, drug, ...}
                              </universe>
      </sort>
      <sort>
                  <title>
                  "type"
                  </title>
                  <type>
                        String
                  </type>
                        <universe>
                  {Space, Medicine,...}
                        </universe>
      </sort>
      <sort>
                  <title>
```

```
                    "description"
                    </title>
                    <type>
                          String
                    </type>
                          <universe>
                    {...}
                          </universe>
        </sort>
        <sort>
                    <title>
                    "list of mechanisms"
                    </title>
                    <type>
                          Power: DiscoveryMechanism
                    </type>

</sort>
<function>
<title>
"determine mechanism"
</title>
<type>
Constant: DiscoveryMechanism
</type>
</function>
<function>
<title>
"addDiscoveryType"
</title>
<type>
            name?→type?→DiscoveryType→DiscoveryType
</type>
<description>
addDiscoveryType: DiscoveryType' = DiscoveryType ∪ {name?, type?}
</description>
</function>
<function>
<title>
"removeDiscoveryType"
</title>
<type>
            name?→type?→DiscoveryType→DiscoveryType
</type>
<description>
<pre>
{name?,type?} ε DiscoveryType
</pre>
<body>
addDiscoveryType: DiscoveryType = DiscoveryType' ∪ {name?, type?}
</body>
```

```
</description>
</function>
</sort>

...

</pattern>
```

4.2.14 Business Issues

4.2.14.1 Business Rules

One can easily find business rules (Sidebar 4.2) in both application development and business management life cycles. Application architects, analysts, and developers base the majority of their work on basic business rules and constraints. Business rules are an excellent tool for automated processes with multiple decision points. Connecting business rules to business processes creates a direct correlation between company policy and business operations. Although such processes do not change very often, business rules keep changing continuously, as managers adjust to internal and external market environments. Business rules differ in complexity. In some instances, a business rule can be very simple, described maybe in one sentence. To make the rules most efficient, each rule should be independent of procedures and work flows (Perry and Kaminski 2008).

Mapping business rules to discovery and finding an instance of discovery with strong evidence ensure the stability of this concept in all areas where discovery is utilized.

The class diagram provides visual illustration of all the classes in the model along with their relationships with other classes. Description of the business rules based on discovery pattern's class diagram is shown below.

- AnyParty and/or AnyActor uses the discovery process for achieving AnyDiscovery (BO) by defining a zero or more than zero set of AnyCriteria (BO) that influence the selected AnyMechanism (BO).
    - *AnyParty.* This refers to any legal user involved with AnyDiscovery, such as scientists, discoverers, assistants of discoverers, any user of the discovery, and so on.
    - *AnyMechanism or discovery process.* Any techniques used to make a specific or particular discovery, such as experimentation and observation.
    - *AnyCriteria.* Any legal user's defined criteria (requirements, constraints, etc.) over what the pattern has.
    - *AnyDiscovery.* Any specific and particular discovery, such as vitamin K discovery, phone discovery, and a particular drug discovery.
- Discovery is the EBT of this model. Discovery (EBT) must have one or more than one AnyType (BO).
    - *AnyType.* Specifies the type of discovery based on domain, such as medical, social, drug, engineering, life, and planetary, or based on criticality, such as major or minor.
- AnyType (BO) represents all the domains in which one can carry out the discovery (EBT). AnyType (BO) can be represented as the discovery of a star or some other innovations like finding another planet that has life.
- AnyType (BO) plays an important role in determining the name and the numbers of AnyEntity (BO).

  - *AnyEntity*. AnyEntity used as part or required for the discovery
- Devising AnyMechanism (BO) specific to the discovery in context involves discovery (EBT) too.
- AnyCriteria influences this AnyMechanism (BO), which is defined by AnyParty (BO) or AnyActor after thorough analysis. Depending on the domain of discovery, AnyMechanism (BO) varies and can range from simple examination to full-fledged experimentation.
- AnyMechanism (BO) must lead to creation of AnyEvidence (BO), because without demonstrable and confirmable evidence discovery cannot be complete.
  - *AnyEvidence*. This refers to patentable information about a particular discovery, such as data and process.
- The presence of AnyEvidence (BO) indicates that AnyDiscovery (BO) qualifies for being called discovery (EBT) as it now meets the criteria.
- AnyDiscovery may utilize zero or more of AnyEntity within zero or more of AnyEvent on AnyMedia.
  - *AnyEvent*. This discusses the events of a particular discovery
  - *AnyMedia*. This describes the media used in a particular discovery, such as computers, lands, nature, air, labs, and the Internet

4.2.14.2 Business Integration

The stability model extracts the core concepts involved in the problem. This *makes it easier to* extend it to fit the needs of any application. The BOs act as the extension points, where the IOs for the particular application can be hooked to make a final product. Hence, it is much easier to integrate the pattern in any business model.

4.2.14.3 Business Enduring Themes

Discovery represents the goal of the business. It answers the question, What is the main and unique goal of the pattern? The system is used to model the concept of discovery.

This pattern can be used in any domain that involves the concept of discovery. The pattern models the concept of discovery in a stable way, so that it can be used in many applications.

4.2.15 Known Usages

The discovery pattern can be used to create the following tools:

1. *Unified data mining engine* (*UDME*). This tool is used to discover various trends from vast amounts of data. UDME is a generalized tool that is usable on any kind of database, as well as on any type of data. It is built by using the knowledge map and discovery is one of the patterns in the knowledge map.
2. *Unified performance evaluation engine* (*UPEE*). This is again a generalized tool that is usable to evaluate the performance of any entity. It again uses discovery stable patterns in its knowledge map.
3. *Electronic discovery* (*e-discovery*). This tool represents a discovery process that uses any form of e-content such as email messages, instant messages, files, databases, and other electronic content that may be stored on a variety of platforms. It is becoming much more important in the context of civil litigation—for example,

roughly three out of four discovery orders today require an email message to be produced as part of the discovery process.

4. *Web services dynamic discovery* (*WS-Discovery*) Version 1.1, retrieved January 6, 2012. WS-Discovery is a technical specification defining a multicast discovery protocol to locate services on a local network. BEA Systems, Canon, Intel, Microsoft, and WebMethods developed WS-Discovery. As the name suggests, the actual communication between nodes is done using web services standards, notably SOAP. Various components in Microsoft's Windows Vista operating system, such as the "People Near Me" contact location system, use WS-Discovery.

4.2.16 Tips and Heuristics

- Describing patterns is a hard task and it requires careful and calibrated work.
- The metamodel is very different compared to a stable model, and it is a traditional model.
- Pattern design must be generic, so that it can be applied to applications spread across various domains.
- EBT must represent the goal of the pattern.
- Intuition and experience are required in order to find the correct EBT for the pattern.
- BOs provide capabilities to achieve the goal of the pattern. Identification of BO requires spending some time in thinking and coming up with correct BOs.
- BOs provide hooks to which specific IOs can be plugged and for getting varied applications in diverse domains. This reduces the cost by encouraging reusability.

SUMMARY

The discovery stable analysis pattern demonstrates or exhibits the entire discovery process in a concise, lucid, and clear manner. The domain can be scientific, geological, engineering, medical, or any other. Similarly, one can use the patterns to implement different discovery mechanisms. The discovery analysis pattern is developed by using software stability paradigm and stable pattern language concepts. These paradigms also help us to develop a pattern that is reusable and stable in nature.

This discovery model can be used for different domains, and IOs can be extended according to the application. The model represents the core knowledge of the pattern in different applications and is presented as EBTs and BOs. The model is explained with three special and specific applications that perform well based on this model.

The correct identification of EBTs and BOs is the most challenging and tedious task, and it requires previous experience and skills. After identifying EBTs and BOs correctly, your next critical challenge is to determine the relationship between EBTs and BOs, so that the discovery pattern can hold true in any context of usage for discovering. Once you perform this task, depending on the nature of the application, the IOs are attached to the hooks provided by BOs. Thus, using the discovery pattern as a basis, an infinite number of applications can be built by just plugging in the application-specific IOs to the pattern. This results in reduced cost and effort and a stable solution. Hence, the discovery design pattern is very useful.

OPEN RESEARCH ISSUES

The following are some of the open research issues that need to be examined and require future work and experimentation.

1. *Software stability and knowledge maps.* Software stability and knowledge maps lead to many knowledge discoveries, which include the generation of problem space patterns (analysis) and ultimate solution patterns (design), redefine knowledge, and discover many possibilities of architectural patterns that are generated from the knowledge map of any domain and used as foundation bases of millions of applications. Software stability and knowledge maps allow the development of meaningful patterns. Developing meaningful patterns is a thing of art and a system of perfect skills; improving the overall quality of patterns is never easy and quick; more often, developers take an inordinately long time to design perfect and meaningful patterns. To develop meaningful and robust patterns, a developer may need to design them in a phased manner. The most important and critical of all these phases is the diagnostic phase, using which one can understand and comprehend the main problems that come in the way of development of today's patterns. Once a pattern developer identifies and notes all the bottlenecks, it becomes very easy to explore the causes of pattern immaturity and their subsequent usability. In addition, software stability and knowledge maps provide simple and clear guidelines for choosing the appropriate patterns from a large inventory of alternatives and distinguishing clearly between analyses, design, and architectural patterns.
2. *e-discovery.* Utilizing the concurrent software development model or knowledge map methodology is a way for developing an e-discovery engine. Building this engine by using traditional development approaches is not an easy exercise, specifically when several factors can undermine their quality success, such as cost, time, and lack of systematic approaches.
3. *Unified programmable dynamic discovery engine (UPDDE).* Utilizing the concurrent software development model or knowledge map methodology is a way for developing UPDDE. The engine mainly focuses on several patterns: dynamism, discovery, adaptability, extensibility, customizability, and so on. The proposed solution attempts to extract out the commonality from all the domains and represent it in such a way that it is applicable to a wide range of domains without trivializing or generalizing the concepts. The engine is a stable structural pattern, and it provides a generic engine that can be applied and/or extensible to any application by plugging application-specific features.
4. *Discovery informatics.* Utilizing the concurrent software development model or knowledge map methodology is a way for developing a unified discovery informatics enterprise framework that facilitates the drug discovery process at any enterprise level. The unified framework includes chemical structure handling (e.g., editing, database storage, and database searching), biological data handling (e.g., database storage, searching, and data reduction), structure–activity relationship handling, e-discovery, dynamic discovery, e-scientific, and chemical inventory management. The unified framework is very generic and can be applied in or easily extended to cheminformatics, bioinformatics, chemoinformatics, medicinal chemistry, computational chemistry, drug discovery innovation,

structure-based drug design, screening, docking, structural biology, predictive toxicology, predictive ADME, chemogenomics, molecular modeling pharmaceutical, and so on.

REVIEW QUESTIONS

1. What do you mean by the term *discovery*? Can you use the term *discovery* in any other context than what you thought of?
2. Find out all such terms that mean exactly same as discovery and can be used interchangeably.
3. Examine the functional requirements of discovery pattern—Are there any missing requirements? Discuss them.
4. Examine the nonfunctional requirements of discovery pattern—Are there any missing requirements? Discuss them.
5. Examine the challenges and add two more challenges to the existing list.
6. Examine the constraints and add five more constraints to the existing list.
7. What are the capabilities to achieve discovery? Describe each of them.
8. What is the trade-off of using this pattern?
9. Present the sequence diagram for applicability of the discovery stable analysis pattern in the drug discovery domain.
10. What are the possible design issues for the discovery EBT when linked to the design phase?
11. What do you think are the implementation issues for the AnyEvidence BO when used in the discovery stable analysis pattern?
12. What do you think are the implementation issues for the AnyMedia BO when used in the discovery stable analysis pattern?
13. List a couple of advantages of using the stable analysis pattern for discovery.
14. List two scenarios that will not be covered by the discovery analysis pattern.
15. Describe how the developed discovery analysis pattern would be stable over time.
16. List some of the lessons learnt from the use of the stable analysis pattern for discovery.
17. List some of the testing patterns that can be applied for testing the discovery stable analysis pattern.
18. List three test cases to test the class members of the discovery pattern.
19. List some of the related design patterns used in formulating the discovery stable analysis pattern.
20. Briefly explain how the discovery stable analysis pattern supports its objectives.
21. Assess two different quantitative measures on the discovery traditional model and discovery stable analysis patterns, and explain the differences between each of the measures.
22. Examine the CRC cards and add two new operations if possible to the EBTs and BOs.
23. Try to create a use case and interaction diagram for each of the scenarios you thought of in the above question.
24. Name two more qualitative metrics and utilize them to measure discovery pattern.
25. Compare the traditional model and stable pattern of discovery using the following adequacies:

a. *Descriptive adequacy.*
   It refers to the ability to visualize objects in the models. Every defined object should be browsable, allowing the user to view the structure of an object and its state at a particular point in time. This requires skill of understanding and extracting metadata about objects that will be used to build a visual model of objects and their configurations. This visual model is domain dependent, that is, based on domain data and objects' metadata. Descriptive adequacy requires that all of the knowledge representation be visual as follows:
   i. Visual models are structured to reflect natural structure of objects and their configurations.
   ii. All the visual knowledge (data and operations) in the visual model is localized.
   iii. Relationships among objects in the visual model are well defined.
   iv. Interactions among objects in the visual model are limited and concise.
   v. The visual model must transcend objects and instead highlight crosscutting aspects.
b. *Understanding adequacy.*
   It relates to be easy to understand.
c. *Simplicity adequacy.*
   It relates to how simple your models will be.
d. *Extensibility adequacy.*
   It relates to the degree of extensibility, adaptability, customizability, and configurability of your models.

26. Compare the traditional model and stable pattern of discovery by using the following modeling essentials (Fayad and Laitinen 1998) as comparative criteria:
    a. *Simple.* This property covers those attributes of the object-oriented model that present modeling aspects of the problem domain in the most understandable manner.
    b. *Complete* (*most likely to be correct*). This property determines if the object-oriented model provides internal consistency and completeness of the model's artifacts. The model must be able to convey the essential concepts of its properties.
    c. *Stable to technological change.* The model should be stable enough to technological changes and it cannot require any changes with a change of technology, such as change of the media or the mechanisms.
    d. *Testable.* To be testable, the model must be specific, unambiguous, and quantitative wherever possible, such as we can run an infinite number of scenarios with the context of the pattern.
    e. *Easy to understand.* In addition to the familiarity of the modeling notations, the notational aspects, design constraints, and analysis and design rules of the model should be simple and easy to understand by the customers, users, and domain experts.
    f. *Visual or graphical.* A picture is worth a thousand words. As a user, you can visualize and describe the model. The graphical model is essential for visualization and simulation.
27. Discuss the benefits of using the discovery analysis pattern to generate business rule.

28. Give some examples of applications, where discovery pattern are currently used.
29. What are the lessons learnt by you from studying the discovery pattern.

EXERCISES

1. Think of a few scenarios where discovery pattern is applicable and come up with corresponding class diagram, use case, and sequence diagram as shown in the solution and applicability sections for each of the scenarios.
2. Draw a sequence diagram of the case study to *discover vitamin K use (case study 1)*.
3. *Research and development.* New product design and development is more than often a crucial factor in the survival of a company. In an industry that is fast changing, firms must continually revise their design and range of products. This is necessary due to continuous technology change and development as well as other competitors and the changing preference of customers. A system driven by marketing is one that puts the customer needs first and only produces goods that are known to sell. Market research is carried out, which establishes what is needed. If the development is technology driven, then it is a matter of selling what it is possible to make. The product range is developed so that production processes are as efficient as possible and the products are technically superior, hence possessing a natural advantage in the market place (Ortega-Argiles et al. 2011).
   Utilize the discovery pattern as an application of a research and development:
   a. Draw a class diagram based on the discovery pattern to show the application of research and development.
   b. Document a detailed and significant use case as shown in case study 1.
   c. Create a sequence diagram of the created use case of b.
4. *Planetary research.* Planetary systems are generally believed to form as part of the same process that results in star formation. Some early theories involved another star passing extremely close to the Sun, drawing material out from it which then coalesced to form the planets. However, the probability of such a near collision is now known to be far too low to make this a viable model. Accepted theories today argue that a protoplanetary disk forms by gravitational collapse of a molecular cloud and then evolves into a planetary system by collisions and gravitational capture (see Darling 2004).

   Some planetary systems may form differently, however. Planets orbiting pulsars—stars which emit periodic bursts of electromagnetic radiation—have been discovered by the slight variations they cause in the timing of these bursts. Pulsars are formed in violent supernova explosions, and a normal planetary system could not possibly survive such a blast—planets would either evaporate or be pushed off of their orbits by the masses of gas from the exploding star, or the sudden loss of most of the mass of the central star would see them escape the gravitational hold of the star. One theory is that existing stellar companions were almost entirely evaporated by the supernova blast, leaving behind planet-sized bodies. Alternatively, planets may somehow form in the accretion disk surrounding pulsars (see Darling 2007; Podsiadlowski 1993).

a. Planetary systems, formation of, David Darling, entry in *The Internet Encyclopedia of Science*, accessed online September 23, 2007.
b. Planet formation scenarios (Podsiadlowski 1993)

Utilize the discovery pattern as an application of a planetary research and

i. Draw a class diagram based on the discovery pattern to show the application of planetary research.
ii. Document a detailed and significant use case as shown in case study 1.
iii. Create a sequence diagram of the created use case of ii.

5. *Drug discovery* (Paul et al. 2010; Warren 2011). In medicine, biotechnology, and pharmacology, *drug discovery* is the process by which drugs are discovered and/or designed. In the past, most drugs have been discovered either by identifying the active ingredient from traditional remedies or by serendipitous discovery. A new approach has been to understand how disease and infection are controlled at the molecular and physiological level and to target specific entities based on this knowledge.

   The process of drug discovery involves the identification of candidates, synthesis, characterization, screening, and assays for therapeutic efficacy. Once a compound has shown its value in these tests, it will begin the process of drug development prior to clinical trials.

   Utilize the discovery pattern as an application of drug discovery and

   a. Draw a class diagram based on the discovery pattern to show the application of drug discovery.
   b. Document a detailed and significant use case as shown in case study 1.
   c. Create a sequence diagram of the created use case of b.

6. *Biomarker discovery* (Jacobs et al. 2005). *Biomarker discovery* is the process by which biomarkers are discovered. It is a medical term.

   Many commonly used blood tests in medicine are biomarkers. The way that these tests have been found can be seen as biomarker discovery. However, their identification has mostly been a one-at-a-time approach. Many of these well-known tests have been identified based on clear biological insight, from physiology or biochemistry. This means that only a few markers at a time have been considered. One example of this way of biomarker discovery is the use of injections of insulin for measuring kidney function. From this, one discovered a naturally occurring molecule, creatinine, that enabled the same measurements to be made easily without injections. This can be seen as a serial process.

   The recent interest in biomarker discovery is because new molecular biologic techniques promise to find relevant markers rapidly, without detailed insight into mechanisms of disease. By screening many possible biomolecules at a time, a parallel approach can be tried. Genomics and proteomics are some technologies that are used in this process. Significant technical difficulties remain.

   There is considerable interest in biomarker discovery from the pharmaceutical industry. Blood test or other biomarkers could serve as intermediate markers of disease in clinical trials and also be possible drug targets.

   Utilize the discovery pattern as an application of biomarker discovery and

   a. Draw a class diagram based on the discovery pattern to show the application of biomarker discovery.
   b. Document a detailed and significant use case as shown in case study 1.
   c. Create a sequence diagram of the created use case of b.

PROJECTS

Develop the following systems using the discovery analysis pattern:

1. *Discovery informatics.* It is the field of computing that facilitates the drug discovery process at the enterprise level. Typical software tools in this space include chemical structure handling (e.g., editing, database storage, and database searching), biological data handling (e.g., database storage, searching, and data reduction), structure–activity relationship handling, electronic scientific notebooks, and chemical inventory management.
2. *Discovery science* (*Chen et al. 2005*). Also known as discovery-based science is a scientific methodology that emphasizes analysis of large volumes of experimental data, with the goal of finding new patterns or correlations, leading to hypothesis formation and other scientific methodologies.

   Discovery-based methodologies are often viewed in contrast to traditional scientific practice, where hypotheses are formed before close examination of experimental data. However, from a philosophical perspective where all or most of the observable *low-hanging fruit* has already been plucked, examining the phenomenological world more closely than using the senses alone (even augmented senses, e.g., via microscopes, telescopes, and bifocals) opens a new source of knowledge for hypothesis formation.

   Data mining is the most common tool used in discovery science and is applied to data from diverse fields of study such as DNA analysis, climate modeling, nuclear reaction modeling, and others. The use of data mining in discovery science follows a general trend of increasing use of computers and computational theory in all fields of science. Further following this trend, the cutting edge of data mining employs specialized machine learning algorithms for automated hypothesis forming and automated theorem proving.
3. *E-discovery* (*Adam and Lender 2011; Various 2009*). It refers to discovery in civil litigation that deals with information in *electronic format* also referred to as electronically stored information (ESI). In this context, *electronic form* is the representation of information as binary numbers. Electronic information is different from paper information, because of its intangible form, volume, transience, and persistence. In addition, electronic information is usually accompanied by metadata, which is never present in paper information unless manually coded (see below). Metadata is the data about the data, or the information that is kept about the electronic files, that is, who the author was, when the file was created, and so on. It is descriptive information that cannot be changed unless spoliation occurs. E-discovery poses new challenges and opportunities for attorneys, their clients, technical advisors, and the courts, as electronic information is collected, reviewed, and produced.

   Examples of the types of data included in e-discovery are e-mail, instant messaging chats, documents (such as MS Office or OpenOffice files), accounting databases, CAD/CAM files, websites, and any other ESI which could be relevant evidence in a lawsuit. Also included in e-discovery is the *rawdata* which forensic investigators can review for hidden evidence. The original file format is known as the *native* format. Litigators may review material from e-discovery in one of several formats: printed paper, *native file*, or TIFF images. If the native file, for example, a Microsoft Word document, contains 10 pages, then an e-discovery vendor will convert it into 10 TIFF images for use in a discovery review database. Documents that are produced are numbered using Bates numbering. Individuals working in the field of e-discovery commonly refer to the field as litigation support.

4. *Legal discovery* (*Kyckelhahn and cohen 2008*). In law, *discovery* is the pretrial phase in a lawsuit, in which each party through the law of civil procedure can request documents and other evidence from other parties or can compel the production of evidence by using a subpoena or through other discovery devices, such as requests for production of documents and depositions.
   a. Name two to three ultimate goals of each of the above discoveries.
   b. List all the functional requirements and nonfunctional requirements of each of the ultimate goals.
   c. List five challenges for the two or three ultimate goals combined for each area.
   d. Name 10 different applications for each of the goals.
   e. Name five different applications for the two or three ultimate goals combined.

SIDEBAR 4.1 Knowledge Discovery

Knowledge discovery is a unique and special concept in the realm of computer science that explains the process of automatically or mechanically searching or seeking large streams of data, for set and convenient patterns that can be considered as pertinent and essential set of knowledge about the data in question. It is about deriving special knowledge from the available set of input data. This complex but defining topic can be classified according to the type of *data* being searched and in what form or type is the result of the data search tabulated. The most famous area of knowledge discovery is the data mining, which is also known as knowledge discovery in databases (Bozdogan 2004).

REFERENCE

Bozdogan. H., (ed.). *Statistical Data Mining, and Knowledge Discovery*, Boca Raton, FL: CRC Press LLC, 2004.

SIDEBAR 4.2 Business Rules

Business architecture, in its simplest explanation, can be viewed as a set of resources that interact under well-defined rules through a set or collection of well-calibrated processes to achieve certain goal(s). These rules are known as *business rules*. This simple view of business architecture does not indeed clarify what business rules are. In fact, there exists no formal or standard definition for business rules; nonetheless, several definitions have evolved over the last decade. In the following paragraphs, we give some of such definitions. A business rule can be defined as

- Units of business knowledge (Odell 1998).
- A statement that defines or constrains certain aspects of a business (Halle 2001).
- Declarations of policies or conditions that must be satisfied (OMG 1992).

It is a statement that defines or constrains some aspect of the business. It is intended to assert business structure or to control or influence the behavior of the business (Morgan 2002). We will use this definition of business rules throughout this book.

- Business rules are abstractions of the policies and practices of a business organization. The *business rules approach* is a development methodology, where rules are in a form that is used by but does not have to be embedded in business process management systems.
- The business rules approach formalizes an enterprise's critical business rules in a language that managers and technologists understand. Business rules create an unambiguous statement of what a business does with information to decide a proposition. The formal specification becomes information for process and rules engines to run.

Ronald Ross (2003) describes several basic principles of what he calls *the business rule approach.* He believes that rules should

- Be written and made explicit.
- Be expressed in plain language.
- Exist independent of procedures and work flows (e.g., multiple models).
- Build on facts, and facts should build on concepts as represented by terms (e.g., glossaries).
- Guide or influence behavior in desired ways.
- Be motivated by identifiable and important business factors.
- Be accessible to authorized parties (e.g., collective ownership).
- Be single sourced.
- Be specified directly by those people, who have relevant knowledge (e.g., active stakeholder participation).
- Be managed.

REFERENCES

Halle, B. V. *Business Rules Applied: Building Better Systems Using the Business Rules Approach.* New York, NY: Wiley, 2001.

Morgan, T. *Business Rules and Information Systems.* Boston, MA: Addison-Wesley Publishing, 2002.

Odell, J. *Advanced Object-Oriented Analysis and Design Using UML.* New York, NY: SIGs Books, 1998.

OMG. *Analysis and Design Reference Model.* Framingham, MA: OMG, 1992.

Ross, R. G. *Principles of Business Rule Approach.* Reading, MA: Addison-Wesley Publishing, 2003.

5 The Knowledge Stable Analysis Pattern

> When you stop learning you might as well be dead (Power 2004), for it is like starving the brain. Knowledge is the food of the mind; and without knowledge the mind must languish (Sanford 1846). Knowledge is the seed that is planted and will develop into a beautiful blossoming tree also known as that *idea*. When the brain is fed it grows and when it grows it become more peaceful, confident, and comfortable with its surroundings.

A seed is planted and it eventually grows into maturity to produce a beautiful flower. That is the actual story of our mind. A mind is fed manure called *knowledge*, and in return, it grows, nurtures itself, and prospers by forming a number of ideas, plans, and concepts. From those ideas and concepts, the world grows, adapts, and eventually becomes a better place to live, for with knowledge comes inner peace and contentment.

Without the right kind of knowledge, we are just little more than an empty shell that breathes. In essence, knowledge is the basic foundation, upon which we as humans grow and enrich our lives. Without knowledge, there is always a fear of the unknown, and when there is fear or scare, there will be a lack of trust, and when there is a lack of trust, there is a greater risk for conflict. Knowledge can set you free. In fact, it can set the whole world free. For, when we have knowledge, we can easily nourish our mind, body, and soul for knowledge definition, refer to Sidebar 5.1.

5.1 INTRODUCTION

One can think of *knowledge* as the mirror of experience that is gained by practice and/or study of a particular discipline. Once this knowledge is acquired by some individuals, it can be used by them to avoid experienced pitfalls and sloppy actions those were experienced in the past. Additionally, it will also allow them to either create new environments/ambience or streamline previously addressed/tackled ones, based on the acquired knowledge. So, the question now is how one can represent such knowledge in a straightforward and coherent manner, so that individuals can use it repeatedly to solve recurrent problems.

Knowledge is the root of the human mind, body, and spirit, for without it, we are nothing but an empty shell. Without knowledge, we fear and when we fear, eventually evil will become us. Knowledge is power, said Francis Bacon, for from knowledge, we grow and prosper and eventually blossom into more compassionate and complete human beings. Knowledge is what we know now. It is not about what is right or wrong, true or false, and good or evil. It is what we have been exposed to, and thus what becomes our knowledge, our reality, and eventually the center of our universe. No two people will or can share identical knowledge, because no two people can ever share identical life experiences, which create knowledge, although we will all share common knowledge.

Software development must attempt to take all those variations that make up the human mind and streamline a product, so that it becomes a product based on common knowledge. For example, we would all agree that to enter some data, we must type something. That is common knowledge.

In order for software development to be both successful and profitable, a less complex methodology must be developed that focuses on common knowledge. From that common knowledge or application, each user would build and fine-tune a product that was specific and special to their needs.

When a software application package is built upon common knowledge, we would reduce the costs significantly, because you are no longer attempting to develop a software application that is different for hundreds of thousands of users. Not only is software development extremely expensive, it is also extremely labor intensive, and spending too much time trying to please the world might result in an inferior and untimely product that is most likely to hit the market well past its target release date.

Instead of focusing more on knowledge and implicitly common knowledge, which is shared knowledge, we will not be continuing to reinvent what has already been invented; instead, we can focus on a common platform that can be expanded on. As a result, a build can occur in a fraction of the time; because of the flexibility of the product, we target a much larger audience.

5.2 PATTERN DOCUMENTATION

5.2.1 Pattern Name: Knowledge Stable Analysis Pattern

The name should be right and appropriate, for knowledge creates stability and with stability comes the power for analysis and subsequently the creation of reoccurring patterns. Thus, this is an excellent choice of title. The name is appropriate and fitting to the analysis that follows in the evaluation of the problem of how knowledge can be stabilized, so that the wheel is not reinvented every other time.

5.2.2 Known As

Many times, acquisition of knowledge is interpreted as gaining information, and hence, information is used many times in the context of knowledge. Although gaining information about any entity is certainly a part of acquiring knowledge, that information must be investigated, analyzed, assimilated, disseminated, and properly used. Thus, knowledge is much more than just acquiring information, and thus, the term *information* cannot be used in place of *knowledge* to represent certain context.

Another issue that is often misinterpreted to mean knowledge is education. People always think that they can gain knowledge through education. This is totally inaccurate and wrong, as education is just one of the means of gaining knowledge. In order to acquire knowledge, one needs to have a keen and intense interest in learning and understanding things. Reading, education, learning, analyzing, thinking, and so on are just a number of means to gain knowledge. Thus, education cannot be used interchangeably with knowledge.

However, cognition can be used interchangeably with knowledge. This is because both knowledge and cognition represent the psychological result of perception, learning, and reasoning. Thus, the stable knowledge pattern can be used as a solution pattern for cognition too.

5.2.3 Context

Knowledge can be gained through experience or study. It represents a collection of facts, rules, tips, or lessons learned with respect to anything that must be synthesized to create knowledge. Sometimes, it might not be possible to obtain complete knowledge about a subject, and it results in partial knowledge. As a result, this partial knowledge needs to be used to solve a problem. The knowledge pattern will be used to represent knowledge synthesis and acquisition.

For example, in autonomic computing, knowledge is a collection of information acquired through examinations of log files, as well as other types of files, located in local or remote repositories (i.e., servers/PCs in local networks or in wide area networks). Another example of knowledge application is encountered in customer relationship management (CRM) systems. In CRMs, knowledge is collected to learn and understand customers' buying behavior, so that they can offer better services and increase their sales. Knowledge is obtained by means of recording the customers' navigational behavior (clicks) during their shopping session. Knowledge may also be represented in the form of symbols or write-ups or just passed on by word of mouth. Thus in this age, storing and dissemination knowledge is essential to gain competitive edge. As a result, knowledge management is important and critical for any organization.

5.2.4 Problem

There is a general tendency to think of knowledge as individual facts, or relevant and specific information that your system collects or requires. This perception makes implementation or application of knowledge different every time it is used, because each system may handle a different subject and with a set of different information and constrains. Why do we need to reinvent the wheel every time we are dealing with a new type of knowledge? This is a significant problem that we must overcome and tackle. Therefore, we need to answer the question: how do we encapsulate the main component of knowledge as an aspect, regardless of its context of applicability and prevent from reinventing the wheel?

For example, a software system solution may be required to use both topic maps from artificial intelligence and implement them in an autonomic software solution. In this case, both topic maps and autonomic computing might view knowledge from different angles and structures. So, their integration may be tricky and tedious, especially when dealing with different representations of knowledge. Hence, the question is restructured in the following way—how can one abstract knowledge characteristics and behavior that are common by different knowledge representations and later generate a single representation of knowledge that can even knowledge acquisition among different software solutions?

5.2.4.1 Functional Requirements

1. *Domain specificity*. What is knowledge of the specific domain or any domain? How is knowledge acquired within the specific domain or any domain and hence, its subjects? What is the knowledge of a specific subject or any subject within any domain? Where the knowledge of any domain obtains one or more subjects matters? Knowledge is defined in the *Oxford English Dictionary* (2011) as (1) expertise and skills acquired by a person through experience or education, the theoretical or

practical understanding of a subject; (2) what is known in a particular field or in total, facts and information; or (3) awareness or familiarity gained by experience of a fact or situation.

2. *Criteria*. Criteria consist of the characteristics of the knowledge and the constraints that are imposed on knowledge, the characteristics, constraints of the knowledge, within AnyDomain. There are specific criteria imposed on the knowledge of AnyDomain and these criteria influence the means for achieving knowledge and lead to change in state. Thus, the pattern should be able to accommodate all these changes with the change in AnyCriteria.
3. *Context*. The broad applicability of the produced or AnySubject defined knowledge.
4. *Type*. Knowledge can be classified based on different factors, such as definition, field or domain, description, or production of knowledge. Usually, AnySubject knowledge has one or more types that are used to describe knowledge, such as meta-knowledge and procedural knowledge. Knowledge can also be classified based on its production, such as actual knowledge, constructive knowledge, or/and imputed knowledge.
5. *Structure*. Structure describes the knowledge structure.
6. *Knowledge*. Knowledge treated as a model has one or more views of it. Knowledge also has a scope.
7. *Mechanisms such as knowledge acquisition*. Knowledge acquisition involves complex cognitive processes: perception, learning, communication, association, and reasoning. The term *knowledge* is also used to mean the confident understanding of a subject with the ability to use it for a specific purpose if appropriate.
8. *Knowledge handlers*. Who possesses the knowledge? AnyActor, such as individual, hardware, software, and/or creatures, and AnyParty, such as individuals, organizations, countries, political parties, and/or a combination of some or all of them as the holder, the user, and/or the creator of knowledge. What does AnyParty know? How do we know what we know?
9. *AnyEntity*.
   a. The knowledge of AnyEntity is gained through logs, which are stored on AnyMedia, which helps us to access AnyEntity. So, the spectrum of AnyEntity should be well understood and well covered.
   b. Knowledge about entities should be available after the criteria hold true. If the criteria fail, knowledge should be restructured.
   c. AnyEntity makes use of AnyMedia. So, the mapping between AnyEntity and AnyMedia should be clear.
   d. The knowledge of AnyEntity should be general enough to make it fit into any application.
10. *Media*.
    a. Media that are used to store and gain access of knowledge, which should be well identified.
    b. Media can be of different types and usage of each may vary.
11. *Logs*.
    a. Logs, such as files, disks, tapes, papers, temples, tomb walls, and/or papyrus, are used to store and gain access to knowledge, which should be well identified.
    b. Logs can be of different types and usage of each may vary.

5.2.4.2 Nonfunctional Requirements

1. *Completeness*. The pattern should be complete in the sense that it should be able to present all the meanings of knowledge and the areas where it can be applied. This means that the pattern is applicable to many areas and has different meanings everywhere. So, the pattern should possess the quality of inferring correct meaning in different contexts and should define the state, rules, assessment, and type according to the context in which it is applied.
2. *Accuracy*. It is the state of being accurate, and the knowledge conforms to known rules and facts or recognized standards. In the domains of applied and pure science, engineering, industry, and statistics, the term *accuracy* is defined as the degree and extent of closeness of a measured, estimated, or calculated quantity to its actual, real (true) value. Accuracy is closely related to the degree of precision, also called the factors of reproducibility or repeatability, the extent to which further or advanced measurements or calculations show or repeat the same or similar results. The term *accuracy* is also defined as the degree to which a given quantity is correct and free from error (Wolfram 2002).
3. *Comprehension*. It is a measurement of the understanding of knowledge and the totality of knowledge, that is, properties or qualities that knowledge possesses. Other nonfunctional requirements of knowledge are *manageability*, *produceability*, and *awareness*.

5.2.5 Challenges and Constraints

- *Challenges*
  - Challenges encapsulate the default structure and behavior that will be shared among different knowledge implementations.
  - They allow canonical handling for a different set of subjects that may be selected during knowledge acquisition.
  - They robustly cope with a set of distinct actors per application on an on-demand basis.
  - They synchronize the provided skills of the actors with the capabilities provided by different mechanisms during knowledge acquisition.
  - Actors must take respective actions according their abilities in varied and uncertain situations.
- *Constraints*.
  - The domain of interest must be selected before any other aspect or subject selection.
  - Subject's definition must wrap, surround, and delegate the default and public capabilities of the mechanism aspect.
  - Before utilizing a specific actor, their skills must be assessed with respect to the selected subject.
  - The knowledge acquisition process is a synchronized and well-tuned process.
  - There must be at least one actor understanding this knowledge.
  - Knowledge can be shared or it can belong to multiple domains.
  - A single domain can contain at least one subject of interest. A subject can be either atomic (does not contain subsubjects) or composed (contains at least one subsubject).

- Knowledge involves a set of mechanism. At least one mechanism needs to be associated with a subject of interest.
- This mechanism streamlines one or more skills of a determined actor.

5.2.6 Solution: Pattern Structure and Participants

The solution is divided into the pattern's structure and its participants.

5.2.6.1 Structure

The structure of this pattern is illustrated by a class diagram.

As seen in Figure 5.1, the knowledge stable analysis pattern requires a determined domain, as the main input for perception, reasoning, and learning. This main input will be broken down into a list of subjects relevant to the main input. These subjects were filtered and reported by the utilization of different mechanisms bound to specific skills of a particular actor. This will result in an understanding of interrelated facts or subjects belonging to a specific domain.

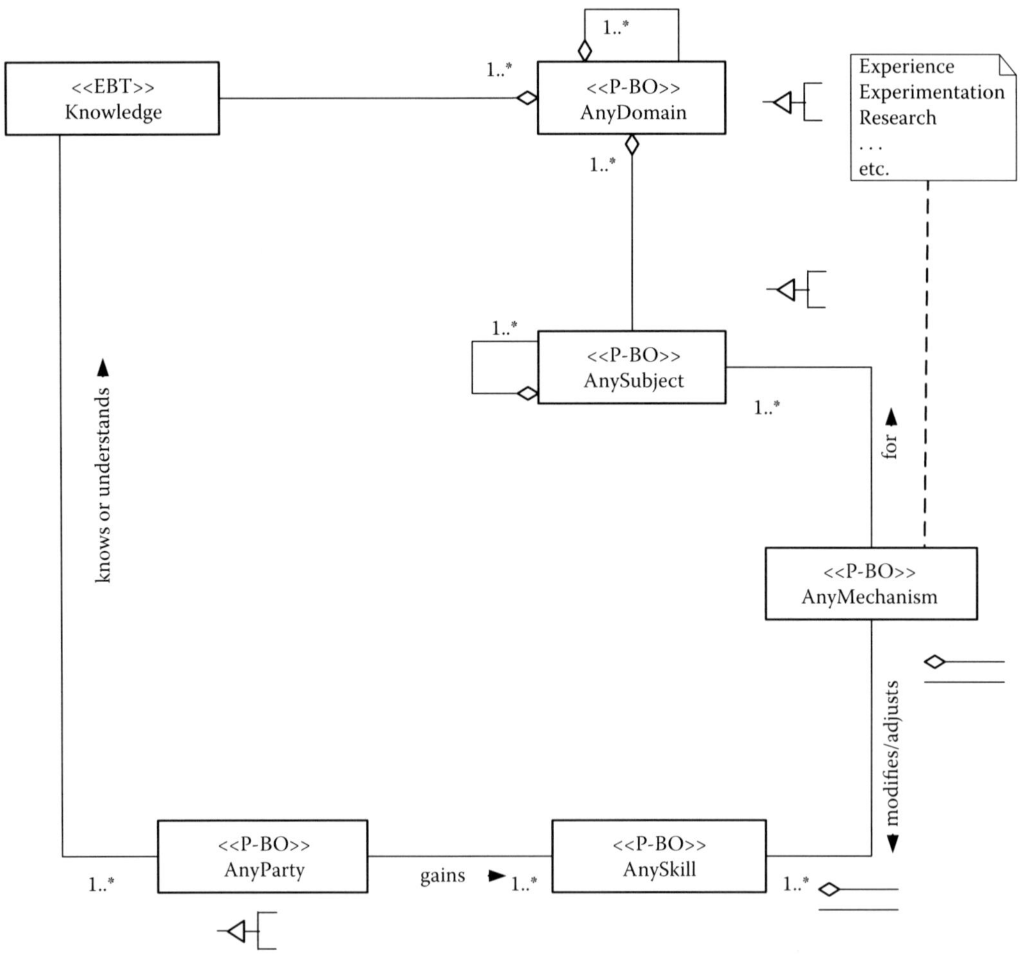

FIGURE 5.1 The structure of the knowledge stable analysis pattern. Some scenarios may require other business objects within knowledge pattern, such as AnyParty, AnyCriteria, AnyType, AnyStructure, AnyEntity, AnyEvent, AnyLog, and AnyMedia.

5.2.6.2 Participants

The pattern's participants are classified into two types: classes and patterns. Classes are individual object constructs that appear in any traditional class diagram. Patterns encapsulate a collection of classes and the associations that stem from these classes.

- *Classes*
    - *Knowledge.* It encapsulates the shared behavior and properties between a set of different types of knowledge.
- *Patterns*
    - *AnyDomain.* It represents the container of a set of distinct and interrelated subjects that have a significant role in a domain's understanding and abstraction.
    - *AnyMechanism.* It encapsulates the canonical (shared) behavior and properties that occur in different types of mechanisms, regardless of their context of applicability.
    - *AnyCriteria.* It represents the requirements and constraints that should hold true in order to access knowledge of some entity or for generating knowledge.
    - *AnyActor.* It represents the set of users or stakeholders, with certain skills, that acquire or understand knowledge.
    - *AnyParty.* It represents a party, an organization, governmental organization, or a country. They may request certain knowledge or generate knowledge about some entity.
    - *AnyStucture.* It states the knowledge format or layout and/or the knowledge structure.
    - *AnyType.* It represents different types of knowledge, when applied to different contexts.
    - *AnyEntity.* It represents the entity that has knowledge. The entity defines the media through which it can be found.
    - *AnyMedia.* It represents the media through which the knowledge of an entity exists or stored. It is possible to have multiple mediums to store knowledge.
    - *AnyLog.* It represents the log, such as record, file, stone, tape, and disk, where the knowledge is stored.

5.2.6.3 CRC Cards

Knowledge (Knowledge) EBT

Responsibility	**Collaboration**	
	Client	**Server**
Represents any type of knowledge that your system requires or collects	1. AnyParty, {OR} 2. AnyActor 3. AnyMechanism 4. AnyType	1. acquire() 2. convey() 3. peek() 4. assess() 5. relatedTo()

Attributes: knowledgeId, KnowledgeArea, KnowledgeProperties, createdBy, context, knowledgeConstraints, knowledgeType

AnyParty (AnyParty) BO

Responsibility	Collaboration	
	Client	**Server**
To use or generate knowledge	1. Knowledge	1. declaresCriteria()
	2. AnyCriteria	2. assesses()
		3. possesses()
		4. defines()
		5. operates()
		6. generates()

Attributes: name, designation, skills, workHrs, location, popularity, noOfMembers, fieldOfOperation

AnyActor (AnyActor) BO

Responsibility	Collaboration	
	Client	**Server**
To use or generate knowledge	1. Knowledge	1. performs()
	2. AnyCriteria	2. succeeds()
		3. accomplishes()
		4. doesTask()
		5. implementsRules()
		6. achieves()
		7. prospers()

Attributes: name, birthdate, designation, skills, qualification, address, status, workingLocation, workingHours

AnyType (AnyType) BO

Responsibility	Collaboration	
	Client	**Server**
To name the type of knowledge	1. Knowledge	1. categorizes()
	2. AnyDomain	2. classifies()
		3. describes()
		4. sorts()
		5. seperates()
		6. makesClass()
		7. organizes()

Attributes: name, status, number, basis, ruleForClassification, parameterUsed, factor, inspirationForClassification

AnyDomain (AnyDomain) BO

Responsibility	Collaboration: Client	Collaboration: Server
To specify domain where knowledge is required	1. AnyType 2. AnyEntity	1. enforces() 2. functions() 3. contributes() 4. comprises() 5. involves()

Attributes: name, type, history, arena, subject, value, peopleInvolved, knowledge

AnyEntity (AnyEntity) BO

Responsibility	Collaboration: Client	Collaboration: Server
To specify entity where knowledge is required	2. AnyMedia	1. exists() 2. maintains() 3. states() 4. demands() 5. hasValue() 6. needs() 7. represents() 8. symbolizes()

Attributes: name, location, type, quintessence, value, history, built, status

AnyCriteria (AnyCriteria) BO

Responsibility	Collaboration: Client	Collaboration: Server
To specify user-defined requirements and constraints for knowledge	1. AnyParty 2. AnyActor 3. AnyMechanism	1. influences() 2. imposes() 3. specifiesStandard() 4. restricts() 5. confines() 6. constraints()

Attributes: typeOfCriteria, stateOfCriteria, specifiedBy, numberOfCriteria, effectOfCriteria, purpose, description

AnyMechanism (AnyMechanism) BO

Responsibility	Collaboration	
	Client	**Server**
To incorporate means/methods for knowledge	1. AnyCriteria 2. Knowledge 3. AnyStructure	1. presentKnowledge() 2. generateKnowledge() 3. classifyKnowledge() 4. findKnowledge() 5. sortKnowledge()

Attributes: nameOfMechanism, criteriaForMechanism, wayOfFunction, usedBy, numberOfMechanism, mechanismDescription, status, application, components, context

AnyStructure (AnyStructure) BO

Responsibility	Collaboration	
	Client	**Server**
To specify the knowledge structure	1. AnyMechanism 2. AnyLog	1. formatKnowledge() 2. stateModel() 3. describeKnowledge() 4. relateKnowledge() 5. IllustrateKnowledge()

Attributes: structureId, structureName, parameterUsed, notationName, modelName, aspectName

AnyMedia (AnyMedia) BO

Responsibility	Collaboration	
	Client	**Server**
To provide media for storage and illustration	1. AnyEntity 2. AnyLog	1. store() 2. formsMedia() 3. used() 4. connect() 5. helpsToAccess() 6. capture() 7. access() 8. navigate() 9. format()

Attributes: usedFor, usedBy, purposeFor, mediaName, mediaType, capability, entry, securityLevel, status, sector

AnyLog (AnyLog) BO

Responsibility	Collaboration	
	Client	**Server**
To record and edit knowledge	3. AnyStructure 4. AnyMedia	1. record() 2. format() 3. open() 4. close() 5. edit() 6. modify() 7. cut() 8. past() 9. add() 10. delete()

Attributes: logId, logName, usedFor, usedBy, purposeFor, logType, capability, entry, securityLevel, status, logSpecifications

AnyDomain (Domain) (BO)

Responsibility	Collaboration	
	Client	**Server**
Serves as the bridge between knowledge acquisition and subject understanding	AnyType, AnyEntity	define(), explore(), localize(), scale(), constrain(), disclose()

Attributes: domainId, domainName, subDomains, domainProperties, domainConstraints

AnyMechanism (Mechanism) (BO)

Responsibility	Collaboration	
	Client	**Server**
Adjust the available processing or ability per actor's skill	Knowledge, AnyCriteria, AnyStructure	bind(), invokes(), cancel(), adjust(), report()

Attributes: mechanismId, mechanismName, mechanismType, mechanismList, mechanismProperties, mechanismParameters

AnySkill (Skill) (Type: BO)

Responsibility	Collaboration	
	Client	**Server**
Represents the abilities an actor can take per the synthesis of new knowledge	AnyMechanism AnyActor	isAble(), obtain(), perform(), indicate(), denoteHost()

Attributes: mechanism, List; actor, AnyActor

AnyActor (Actor) (Type: BO)

Responsibility	Collaboration	
	Client	**Server**
Represents the actors acquiring knowledge matching a subject of interest	Knowledge AnySkill	act(), demonstrate(), understand(), gain()
Attributes: skills, List; knowledge, Knowledge		

Knowledge acquisition will initiate when a determined actor localizes and explores a domain of particular interest. Then, the actor will break down the domain into a list of subjects to ease the actor's understanding toward the domain of interest. This breaking will be carried out by using certain well-defined mechanisms. The actor will then process the explored subjects and gain important skills, which are the implementation of the gained knowledge toward the actor's benefits.

5.2.7 Applicability with Illustrated Examples

The following two scenarios provide two possible uses for this pattern. For the sake of simplicity, we did not include the complete pattern's model.

5.2.7.1 Scenario 1—Autonomic Computing Context

Briefly, autonomic computing relies on the idea of the creation of self-governing that can adapt and manage themselves in accordance with stakeholders' interest (i.e., automatic system installation, etc.). Self-governing systems will use gained knowledge acquired from detecting and analyzing log files, system configuration, and so on to detect and solve local problems related to bugs or logical errors. Using the gained knowledge about these possible failures, the self-governing system will look into a specific repository for the right patches (if available) and fix the problems.

5.2.7.1.1 Class Diagram

The autonomic computing application, Figure 5.2, will consist of an agent that will be responsible to determine by experience and context which task to perform during its installation and deployment in a determined environment. This agent will analyze the context, where it is trying to be installed, via the checking of logs files. This information will be filtered and stored in a database for posterior use during its deployment. If a problem arises, this information will be retrieved, so that the agent will know what to do to guarantee a successful deployment and installation.

5.2.7.1.2 Use Case

Use Case Title: Use Autonomic Computing

Actors	Roles
AnyActor	Agent

Classes	Type	Attributes	Operations
Knowledge	EBT	id	presentFact()
		area	
AnyActor	BO	id	learn()
		actorName	participate()
		type	
		role	
AnyDomain	BO	id	storeKnowledge()
		name	
		type	
AnyType	BO	id	report()
		type	
		domainName	
AnyMechanism	BO	name	provideAutonomicComputing()
		type	
		interface	
AnyCriteria	BO	name	adjustMechanism()
		type	
		value	
AnyEntity	BO	entityId	exist()
		entityType	
		entityName	
AnyMedia	BO	mediaId	store()
		mediaName	
		mediaType	
		interface	
AnyStructure	BO	structureId	layoutKnowledge()
		structureType	
		structureName	
Agent	IO	name	retrieveData()
		type	gainSkill()
			understandKnowledge()
Deployment	IO		
Database	IO	type	storeData()
		size	
SolutionSearch	IO	criteria	searchForSolution()
		mechanismType	
InfoManagement	IO	dataType	manageData()
		mechanismType	
Activity	IO		
ActivityProduct	IO		
Installation	IO		
ActivityFactory	IO		

BO, business object; EBT, enduring business theme; IO, industrial object.

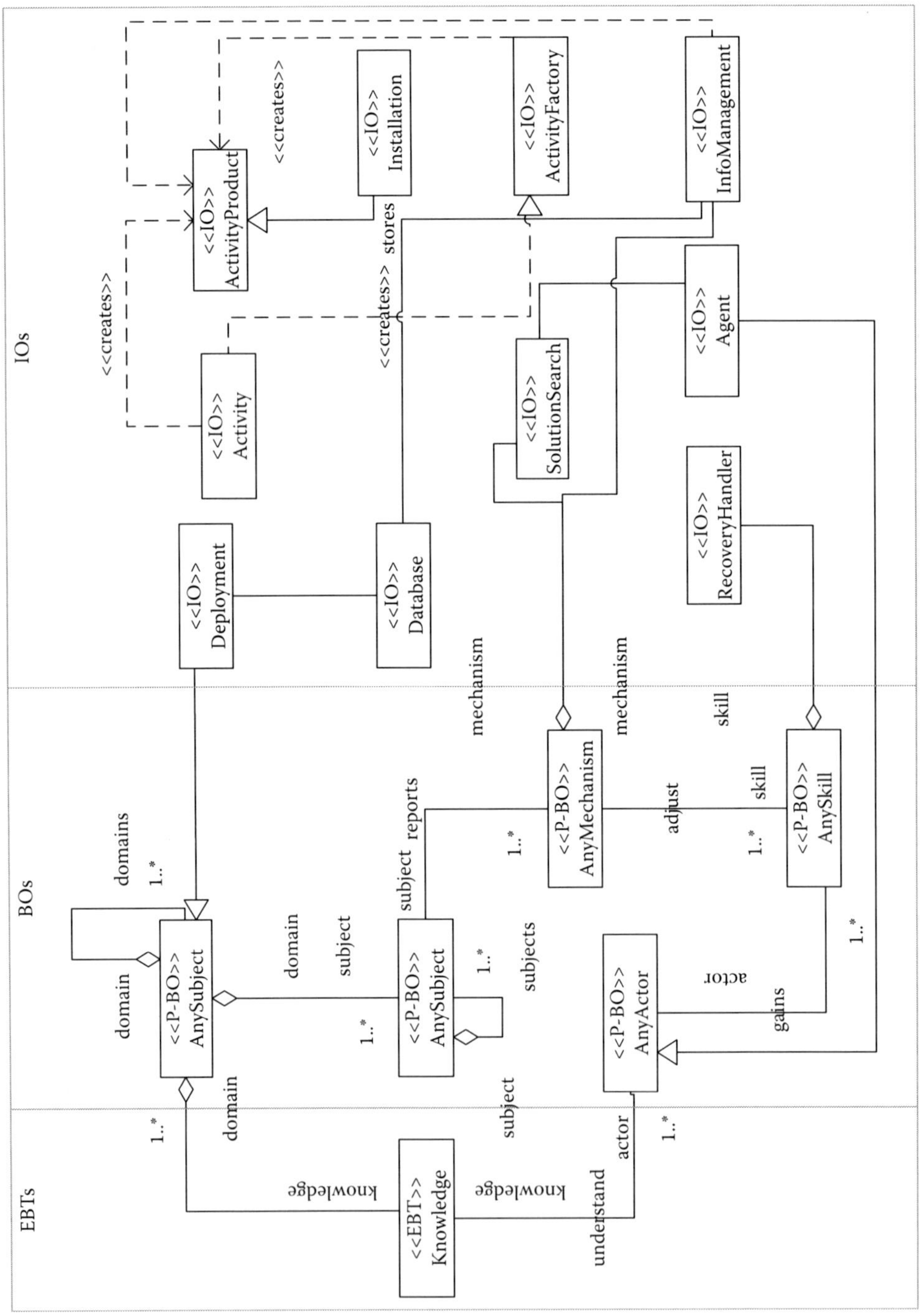

FIGURE 5.2 The autonomic computing context scenario.

Use Case Description

1. AnyActor (Agent) understands knowledge (enduring business theme [EBT]) of AnyDomain (Deployment) by detecting and analyzing log files, system configuration, and so on. What are the other techniques available to AnyActor (Agent) for understanding knowledge (EBT) of AnyDomain?
2. AnyDomain (business object [BO]) under consideration can again be a part of another AnyDomain (BO). What happens if AnyDomain (BO) is a part of AnyDomain (BO)?
3. AnyActor (Agent) gains AnySkill (RecoveryHandler) through understanding of knowledge (EBT). How does AnyActor (Agent) gain AnySkill (RecoveryHandler)?
4. This AnySkill (RecoveryHandler) adjusts AnyMechanism (BO). How is AnyMechanism (BO) adjusted through AnySkill (RecoveryHandler)?
5. AnySubject (BO), which is a part of AnyDomain (Deployment), reports the use of AnyMechanism (BO) too. In what way AnyDomain (Deployment) reports the use of AnyMechanism (BO)?
6. AnyActor (Agent) uses AnyMechanism (SolutionSearch) to give the desired results. How is AnyMechanism (SolutionSearch) used by AnyActor (Agent)?
7. In addition, AnyMechanism (InfoManagement) is also available for storing the filtered data in a database for posterior use during deployment (industrial object [IO]). If a problem arises, this information will be retrieved, so that the AnyActor (Agent) will know what to do to guarantee a successful deployment and installation. What is the structure and type of database? Who stores data in database and how? How does AnyActor (Agent) retrieve the information upon failure?

5.2.7.1.3 Sequence Diagram

The flow diagram in Figure 5.3 represents the flow of messages, when AnyActor (Agent) understands knowledge (EBT) of AnyDomain (Deployment) under consideration by use of various strategies. By understanding the knowledge (EBT), AnyActor (Agent) gains AnySkill (RecoveryHandler). This AnySkill (RecoveryHandler) is used to adjust AnyMechanism (SolutionSearch). AnyMechanism (SolutionSearch) is reported by AnySubject (BO). This AnySubject (BO) is a part of AnyDomain (BO) and is made available to AnyActor (Agent). AnyActor (Agent) also uses AnyMechanism (InfoManagement), which stores the data for later retrieval in database (BO). Data in the database (BO) are retrieved by AnyActor (Agent) when needed.

5.2.7.2 Scenario 2—CRM System

In B2B, C2B, or any other systems within the e-commerce's realm, the acquisition of customers' knowledge is the cornerstone, especially understanding these customers including their needs, aims, and wants. Therefore, we need a system that facilitates the proper understanding of the customers' knowledge and allows businesses to align their processes, products, and services to build good customer relationships and increase the benefits of the businesses.

5.2.7.2.1 Class Diagram

The CRM application (Figure 5.4) will consist of an efficient catalog subsystem that will be responsible to analyze customer data in order to analyze customer behavior. The product search and purchase activity of the customers is tracked. These data mined for the customers will be stored in the database, via a customer mining application. Using these data, the subsystem will collect and predict the behavior of the customers.

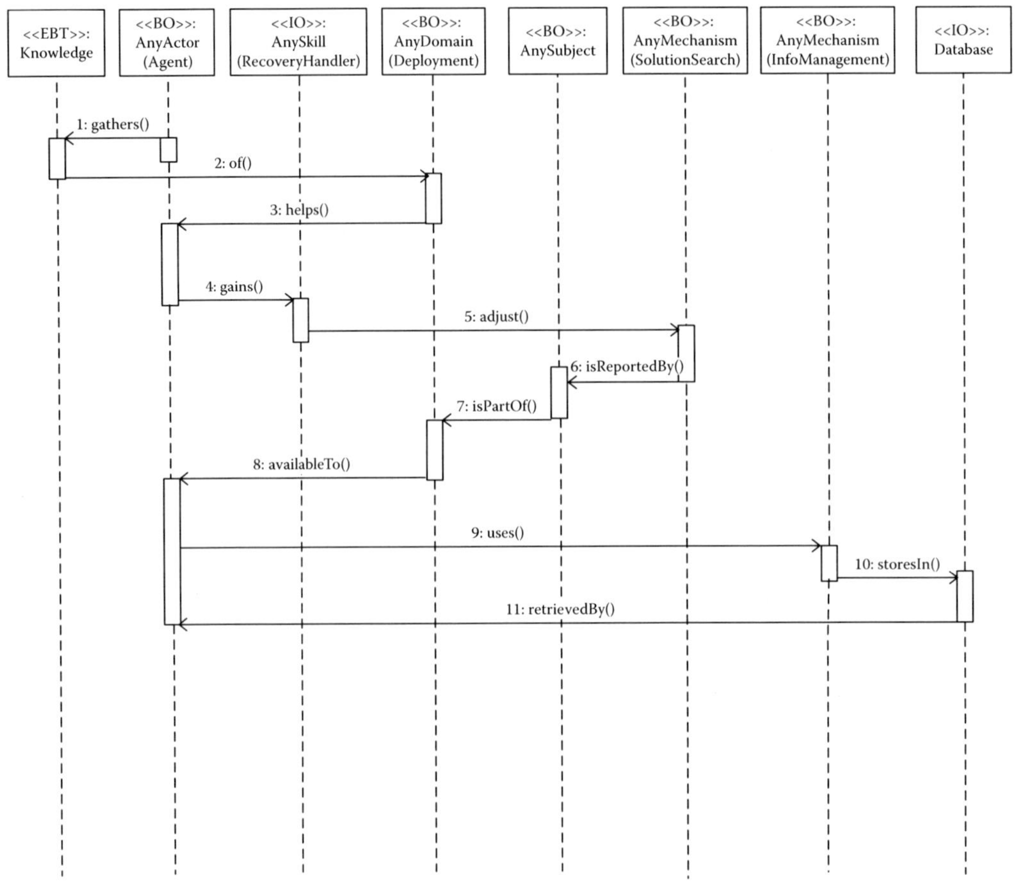

FIGURE 5.3 Sequence diagram for autonomic computing.

5.2.7.2.2 Use Case

Use Case Title: Maintaining Customer Relationship

Actors	Roles
AnyActor	Catalog subsystem

Classes	Type	Attributes	Operations
Knowledge	EBT	id area	presentFact()
AnyActor	BO	id actorName type role	learn() participate()

(Continued)

Classes	Type	Attributes	Operations
AnyDomain	BO	id name type	storeKnowledge()
AnySubject	BO	id type domainName	report()
AnyMechanism	BO	name type interface	provideAutonomicComputing()
AnySkill	BO	name type value	adjustMechanism()
CatalogSubsystem	IO	name type	retrieveData() understandCustomerBehavior()
Purchase	IO		
Database	IO	type size	storeData()
ProductSearch	IO	criteria mechanismType	searchForProduct()
BehaviorCollection	IO	dataType mechanismType	collectCustomerbehavior()
BehaviorPrediction	IO		predictCustomerBehavior()
CustomerMining	IO		analyzeData() inferTrends()

Use Case Description

1. AnyActor (CatalogSubsystem) understands knowledge (EBT) of AnyDomain (CustomerMining) by detecting and analyzing customer-related data. What are the other techniques available to AnyActor (CatalogSubsystem) for understanding knowledge (EBT) of AnyDomain?
2. AnyDomain (BO) under consideration can again be a part of another AnyDomain (BO). What happens, if AnyDomain (BO) is a part of AnyDomain (BO)?
3. AnyActor (CatalogSubsystem) gains AnySkill (OrganizationHandler) through understanding of knowledge (EBT). How does AnyActor (CatalogSubsystem) gain AnySkill (OrganizationHandler)?
4. This AnySkill (OrganizationHandler) adjusts AnyMechanism (BO). How is AnyMechanism (BO) adjusted through AnySkill (OrganizationHandler)?
5. AnySubject (BO), which is part of AnyDomain (CustomerMining), reports the use of AnyMechanism (BO) too. In what way AnyDomain (CustomerMining) reports the use of AnyMechanism (BO)?
6. AnyActor (CatalogSubsystem) uses AnyMechanism (BehaviorCollection and Behavior-Prediction) to give the desired results. How is AnyMechanism (BehaviorCollection and BehaviorPrediction) used by AnyActor (CatalogSubsystem)?
7. In addition AnySubject (BehaviorProduct) is also available for storing the filtered data in a database for posterior use during CustomerMining (IO). If a problem arises, this

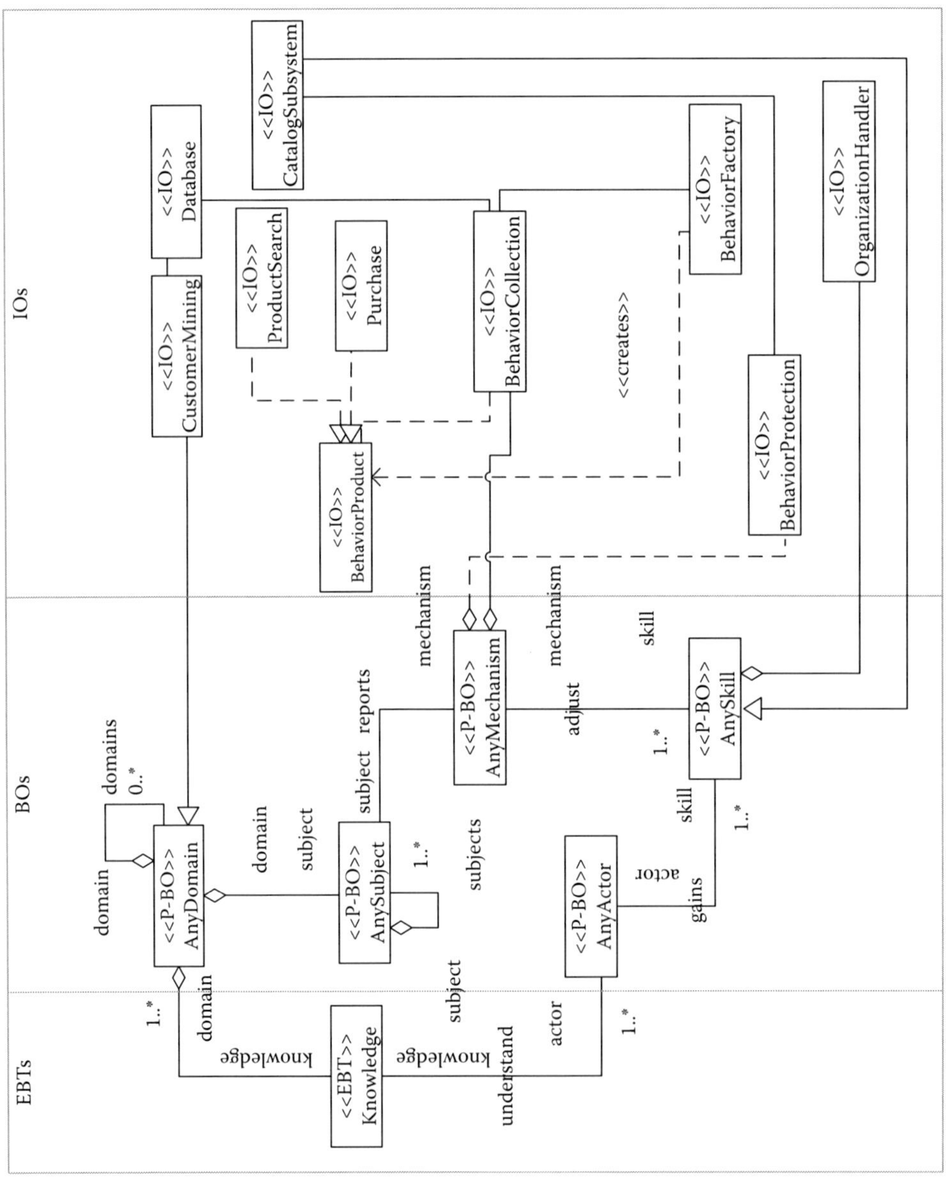

FIGURE 5.4 The CRM system scenario.

information will be retrieved, so that the AnyActor (CatalogSubsystem) will know what to do to guarantee a successful knowledge of the customer's behavior. What is the structure and type of a database? Who stores data in a database and how? How does AnyActor (CatalogSubsystem) retrieve the information upon failure?

5.2.8 Related Patterns and Measurability

Knowledge for stability pattern is generalized enough to allow its applicability in diverse application domains. This pattern includes EBTs and BOs as objects, so its applicability in other related domains only requires attachment of IOs on peripheral boundaries. Even this pattern is complex in design and requires deeper analysis to identify key EBTs and BOs, but it greatly enhances pattern reusage and effectiveness to find solution.

Traditional model is based on IOs. IOs are the physical objects and are unstable. The traditional model caters to the current requirements. The traditional model is hard to reuse if the requirements change. Any changes in the requirements cause complete reengineering of project. Thus, the traditional model involves high maintenance cost in terms of time, labor, and money. The system built by using the traditional model cannot be extended or adapted.

Knowledge traditional model is based on IOs, which are nonenduring and nonadaptable objects. Change in single IO may initiate a cascade of changes throughout other IOs, thus making it highly unstable. So, this model cannot remain stable for longer time span, whereas knowledge stable pattern is based on enduring concepts, that is, EBTs and BOs, which are adaptable and durable. This confirms its applicability forever.

Applicability of traditional model is limited to a particular domain area. In case of knowledge traditional model, it is tied to knowledge related to area, such as engineering, science, and so on. On the other side, a stable model on knowledge is applicable to a number of domain areas having core themes in common. Hooks can be easily used to extend and reuse this stabilized model.

The identification of objects involved in the traditional model just requires brief knowledge and documented data about specific domain. These objects can be easily found in problem statement. But, in stable model, one requires deeper study, experience in domain, and intuition to come up with useful set of EBTs and BOs.

5.2.8.1 Measurability

5.2.8.1.1 Quantitative Measurability

- *Number of Behaviors*
    - The number of behaviors in the traditional model is far more when compared to the stability model. All the classes have to be changed in a traditional model for each application, where as the stability pattern designed here will have the same EBTs and BOs. For any application, we only have to choose the IOs accordingly, which will obviously result in less number of classes and with no rework.

5.2.8.1.2 Qualitative Measurability

- *Knowledge Imparted*
    - A traditional model requires the knowledge of a specific domain; thus, the knowledge imparted by the traditional model is limited, whereas the stability

model, which covers the whole domain, requires a considerable amount of knowledge to understand.

- *Usability*
    - Because the stability model is modeled by using the ultimate goal, it is usable for all kinds of similar problems or applications, whereas a traditional model defeats the purpose of usability, because it is only used for the application it is designed for.

5.2.9 Known Usages

The knowledge analysis pattern can be applied in any industry, where prior knowledge is useful. Some of the industries where it is most effective are listed as follows:

- Manufacturing
- Retail industry
- Media and marketing
- Software industry

Listed below are some of the applications where knowledge is being currently used. However, the need for knowledge is different in each case and so is the mechanism used to achieve knowledge, but the core concept remains the same.

1. E-learning tools that obtain the knowledge level of the user and then provide them accordingly with the learning aids.
2. Amazon, where the system gains knowledge of the user's preferences from the items that he or she buys and provide recommendations of items that are of interest.

SUMMARY

One can quickly understand that for the software industry to remain vital and healthy, specific methodology changes or modifications must occur that might result in reduced developmental cost. Currently, these developmental costs continue to skyrocket out of control, thereby threatening the very existence of the industry.

The most important change that must occur is for software to be built on a common knowledge platform that is shared by all and then to allow the user flexibility to expand the software according to their own needs and requirements. Only then, will software development become a profitable venture.

In this chapter, we saw how the CRM system collected information and how it could help distinguish their buyers' buying habits. The pattern allows for better targeting of the consumer and thus the ability to identify areas that can be improved to attract that buyer and to increase the amount of purchases. The end result means a decrease in expenses incurred for wasted marketing and adverting dollars and an increase in sales and profits because of better targeting.

In conclusion, the knowledge pattern is a powerful and efficient tool that will represent both synthesis and acquisition of collecting information, by allowing for the collection of information in a variety of files, which can then be examined and used for decision-making processes, thus allowing for a much better targeting of one's market and reducing expenditures significantly and increasing profit margins. There are few companies that would not jump at the chance for increased revenues and the knowledge pattern is able to meet that demand at a reasonable cost to the company.

OPEN RESEARCH ISSUES

No formal process or tools for testing are available for testing knowledge pattern use for making an infinite number of applications. Thus, one needs to come up with formal documents and tools for testing the same.

REVIEW QUESTIONS

1. What do you mean by the term *knowledge*? Can the term *knowledge* be used in any other context other than what you thought of?
2. Find out all such terms that mean exactly the same as knowledge and can be used interchangeably?
3. What are the capabilities to achieve knowledge? Describe each of them.
4. Draw and describe the class diagram for stable knowledge pattern?
5. Come up with two scenarios other than those given in this chapter, where knowledge can be used. Try to fit these scenarios with the knowledge pattern.
6. Try to create a use case and interaction diagram for each of the scenarios you thought of in the above question.
7. List differences between the knowledge pattern described here and the traditional knowledge pattern.
8. List some design and implementation issues faced when implementing knowledge pattern. Explain each issue.
9. Give some applications where the knowledge pattern is being used.
10. What are the lessons learnt by you by studying the knowledge pattern.
11. Define the knowledge stable analysis pattern.
12. Is the following statement true or false? The knowledge stable analysis pattern can be applied and extended to any domain.
13. List some of the domains where the knowledge analysis pattern can be applied.
14. List any four new applications of the knowledge stable analysis pattern.
15. List three challenges in formulating the knowledge analysis pattern.
16. List 10 different constraints in the knowledge analysis pattern.
17. What are the classes and patterns involved in defining the stable pattern for knowledge?
18. Illustrate using a class diagram of knowledge patterns of each of the new applications of question 14.
19. Illustrate with a class diagram how AnyParty or AnyActor BO has a second abstraction level.
20. Document the CRC card for the knowledge EBT.
21. Is the knowledge pattern incomplete without the use of other patterns? Explain briefly.
22. What is the trade-off of using this pattern?
23. Present the sequence diagram for applicability of the knowledge stable analysis pattern in the e-commerce domain.
24. What are the possible design issues for the knowledge EBT, when linked to the design phase?

25. What do you think are the implementation issues for the AnySkill BO when used in the knowledge stable analysis pattern?
26. List a couple of advantages of using the stable analysis pattern for knowledge.
27. List two scenarios that will not be covered by the knowledge analysis pattern.
28. Describe how the developed knowledge analysis pattern would be stable over time.
29. List some of the lessons learnt from the use of the stable analysis pattern for knowledge.
30. List some of the testing patterns that can be applied for testing the knowledge stable analysis pattern.
31. List three test cases to test the class members of the knowledge pattern.
32. List some of the related design patterns used in formulating the knowledge stable pattern.
33. Briefly explain how the knowledge stable analysis pattern supports its objectives.
34. Assess two different quantitative measures on the knowledge traditional model and knowledge stable analysis patterns and explain the differences between each of the measures.

EXERCISE

Think of a few scenarios where knowledge pattern is applicable and come up with corresponding class diagram, use case, and sequence diagram, as shown in the solution and applicability sections for each of the scenarios.

PROJECTS

1. *Knowledge retrieval* (Martin and Eklund 2000; Yao et al. 2007). It is a field of study that seeks to return information in a structured form, consistent with human cognitive processes as opposed to simple lists of data items. It draws on a range of fields including epistemology (theory of knowledge), cognitive psychology, cognitive neuroscience, logic and inference, machine learning and knowledge discovery, linguistics, information technology, and so on.
2. *The knowledge economy* (Powell and Snellman 2004). It is a term that refers either to an *economy of knowledge* focused on the production and management of knowledge in the frame of economic constraints or to a *knowledge-based economy.* In the second meaning, more frequently used, it refers to the use of knowledge technologies (such as knowledge engineering and knowledge management) to produce economic benefits. The phrase was popularized, if not invented, by Peter Drucker (1969), as the title of Chapter 12 in his book *The Age of Discontinuity.*
3. *Knowledge acquisition* (Potter 2014). It is the transformation of knowledge from the forms in which it exists into forms that can be used in a knowledge-based system.
4. *Knowledge ecosystem* (Bahrami and Evans 2005). As an extension of knowledge management ideas, a *knowledge ecosystem* fosters the dynamic evolution of knowledge interactions between entities. This bottom-up approach seeks to provide a more resilient approach. Within certain contexts (e.g., turbulent

environments), a top-down knowledge management is viewed as indeterminate; hence, the intention of creating a knowledge ecosystem to improve decision making and innovation through improved evolutionary networks of collaboration.

a. Name two to three ultimate goals of each of the above knowledge areas.
b. List all the functional requirements and nonfunctional requirements of each of the ultimate goals.
c. List five challenges for the two or three ultimate goals combined for each area.
d. Name 10 different applications for each of the goals.
e. Name five different applications for the two or three ultimate goals combined.

SIDEBAR 5.1 Knowledge Definition

Knowledge is what *I* know.
Information is what *we* know.

In the world of philosophy, knowing and understanding that *something* is a likely scenario assumes that what is known is always true and 100% correct. Is it possible to announce that someone really knows something but it is not exactly true and correct? It is almost impossible to say or ascertain that something is true unless we are successful in demonstrating that *that something* is also true and correct (Foskett 1982).

So, what exactly is knowledge? Is it possible to define or explain it in very simple and easy to understand terms?

According to Merriam Webster's online dictionary, knowledge is defined as

1. The fact or condition of knowing something with familiarity gained through experience or association.
2. Acquaintance with or understanding of a science, art, or technique; the fact or condition of being aware of something.
3. The range of one's information or understanding (e.g., answered to the best of my *knowledge*); the circumstance or condition of apprehending truth or fact through reasoning.

The definition of knowledge is a matter of intensive and ongoing debate among philosophers in the field of epistemology. The classical definition, described but not ultimately endorsed by Plato, has it that in order for there to be knowledge, *at least* three criteria must be fulfilled: to count as knowledge, a statement must be justified, should be truthful, and should be believed.

REFERENCE

Foskett, A. C. *The Subject Approach to Information.* Hamden, CT: Linnet Books/The Shoe String Press, Inc., p. 1, 1982.

Section III

Knowledge Map Capabilities

Simply speaking, capabilities are entities that form the heart and soul of a business. Without appropriate capabilities, it may be very tedious for an organization to conduct day-to-day business operations. Capabilities also form the essential components of business process to ensure maximized productivity, stability, durability, and steadfastness of business functions. In essence, capabilities may mean many things to people and in different industries.

Capabilities also complement the goals of a given concept by guiding themselves systematically toward their goal achievement. In fact, they are the enduring business workhorses of the goals. Their behavior is driven forward by a set of enduring business rules, which are encapsulated by a number of goals. Each capability has an ultimate goal responsible for constraining the capabilities' internal behavior toward a nondeterministic outcome without causing strugglers. The main idea of embedding the capability's ultimate goal within the capability's workflow is to ensure and ascertain a business-centric behavior that is usually coherent with the rationale or goals of interest.

Section III discusses capabilities or/and business objects and documents two major capabilities of the knowledge maps as stable design patterns: AnyMap and AnyContext. Section III also contains three chapters and six sidebars.

Chapter 6 is titled "Capabilities: The Heart of Business," and it defines capabilities or business objects and the origin of business objects, discusses the workflows, shows how to deal with capabilities: identification and assessment, the essential properties of business objects, and briefly discusses the capabilities of the knowledge maps. This chapter concludes with a brief summary and open research issues for the future. This chapter also provides a series of review questions, exercises, and projects.

Chapter 7 is titled "AnyMap Stable Design Pattern," and it discusses, models, and documents this pattern by using Fayad's stable pattern documentation template as shown in Appendix A. The chapter also includes three different and distinguishable AnyMap patterns that focus on three different goals: navigation, visualization, and recording. This chapter concludes with a brief summary and numerous open research issues. This chapter also provides review questions, exercises, and projects.

Chapter 8 is titled "AnyContext Stable Design Pattern," and it discusses, models, and documents this pattern by using Fayad's stable pattern documentation template as shown in Appendix A. This chapter concludes with a brief summary and many open research issues. This chapter also provides review questions, exercises, and projects.

Sidebar 6.1 is titled "Business Objects," and it provides different definitions of business objects.

Sidebar 6.2 is titled "Learning about Capability," and it provides different definitions of capability.

Sidebar 6.3 is titled "Work Flow," where the workflow is defined, and it shows how different activities coordinate within the workflow.

Sidebar 7.1 is titled "Genetic or Linkage Map," where a definition has been provided.

Sidebar 7.2 is titled "Site Map," where a definition has been provided.

Sidebar 7.3 is titled "Topographic Map," where a definition has been provided.

6 Capabilities
The Heart of Business

Capabilities are the true understanding of the solution space.

M. E. Fayad
2015

6.1 INTRODUCTION

Capabilities are those entities that form the heart and soul of a business. Without proper capabilities, it may be very tedious for an organization to conduct day-to-day business operations. Capabilities also form the essential components of a business process, to ensure maximized productivity, stability, durability, and steadfastness of business functions. In essence, capabilities mean many things to people in different industries. For business, capability means abilities that are acquired and are applied at their workplaces. It may include a series of knowledge pools and skill sets required to run, manage, and administer numerous business processes. In some domains, capabilities may also include wisdom and knowledge acquired because of years of experience; more often, a person is said to be capable when he or she has the required abilities to perform certain things or actions. Capabilities are always enduring and persisting in their nature; once you acquire capabilities in a given domain, it is almost certain that you will be capable to excel in that domain for life.

Capabilities in software architecture and development system may encompass several things like required knowledge and skills, prior experience in the industry, and expertise required to run several functions. A piece of software system is said to be capable when it has the ability to remain stable for a long time, acquire the intrinsic ability to resist frequent architectural changes or modification, and finally display an ability to perform a series of functions with different variables.

It is possible to acquire capability by developing the capabilities of software systems by specializing in a given domain, or by repeating the same thing in different contexts, or even by developing them in a larger domain with more variables, or in a new domain, or in a totally different domain (Sidebar 6.2).

6.1.1 The Heart of the Business

Capabilities or business objects (BOs) (Sidebar 6.1) are the heart and soul of business operations. They always implement a number of entities that realize an assured and definite business concept (Patel, Sutherland, and Miller 1998). Such a responsibility allows them to specify, execute, and track internal norms or comply with a series of logical steps across a wide spectrum of environments. Like goals, they are the intrinsic parts of the knowledge maps. Their main responsibility, within the knowledge maps' standpoint of view and perspective, is to interpret goal definition, to encapsulate the processing points where this goal is achieved, and to interact with work flow participants. Additionally, capabilities have a major effect on business processing (Lawrence 1997). They are able to execute the logical steps pertinent to their work flow in parallel, thereby reducing substantially the cycle time necessary to handle a particular goal. Thus, for example, a common task that requires 10 logical steps can have 5 of its logical steps executed in parallel (Lawrence 1997). Likewise, capabilities also support or enable the execution of their logical steps manually or in sequence.

Capabilities, like BOs, represent *complete* work flows that are responsible for specifying and managing a stable set of processing tasks. To handle these tasks, the work flows of capabilities are expressed as sets of smaller chunks of functionality that are easier to understand than high-level complex processing tasks (James, Hawick, and Coddington 2000). In essence, one can design capabilities with a high cohesion of their individual instructions. The basic definition of these instructions is completed by means of a systematic process that involves focused role assignation, their respective responsibilities, the services that cope with these responsibilities, and their work flow collaborators (Fayad, Sanchez, and Hamza 2004).

Capabilities (Sidebar 6.2, Fayad 2002a, 2002b) also complement the goals of a given concept by guiding themselves systematically toward their goal achievement. In fact, they are the enduring business workhorses of the goals. These capabilities or workhorses behavior is driven forward by a set of enduring business rules, which are encapsulated by a number of goals. Each capability has an ultimate goal responsible for constraining the capability's internal behavior toward a nondeterministic outcome without causing struggles. The main idea of literally embedding the capability's ultimate goal within the capability's work flow is to ensure and ascertain a business-centric behavior that is usually consistent with the rationale or goals of interest. This idea neatly corroborates and supports two things: a capability without a goal is not a capability at all, and capabilities are recursively related to the goals without losing generality. Together, they form the basis for the generation of many applications across the spectrum of interest, such as transactions and services in service-oriented architectures (Fayad and Kilaru 2005).

Similar to goals, capabilities are the enduring artifacts, but with a minor difference: they are externally adaptable, via extension points or hooks (Fayad, Sanchez, and Hamza 2005). Their adaptable natures can be determined by examining the relationships between the underlying business, and their direct application, and the application of the right *hooking code*. These relationships can be inheritance, aggregation, and associations. Rather than focusing on the generalization and specification problems, the *hooking code* is mainly responsible for weaving business and the actual application together on a real-time basis. An important point here is that BOs (Sidebar 6.1) are not directly adapted by the industrial objects (IOs); in fact, they are not. Rather, hooks create an environment where capabilities are able to attach to any IO without changing the capabilities internal structure and without a bit of chance of a collapse.

Further sections in the book introduce how capabilities form part of the work flow metaphor specification and how they consolidate a complete goal processing. Concretely speaking, they make sure you are properly achieving the goal that you discovered during the analysis phase in a complete and accurate manner. That is, they represent how the goal definition should happen, rather than focusing on what needs to be performed, that is, the solution space. The following section represents such a work flow metaphor and what its role is in a definition of the capabilities and their abilities facing a set of undetermined contexts.

6.1.2 Work Flow Metaphor—Workhorses

Metaphors are useful and beneficial for introducing a new idea or concept to a particular cast (Odell 1998). Their application to BOs is quite common, especially for specifying and managing their work flows (Fayad, Hamu, and Brugali 2000). In the capabilities domain, these work flows or business processes are well defined with a stable nature in mind. Nonetheless, they provide the right type of facilities to cope with determined contexts of applicability. Using work flows, as a metaphor, is always interesting and compelling, because it allows us to capture and understand those essential *high-level* sequences of activities required to fulfill system behavior. In fact, they dramatically allow complex business processing to be streamlined among work flow participants in determined environments (Fayad, Hamu, and Brugali 2000) and ease their execution and management.

Work flow metaphors have a strong effect on business process definition and understanding. They provide the capabilities for constructing enduring business processes. They also ensure that each of the activities they enclose and divulge is an essential activity, a complete one, of the target environment or subject matter. The interactions of the activities, internal and external, cover detailed (enduring) and dynamic (nested) processing according to the target environment's conditions and opportunities. To assure proper execution and management of these activities, BOs or capabilities must contain the right type of services. Using the divine mantra *divide and conquer* as the starting point, we could write and postulate that a proper and focused definition of a capability's services depends on how you distribute the load of each service to achieve the capability's responsibility, hence providing a proper work flow execution.

One can express the work flows of capabilities as a set of smaller chunks of functionalities that are easier to understand than those *high-level* complex processing tasks. This functionality is one to one and is mapped with the services the capability provides. How to define these services will determine how accurate the work flow is and how one perceives the enduring quality along the entire work flow. We will not cover, in this section, the identification of the services of capabilities work flows. However, we will describe the process to identify these capabilities. The next section describes such a process.

6.1.3 Dealing with Capabilities: Identification and Assessment

Capabilities are adaptive concerns that represent the solution space of a problem of interest. They embody knowledge skills, processes, and abilities associated with the execution of a specified course of actions or actions work flow. Due to their extreme adaptive nature, they also ensure a reduced cycle time for coping with a vast number of transient requirements and handling other goals and capabilities in a determined topology of patterns. This peculiarity of capabilities allows both demand adaptation and flexibility against IOs or transient aspects, and on-demand scalability of the environment to expand the abilities needed to achieve a goal. The above behavior of capabilities is, in the end, translated as a faster return of investment (ROI), while still maintaining a high-quality solution. This means that a customer could manage his or her products basic functionality (e.g., add, remove, override, and extend), when facing new requirements without the necessity of reinventing the wheel.

A tendency or trend exists with developers to be more familiar or accustomed with capabilities than with goals. One reason here is that they are usually exposed to capabilities on a daily and consistent basis. However, this *familiarity with capabilities* does not mean that capabilities are obvious to extract whatsoever; in fact, their extraction is a challenging activity due to different reasons. For instance, following are the common difficulties experienced when trying to extract capabilities:

1. Software practitioners do not know or understand if they have discovered a capability or a goal due to the conceptual nature of both. For example, one can see a collection as both a goal and a capability. There is a fuzzy line between the identification of goals and capabilities.
2. Software practitioners have quite a bit of difficulty knowing and understanding, whether they have discovered the right capabilities for a determined goal. For example, what are the capabilities of branding? By reading a problem statement, practitioners will consider position, advertisement, promotion, market, product, brand, and so on. However, they should be looking for the right ones: AnyEntity, AnyBrand, AnyMechanism, AnyParty, and AnyIdentity (Sanchez 2005).
3. Software practitioners may misunderstand the real nature of capabilities. Instead of thinking in terms of the enduring principles that drive their business, they always start thinking in terms of an application's artifacts. From the previous example, we have the following application-centric artifacts, which are part of the marketing field: product, market, promotion, and so on What if we used and employed branding for identifying humans in the United States. It is obvious that we have a different set of application, where we will not have the objects: promotion, market, product, and so on.
4. Belaboring the rareness of a one-to-one mapping between goals and capabilities (Hamza and Fayad 2002), a goal can be mapped into several capabilities. This increases the complexity of extracting the right capabilities.
5. There are also cases or instances, when candidate capabilities are off content and are thus recursively related to other capabilities and not to goals. This is understood by delving into the capability's context of applicability.

Hamza and Fayad (2002) described a process for extracting capabilities or BOs that support a determined goal of interest. This process implicitly conveys the idea that capabilities must be focused with the sole purpose of avoiding practitioners to be bogged down with many irrelevant details (e.g., application-specific details). An important remark here is that this process was addressed from a point where we already knew and understood the goal of the subject but that goal did not have capabilities in reality. In this chapter, we focus especially on *how does the subject do it*, by describing two ways for dealing with capabilities: first, when we have a goal with no capabilities, and second, when we have one capability with no goal at all. One needs to evaluate the capabilities and filter down then by using a set of heuristics, provided in the form of questions.

Before providing the capabilities identification process, we have to step back or retract and understand that two perspectives will drive this process' success. First is that we will have, in most of the cases, a goal with no capabilities. In this case, our important job is to identify those capabilities and the relationships that stem from them and provide that expected and cherished harmony between those capabilities and their goals. The second is that each capability has a second level of abstraction or level of granularity, where it may or may not have a set of collaborators or other capabilities interacting with it to satisfy an internal goal. In this case, our immediate task will be to identify the internal goal and the set of collaborators (capabilities) that will aid in the definition of the capability's second level of abstraction. You can use the process described in Chapter 3 to identify this goal, but now the subject explored will be the selected capability. Other sections of this book will cover the processes of both perspectives.

6.1.3.1 The Impact of BO on Creating Multiple Applications: Generality

BOs, along with goals, serve as an essential framework for building important applications. Because BOs represent the capabilities to achieve goal, they do not always depend on any application in particular. BOs abstract the actual application-specific IOs. Thus, if we have corrected BOs at our disposal, infinite number of applications can be built on top of them easily by just attaching specific IOs.

6.1.3.2 BO = Stable Design Patterns

BOs provide high-level and accomplished designs to which one can plug application-specific IOs. Because BOs are abstract and generalized, we need to arrive here with them, before implementing the actual application; hence, they represent stable design patterns.

6.1.3.3 Essential Properties of BO

A complete description of the BOs essential properties can be found in Fayad (2008):

- Timeless notion patterns
- Working horse of the system patterns
- Adaptability patterns
- True presentation of the solution space patterns
- Management work flow metaphor patterns
- Domain-independent patterns

6.1.4 A Goal with No Capability

Even though the identification of capabilities seems to be a little bit less complicated and complex than the identification of goals, their inherent processes require the same amount and care as do those for the identification of goals. Therefore, we will provide in this section a set of questions and heuristics that will guide you throughout the entire identification process. By following these questions, the success of this identification process in terms of finding the right capabilities will be more tangible. The process of identifying the capabilities of a goal is described herein.

1. Set the context, where all candidate capabilities will need extraction. We determine the context by asking, "What is the subject of interest?"
2. Identify the goal(s) of interest by asking, What is the subject for?
3. Determine a focused problem or subject understanding by asking, Can we divide this subject into smaller chunks of understanding? To support this step, we use the following questions:
   a. How can we approach the underlying goal?
   b. What do we need to fulfill this goal?

   c. What are we looking for with this goal in the determined subject of interest?
   d. Who is going to use the goal?
4. Filter the entire list of found capabilities.
   a. Does the outcome apply to the goal itself, or is it just part of a wish list of the stakeholder? If it is part of a wish list, this is not a good candidate capability.
   b. Can the subject matter exist without this capability? However, is the capability redundant or nonexistent within the subject of interest? If yes, this is not a good candidate capability.
   c. Does the outcome have a physical representation in application-specific environments? If not, this is not a good candidate capability.
5. Use the prefix *Any* as a generality indicator for both atomic and capabilities. We will use the stereotype BO for atomic capabilities, and *Pattern-BO* for nonatomic capabilities. There is an exception for this naming rule. Capabilities that represent types of a unit or element will follow a determined naming convention: the UnitName + Type keyword. For example, you can rename types of resources as *ResourceType*.
6. Evaluate the list of candidate capabilities to assure their accuracy and relevance to the goal of interest. We can use the following questions to support this step:
   a. What is the relationship between the capability and the target goal? Is it purely obscure or confusing? Alternatively, is it clear? If it is clear, this may be a good candidate capability.
   b. Is the capability's execution and management bound to external conditions and abilities related to a specific context? If yes, this is not a good candidate capability.
   c. Does the capability comply with the so-called stateless class definition? If yes, this is not a good candidate capability. A good example of this is Chotin.
7. Model the found capabilities and goals using UML notation.

The next section will mainly concentrate on the second perspective for dealing with capabilities. This specific and fixed way concentrates on providing the right abstraction level for each one of the extracted capabilities. That is, we must delve into the second level of abstraction of certain capabilities and define their ultimate goals and the set of capabilities that support their ultimate goal.

6.1.5 A Capability with No Goal

Once you have identified all the capabilities of a determined goal, your need to take each capability in isolation for immediate perusal or follow-up. During this perusal, we will identify the ultimate goal of the isolated capability. In addition to this, we will also identify, if they exist, the rest of the capabilities that complement the isolated capability's ultimate goal.

Capabilities have their own internal goal to fulfill. The fulfillment of this goal depends on how well software practitioners have examined the isolated capability's structure in search of the pieces that synthesize and realize, as a whole, the rationale of this capability in question.

TABLE 6.1
Scenario Format

Scenario 1: Scenario Name	
Definition	**Business Rules**
Concept of interest definition	List of business rules of the concept

To simplify this examination process, we can provide a set of questions and heuristics that will guide software practitioners throughout the entire examination process. This examination process is described herein.

1. What is the subject of interest? In this case, the selected capability is the subject of interest.
2. What is the subject matter? The process described in Section 3.4 can help one come up with the right goal.
3. What are the core elements of the subject matter? This question is addressed by means of the following steps:
   a. Specify a set of scenarios, where the subject matter is present.
   b. Describe each of the scenarios by identifying the pertinent business issues or rules that drive the subject matter's realization.
   c. Make sure that each of the scenarios is described from a domain-independent perspective. Table 6.1 depicts what we mean and understand by the word *scenarios* and the *business rules* that were extracted from them. The number of scenarios can be almost infinite.
   d. Abstract the business rules or issues shared by all the scenarios
   e. Identify the formal axioms that constrain business rules' application across several scenarios.
   f. Does each one of the capabilities have a physical representation? If yes, these are tentative core elements of the subject matter.
   g. Discuss the results with the technical cast and look for consensus of the core elements of the subject.
   h. Continue with this process, until you have covered all the capabilities exploration.

In order to support the previous process, we will provide an example (in the next section), where we put in action and work all the questions and heuristics previously described.

6.1.6 Identification Process of Capabilities: An Example

Let us select a possible goal that is easy to explore in detail, so that we can start finding its capabilities. We will walk through the steps of the process with the CRC cards example described in Fayad, Sanchez, and Hamza (2004). The following illustrates the application of heuristics and the questions for extracting the capabilities of a goal:

1. We will determine the context by asking, What is the subject of interest? In this case, it will be *CRC cards*.
2. What are CRC cards' goals? One of CRC cards' goals is brainstorming.

3. Can we divide the brainstorming concept into smaller chunks of understanding? Yes, we can. See below.
   a. How can we approach the brainstorming concept? The result: location where a participant will brainstorm, rules for doing brainstorming, time for brainstorming, brainstorming media, and so on.
   b. What do we need to fulfill this brainstorming goal? We need motivation or interest for doing brainstorming, processes for doing brainstorming, forms of brainstorming, the target context where brainstorming is applied, the specific topic of interest, engagement among participants, and so on.
   c. What are we looking for with the brainstorming goal in the CRC cards context? To complete a brainstorming session, to explore a context, and to produce a list of new ideas or candidate classes represented by CRC cards, ideas assessment, and so on.
   d. Who is going to perform brainstorming in the CRC cards context? The results are facilitators, analysts, designers, scribes, and so on.
4. We will also filter all the identified capabilities with the sole purpose of avoiding being constrained with irrelevant details associated with the application side of brainstorming.
   a. Does the outcome (found capabilities) apply to the brainstorming goal itself, or is it just part of a wish list of the stakeholder? The result: new ideas and idea assessment will be removed.
   b. Can the CRC cards approach exist without some of the found capabilities? Alternatively, are some capabilities redundant within the CRC cards approach? The result: the capabilities, such as time for brainstorming, rules for brainstorming, location where brainstorming is carried out, and the brainstorming session, will be removed.
   c. Does the outcome (found capabilities) have a physical representation? The result: this capability is removed: motivation or interest.
5. Use the aforementioned naming conventions with the left capabilities. The result: *AnyMedia*, *AnyForm*, *AnyContext*, *AnyEngagement*, *AnyParty*, *AnyTopic*, and *AnyMechanism*.
6. Evaluate this list of capabilities.
   a. What is the relationship between each one of the capabilities and the brainstorming goal? The result: *AnyTopic* capability will be removed, because its association with the goal is too vague and almost redundant. *AnyContext* is already covering the *AnyTopic* role.
   b. Is each one of the capabilities' execution and management bound to external conditions and abilities related to a specific context? The result: none of the current capabilities is bound to external conditions. Therefore, they will not be removed.
   c. Does each one of the capabilities comply with the so-called stateless class definition? The result: none of the current capabilities is stateless classes.
7. The final list of capabilities: *AnyMedia*, *AnyForm*, *AnyContext*, *AnyEngagement*, *AnyParty*, and *AnyMechanism*.

The second phase is to pick, select, choose, and explore only one capability at a time. We do this with the sole purpose of finding each capability's internal goal, along with other capabilities that may complement the picked capability's internal goal. To support this process, we will use the *AnyMechanism* capability as the subject to be explored.

1. What is the subject of interest? In this case, it is the *AnyMechanism* capability.
2. What is *AnyMechanism* used for? The process described in Chapter 3 is used to come up with the right goal. After following this process, we will come up with the following goal: computing.
3. What are the core elements of *AnyMechanism*? This question involves the following:
   a. Specify a set of scenarios, where *AnyMechanism* is present. We can use the scenario format illustrated in Table 6.1.
   b. Describe the business rules of the *AnyMechanism* capability. Table 6.2 shows the result.
   c. For example, take the business rules or issues shared by all the scenarios. For the sake of simplicity, we did not include all the business rules that we used to extract the atomic structure of *AnyMechanism*. Based on the scenarios, we found the following capabilities: *AnyAlgorithm*, *AnySequence*, *AnyPeriod*, *AnySignature*, *AnyType*, *AnyClass*, and *AnyMedia*.
   d. Define the pertinent axioms of the *AnyMechanism* capability. For example, in order to run the sequence of steps, a period must be defined and described. The aggregation mechanism into the sequence must be ordered.
4. The final list of capabilities, along with the *AnyMechanism* goal, is computing (goal), *AnyMechanism*, *AnyAlgorithm*, *AnySequence*, *AnyPeriod*, *AnySignature*, *AnyType*, *AnyClass*, and *AnyMedia*.

The complete model of *AnyMechanism* can be seen in section on the capabilities of knowledge maps.

TABLE 6.2
***AnyMechanism's* Scenarios**

Definition	Business Rules
Scenario 1: Chemical Reactions	
An atomic process that occurs during a chemical reaction	A systematic sequence of steps or reactions that influences a chemical change
	A process occurs in natural phenomena
	Represents the implementation of an algorithm
Scenario 2: Earth's Sunlight Process	
A process that has occurred on earth due to a chain of causes	A process on the earth that occurs according to a determined period
	Certain parameters determine when the light will be perceived on earth
	A process that consists of other processes, such as Earth orbit, Sun, and nuclear reactions

6.2 CAPABILITIES OF KNOWLEDGE MAPS

This section covers the capabilities of knowledge maps. They are illustrated as stable design patterns (see Chapter 2 for reference). Nine capabilities drive the formation of the knowledge map. Table 6.3 summarizes these capabilities.

6.2.1 Capability 1

- *Name.* AnyMechanism stable design pattern
- *Context.* The pattern is trying to capture the essentials or the core knowledge of any mechanism concept whenever it appears.
- *Problem.* How to model the core abstractions of a concept that spans multiple application domains.
- Solution and participants
  - *Solution* (Figure 6.1)
  - *Participants*
    - *Classes*
      - *AnyMechanism.* It represents or signifies the process or logical steps to perform a determined activity.
    - *Patterns*
      - *AnyMedia.* It identifies and defines the media upon which the mechanism will be executed. It also represents the media, by which the sequence of logical steps will be executed.

TABLE 6.3
Capabilities of Knowledge Maps

Capability	Description	Provided?
AnyMechanism	AnyMechanism represents the process or logical steps to perform a determined activity	Yes
AnyView	It is the view of a model that extracts the essential information relevant to a particular purpose and ignores the remainder of the information	Yes
AnyModel	It visualizes the relevant details of a subject or discipline, while ignoring the irrelevant details	Yes
AnyLevel	It represents the level of abstraction that a concern or concept can be represented	No
AnyContent	This deals with the sum or range of what (patterns) has been perceived, discovered, or learned in a particular discipline	Yes
AnyContext	This defines what elements are/are not part of the problem under discussion	No
AnyArchitecture	This refers to the integration of two or more patterns. Architecture should contain more than one EBT or goal[a]	No
AnyPattern	This is the best solution for a set of recurring problems or events	No
ExtensionPoint	They represent "the slots, knobs, and dials that must adjust in order to adapt the framework to your context"[a]	No

Source: [a]Fayad, M.E. *Stable Design Patterns for Software and Systems.* Boca Raton, FL: Auerbach Publications, 2015.

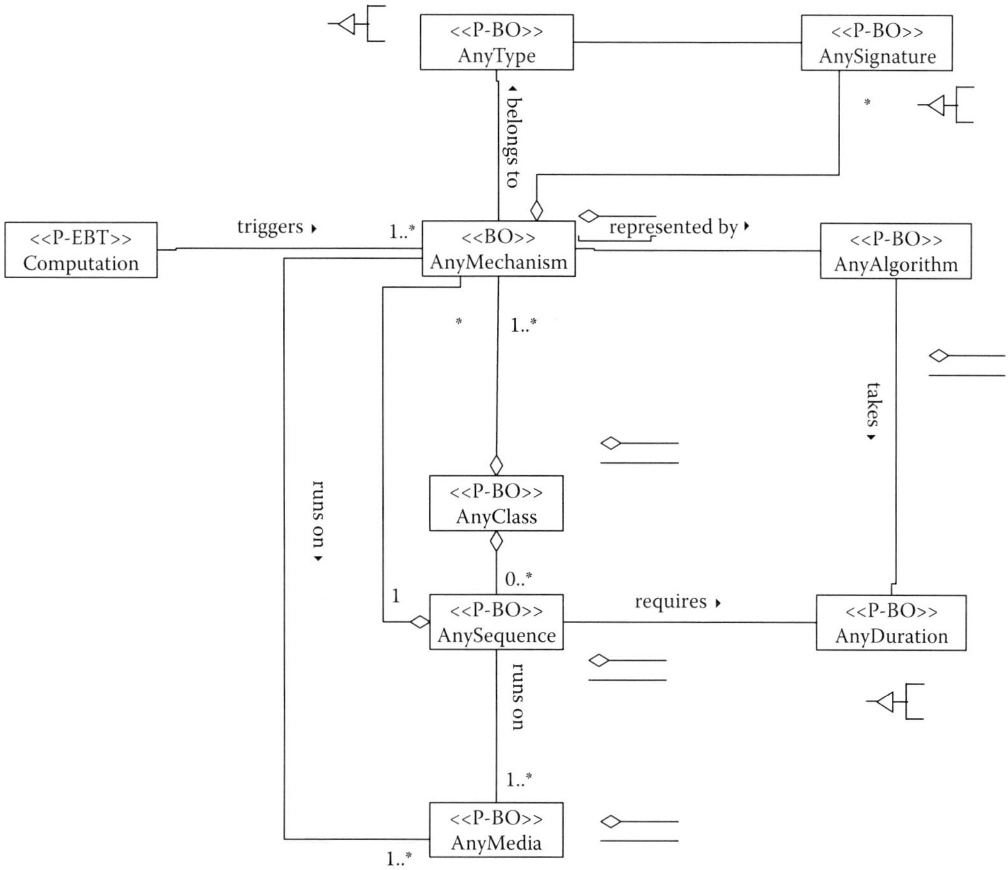

FIGURE 6.1 The AnyMechanism stable design pattern.

- *AnyClass.* It classifies the common properties and behavior for a set of specific kinds of objects.
- *AnyType.* It tags the interfaces that form the mechanisms. The provided signatures of a set of services classify it.
- *AnySequence.* It represents the logical order in which a set of instructions is executed in a determined period.
- *AnyPeriod.* It indicates the amount of time duration required by the sequence of steps to be executed in accordance with certain algorithm.
- *AnyAlgorithm.* It indicates a set of systematic rules that produce a determined outcome or solution.
- *AnySignature.* It represents the specification of an instruction that is part of a mechanism.

6.2.2 Capability 2

- *Name.* AnyView stable design pattern (Fayad, Islam, and Hamza 2003)
- *Context.* It represents the core knowledge for the situations in which a view may occur.
- *Problem.* How can we model the core knowledge of a concept that spans multiple application domains?

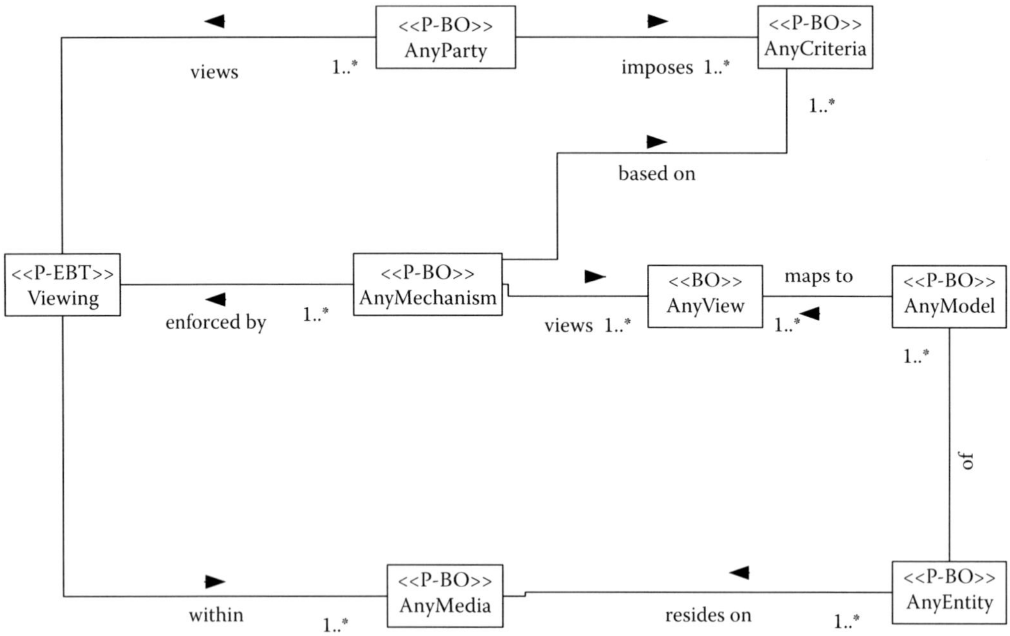

FIGURE 6.2 The AnyView stable design pattern. (From Fayad, M. E. *Software System Engineering*, Lecture Notes, Computer Engineering Department, College of Engineering, San Jose State University, San Jose, CA, 2002–2014.)

- *Solution and participants*
  - *Solution* (Figure 6.2)
  - *Participants*
    - *Classes*
      - *AnyView.* It represents the view of a collection of data (the model).
      - *Viewing.* It describes the viewing process and rules.
    - *Patterns*
      - *AnyParty.* This represents the viewer. This viewer perceived the mapped view of a requested model.
      - *AnyMedia.* It identifies and defines the media upon which the views are mapped and transmitted. It also represents the media by which the views are to be displayed.
      - *AnyCriteria.* It describes the properties that govern specific kinds of views.
      - *AnyEntity.* It describes the entity where models and views are produced.
      - *AnyModel.* It describes the model of any entity.

6.2.3 Capability 3

- *Name.* AnyModel stable design pattern (Fayad, Islam, and Hamza 2003)
- *Context.* A model is a significant representation of a subject. The model concept spans and stretches across multiple application domains.
- *Problem.* How can we capture the reusable essentials or the core knowledge of a model?

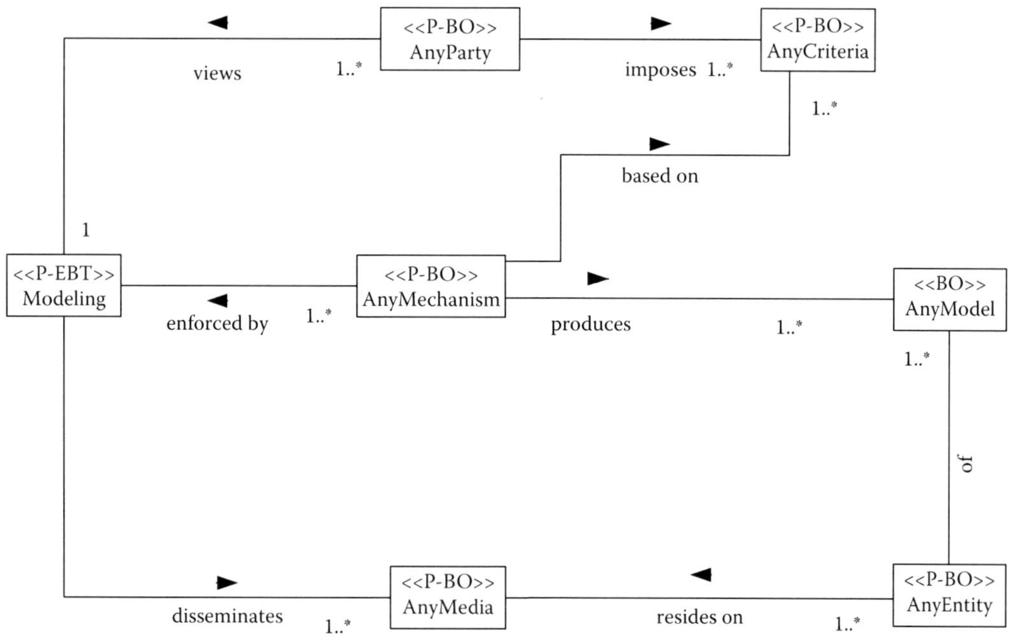

FIGURE 6.3 The AnyModel stable design pattern. (From Fayad, M. E. *Software System Engineering*, Lecture Notes, Computer Engineering Department, College of Engineering, San Jose State University, San Jose, CA, 2002–2014.)

- *Solution and participants*
  - *Solution* (Figure 6.3)
  - *Participants*
    - *Classes*
      - *AnyModel.* It describes the model of a determined application (e.g., plane manufacturing and game development). In other words, a model is a representation of data within a specific application.
    - *Patterns*
      - *Modeling.* It defines the modeling process.
      - *AnyParty.* It represents the modeler. The modeler is responsible for building the data models in an appropriate abstract level.
      - *AnyMedia.* It identifies and defines the media upon which the models are built or exchanged.
      - *AnyCriteria.* It describes the properties that govern specific kinds of models.

6.2.4 Capability 4

- *Name.* AnyContent stable design pattern
- *Context.* From an informational perspective view, content is a special concept that represents any type of information in a digital context, such as web pages' content and files' content.
- *Problem.* How can we capture the core knowledge of any content focusing on the information category?
- *Solution and participants*

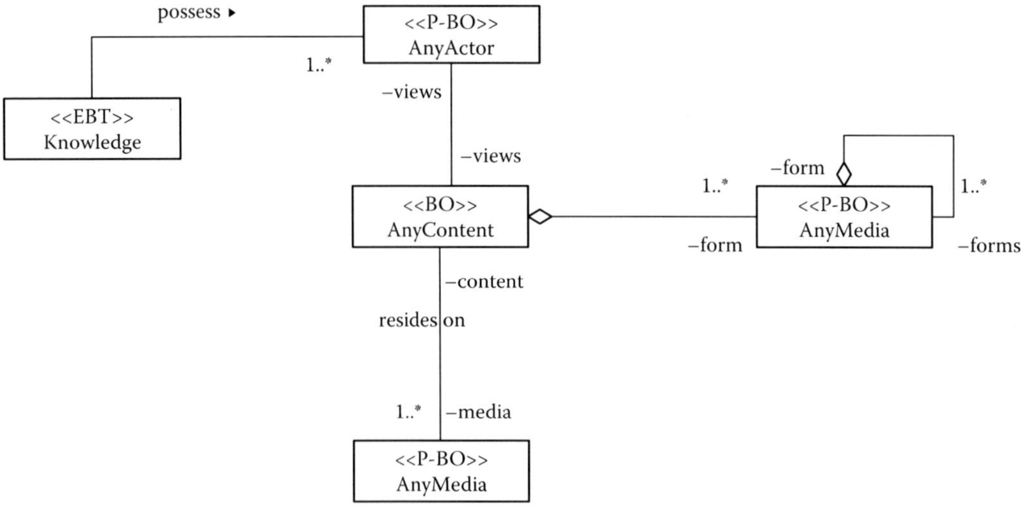

FIGURE 6.4 The AnyContent stable design pattern (From Fayad, M. E. *Software System Engineering*, Lecture Notes, Computer Engineering Department, College of Engineering, San Jose State University, San Jose, CA, 2002–2014.)

- *Solution* (Figure 6.4)
- Participants
  - *Classes*
    - *AnyContent.* It describes the type of information in a digital context.
  - *Patterns*
    - *AnyActor.* The actor who successfully generates the content
    - *AnyMedia.* It identifies/defines the media upon which the contents are generated and exchanged.
    - *AnyForm.* It identifies/defines the different types of content in that digital context, such as text, images, audio, and symbols.
    - *Knowledge.* It represents the gained experience that will be used as the basis for the actor to generate content.

SUMMARY

By applying the software stability concepts approach to the definition of capabilities, we have concluded that capabilities are the true BOs that allow software practitioners to achieve determined goals in a systematic manner. They will always comprise a focused logical processing or work flow consistent with a set of enduring business rules. The following summarizes the information covered in this chapter:

1. Capabilities are knowledge-centric BOs with an ultimate goal in mind and view. A capability without a goal means a capability with an obscure purpose in a determined system.
2. Capabilities are internally stable and externally adaptable. This means that development of capabilities involves both the internal norms and compliance-logical steps forming the capabilities work flow and the comprising of *hooking code* that enables capabilities to adapt to external application-specific environments.

3. Capabilities collaborate with other capabilities, working together to achieve a common goal, which is the extension and adaptation of their periphery (behavior and structure) to maximize their goal's success. We usually focus on the implementation of gluing or gelling points, where the capabilities can associate other capabilities, rather than focusing on the capabilities per se.
4. Capabilities' work flow interoperability is defined with a role in mind, which enables the responsibility of each of the capabilities involved in the goal's resolution. Being aware of the responsibility of each capability will enable a designer to create the right abilities or services of a particular capability.
5. Dynamic change in capabilities' abilities is allowed. Capabilities are embedded with the ability to dynamically change their abilities through extension points that are responsible to add, remove, update, override, and extend functionality on a real-time basis.

In summary, the capabilities of a discipline are important and critical aspects attempting to encapsulate or encompass the business processes and categories of a business-centric theme or goal. When they are directly associated with their goals, they form a synergetic force indented to represent the groundwork of any discipline's understanding: the knowledge maps.

OPEN RESEARCH ISSUES

Capabilities are conceptual concepts, and hence, finding or detecting them requires thorough knowledge and acquiring skills of the domain for which capabilities need to be found. In addition, the goal(s) must be known and identified, before hand, in order to find the correct BOs. Thus, finding correct capabilities comes naturally only through intense and regular practice.

REVIEW QUESTIONS

1. Discuss the following statement: capabilities/BOs are heart of business.
2. What is the responsibility of capabilities from the standpoint of knowledge maps?
3. Justify: capabilities have a major effect on business processing.
4. Explain how BOs represent complete work flow for specifying and managing a stable set of processing tasks.
5. What drives the capabilities behavior?
6. What is the main difference between capabilities and goals?
7. What are the two main ideas corroborated by business-centric behavior of capabilities?
8. Is the following statement true or false? BOs are directly adapted by the IOs.
9. ____________ create an environment, where capabilities are able to attach to any IO without changing the capabilities' internal structure.
10. What is the main function of hooks?
11. What mechanism is used to attach industrial-specific IOs to BOs?
12. Is the work flow metaphor of workhorses apt for capabilities? Explain.
13. Which behavior of capabilities can be translated into faster ROI with high-quality solution?

14. List all the difficulties experienced by software practitioners, when extracting capabilities for a domain.
15. What are the two scenarios under which a software practitioner needs to deal with capability? Explain each one of them.
16. What is the impact of BOs on creating multiple applications?
17. Explain how BOs represent stable design patterns.
18. Enlist essential properties of BOs and explain each one of them briefly.
19. What steps need to be followed for identifying capabilities given a goal?
20. What steps need to be followed for identifying capabilities given a capability with no goal?
21. Compare and contrast the process of extracting capabilities, when goal is given to the process of extracting capabilities, when one capability and no goal are given.
22. Explain all the capabilities of the knowledge map using the standard template given in Section 6.2 as shown in Table 6.3.
23. A capability without a goal means a capability with an obscure purpose in a determined system. Explain.
24. Is the following statement true or false? Capabilities collaborate with other capabilities toward a common goal.
25. Is the following statement true or false? Dynamic change of capabilities' abilities is not allowed.
26. Why is capabilities' work flow interoperability definition defined with a role in mind?
27. What are the differences and similarities between traditional BOs and knowledge map's BOs?

EXERCISES

1. Find out if following are goals or capabilities
   a. Personalization
   b. Dynamism
   c. Secrecy
   d. Collection
   e. Storage
   f. Perfection
   g. Entity
2. Find the rest of the capabilities of the following concepts:
   a. AnyProject
   b. AnyRevenge
   c. AnyClaim
   d. AnyOption
   e. AnyPresentation
   f. AnyAlarm
   g. AnyEvent
   h. AnyTask
   i. AnyHabit
   j. AnyCrime

3. Name the BOs or capabilities of the ultimate goal(s) of the following classic movies:
   a. *Lagaan*
   b. *The Lord of the Rings* trilogy (2001–03)
   c. *Titanic* (1997)
   d. *Toy Story* (1995)
   e. *The Silence of the Lambs* (1991)
   f. *Crumb* (1995)
   g. *The Lion King* (1994)
   h. *Shrek* (2001)
   i. *The Breakfast Club* (1985)
   j. *Speed* (1994)
   k. *Scarface* (1983)
   l. *Fatal Attraction* (1987)
   m. *Ghostbusters* (1984)
   n. *Dirty Dancing* (1987)
   o. *Back to the Future* (1985)
4. Name the BOs or capabilities of the ultimate goal(s) of the following classic books:
   a. *The Road*, Cormac McCarthy (2006)
   b. *Harry Potter and the Goblet of Fire*, J. K. Rowling (2000)
   c. *Beloved*, Toni Morrison (1987)
   d. *The Liars' Club*, Mary Karr (1995)
   e. *American Pastoral*, Philip Roth (1997)
   f. *Mystic River*, Dennis Lehane (2001)
   g. *Cold Mountain*, Charles Frazier (1997)
   h. *Watchmen*, Alan Moore and Dave Gibbons (1986–1987)
   i. *Black Water*, Joyce Carol Oates (1992)
5. Name the BOs or capabilities of the ultimate goal(s) of the following classic TV shows:
   a. *The Simpsons*, Fox (1989–present)
   b. *The Sopranos*, HBO (1999–2007)
   c. *Seinfeld*, NBC (1989–1998)
   d. *The X-Files*, Fox (1993–2002)
   e. *Sex and the City*, HBO (1998–2004)
   f. *Survivor*, CBS (2000–present)
   g. *The Cosby Show*, NBC (1984–1992)
   h. *Friends*, NBC (1994–2004)
   i. *The Oprah Winfrey Show*, Syndicated (1986–present)
   j. *American Idol*, Fox (2002–present)
   k. *Beverly Hills, 90210*, Fox (1990–2000)
   l. *Star Trek: The Next Generation*, Syndicated (1987–1994)
   m. *Miami Vice*, NBC (1984–1989)
   n. *L. A. Law*, NBC (1986–1994)
   o. *Moonlighting*, ABC (1985–1989)
   p. *Planet Earth*, Discovery Channel (2007)
   q. *The Golden Girls*, NBC (1985–1992)
   r. *Prime Suspect*, ITV (1991–2006)

PROJECTS

1. Form a group and discuss other ways to find capabilities.
2. Identify and model two to four ultimate goals, their capabilities, and connect them together for the following domains.
   a. Manufacturing
   b. Modeling
   c. Requirement analysis
   d. Customer relationship management
   e. Database
   f. Project
   g. Kitchen

   Identify the common capabilities in each of the models.
3. Identify and model the class diagrams by using 2–3 ultimate goals and their capabilities for each of the problem statements in Appendix D. Identify the common capabilities in each of the models.

SIDEBAR 6.1 Business Objects

The traditional definition of a business object is that they are objects in an object-oriented computer program that represent the entities in the business domain that the program is designed to support. For example, an order entry program might have a BO to represent each order, line item, and invoice. BOs are sometimes called *domain objects*; a domain model represents the set of domain objects and the relationships between them. BOs often encapsulate all of the data and business behavior associated with the entity that it represents. They do not necessarily need to represent objects in an actual business, though they often do. They can represent any object related to the domain for which a developer is creating business logic. The term is used to distinguish between the objects a developer is creating or using related to the domain, and all the other types of object he or she may be working with, such as user interface widgets and database objects such as tables or rows.

Technically, BOs encapsulate traditional lower-level objects that implement a business process (i.e., they are a collection of lower-level objects that behave as single, reusable units). User interfaces can be thought of as views of large-grained BOs. Databases maintain a record of the *state* of BOs as they change over time (Sutherland 1997).

A BO is an object that is modeled after a business concept, such as a person, place, event, or process. BOs represent real-world things such as employees, products, invoices, or payments. To remain competitive, modern day enterprises need information systems that serve and adapt to their complex needs. Applications designed from the ground upward (not hacked) by using the BO model are better suited to meet the requirements of rapidly evolving businesses Sutherland (1997).

Our definition of BO is completely different from all of the above. Our BOs are capabilities that are used to achieve the business goals, which we call them, enduring business themes (EBTs) (Fayad 2002a, 2002b, 2015; check Chapter 6 for more information).

REFERENCES

Fayad, M. E. "Accomplishing Software Stability." *Communications of the ACM* 45, no. 1 (2002a): 111–115.

Fayad, M. E. "How to Deal with Software Stability." *Communications of the ACM* 45, no. 4 (2002b): 109–112.

Fayad, M. E. *Stable Design Patterns for Software and Systems*. Boca Raton, FL: Auerbach Publications, 2015.

Sutherland, J. "Business objects in corporate information systems." *ACM Computing Survey* 27, no. 2 (1995): 274–276.

Sutherland, J. "The Object Technology Architecture: Business Objects for Corporate Information Systems." In *Business Object Design and Implementation*, Sutherland, J. V., D. Patel, C. Casanave, J. Miller, and G. Hollowell. eds. Springer, 1997.

Sutherland, J. "Business Object and Component Architectures: Enterprise Application Integration Encounters Complex Adaptive Systems" (invited paper). *IEEE Hawaii International Conference on Software Systems*, 2001.

SIDEBAR 6.2 Learning about Capability

The basic term of capability was coined first by Dennis and Van Horn (Dennis and Van Horn 1966). According to them, the basic idea of capability is just like a token that can designate an object and give the program an authority to carry out a specific and unique set of actions on the given object. The token in this context is the capability.

A capability is very similar to the keys on your key ring, such as a car key or house key, or password. It is just like the password to be used to log into your computer system. Just consider this simple example: the password that you use can open only your computer and it is specific to your machine. Anyone who has your password can open your computer without any problems. It means that your password does not know or identify who is holding the password, be it you or anyone else.

Computer login passwords can come in several variations. One common type of password is the computer boot password, which starts or boots your computer, whereas the other one is the data encryption password that can help you open your sensitive files and folders. Though both passwords perform the same action, which is opening up something, the actions performed are entirely different.

Thus, *two capabilities can tag or designate the same type of object, but they will always authorize different set of actions! This is similar to capability-based security*, which is a concept in the design of secure computing systems. A *capability* (known in some systems as a *key*) is a communicable token of authority. It refers to a value that references an object, along with an associated set of access rights (Levy 1984, Miller et al. 2003).

In summary, one can delegate capabilities that mean that you can hand over the capability to anyone that you rely and trust. One can copy a set of capabilities, whereas the other can hand over the capability only after the clause of trust is acknowledged and obeyed. However, one can even change the nature of the capabilities by rescinding them, if the situation compels you to do so.

Our definition of capability is different from all of the other definitions: Dennis and Van Horn (1966); http://www.erosos.org/essays/capintro.html. Capabilities are equal and identical to BOs that are used to achieve the business goals, which we call them EBTs (Fayad 2002a, 2002b). Capabilities are adaptive concerns that represent the solution space of a problem of interest. They embody knowledge skills, processes, and abilities associated with the execution of a specified course of actions or actions work flow (check Chapter 6 for more information).

REFERENCES

Dennis, J. B., and E. C. Van Horn. "Programming Semantics for Multiprogrammed Computations." *Communications of the ACM* 9, no. 3 (1966): 143–55.

Fayad, M. E. "Accomplishing Software Stability." *Communications of the ACM* 45, no. 1 (2002a): 111–15.

Fayad, M. E. "How to Deal with Software Stability." *Communications of the ACM* 45, no. 4 (2002b): 109–12.
http://www.eros-os.org/essays/capintro.html.
Levy, H. M., *Capability-Based Computer Systems*, Digital Equipment Corporation 1984.
Mark S. M., Ka-Ping Yee, and J. Shapiro. *Capability Myths Demolished*, Technical Report SRL2003-02, Systems Research Laboratory, Johns Hopkins University, 2003.

SIDEBAR 6.3 Work Flow

Work flow is a depiction of a sequence of many operations, declared as works of a person, the work of a simple or complex mechanism, work of a group of persons (Fischer 2007), and work of an organization of staff, or machines. It may be seen as any abstraction of real work, segregated in work share, work split, or whatever the types of ordering. For control purposes, work flow may be a detailed view on real work under a chosen or selected aspect (Fischer 2005), thus serving as a virtual representation of actual work.

A work flow is a reliable, trusted, repeatable, and consistent pattern of activities empowered by a systematic and orderly organization of a number of resources, well-defined and specific roles and mass, energy and information flows, into a complete and wholesome *work process* that can be efficiently documented and learned. Work flows are designed to achieve specific processing intentions of some sort, such as physical transformation, service provision, or information processing (Jackson and Twaddle 1997; Sharp and McDermott 2009).

Better work flow will provide a number of benefits and advantages like improved business process efficiency, enhanced business process control, improved consumer service, flexibility and simplicity, and an overall improvement in business processes. The term is widely used in computer programming and designing to seek, develop, capture, and streamline man-to-machine communication. Work flow software scripts try to provide end users with a flexibility to create and design or describe complex processing of data in an easier way to understand visual form, much like flow charts, but without the need to understand computers or programming (Jackson and Twaddle 1997; Sharp and McDermott 2009).

Work flows as indicated at Zhu 2010, like traditional programs, allow you to coordinate the work. In addition, they can be defined as follows:

- Work flows can handle and manage long running work schedules by persisting on a durable store, such as a database, when idle, and loading again once there is a pending work.
- An instance of a work flow can be modified dynamically, while running the event that new conditions require the work flow to behave differently than it did when it was created.
- Work flows are a declarative way of writing programs by linking together predefined activities, rather than an imperative programming model of writing lines of code.
- They allow you to declare business rules that are separated from your code, thus making it easier for you to modify them in the future.
- They support different styles of systems with sequential and state machine work flows.

Work flows will also lead to EBTs (Fayad, Hamu, and Brugali 2000), which are the keys for developing stable object-oriented systems. One should manage and maintain the work flow to streamline the complex interactions between objects found in large-scale object-oriented applications.

Proponents of stable architectures, component-based, and pattern-oriented systems go so far as to suggest that work flow mechanisms should eliminate the need for most application programming in the workplace (Fayad, Hamu, and Brugali 2000).

REFERENCES

Fayad, M. E., D. S. Hamu, and D. Brugali. "Enterprise Frameworks Characteristics, Criteria, and Challenges." *Communications of the ACM* 43, no. 10 (2000): 39–46.

Fischer, L. *Workflow Handbook 2005*. Future Strategies Inc, 2005.

Fischer, L., ed. *BPM and Workflow Handbook*. Lighthouse point, FL: Future Strategies Inc, 2007.

Jackson, M., and Twaddle, G. *Business Process Implementation: Building Workflow Systems*, Addison-Wesley, ACM Press, July 1997.

Sharp, A., and McDermott, P. *Workflow Modeling*, Artech House Publishers, 2009.

Zhu, A. *Microsoft Windows Workflow Foundation 4.0 Cookbook*, Packt Publishing, September 24, 2010.

7 AnyMap Stable Design Pattern

We believe in *map* because in our view, it's what the customer needs and wants: To have access anywhere, through any kind of technology.

Jean-Marie Messier
Neligan 2006

The word *map* ordinarily means pictorial representation of some geographical area, but this word is not limited only to represent some geographical areas. It has wide applicability in many fields, like genetic maps that are used by biologists to analyze the genetic structure of humans in order to cure genetic diseases, nonspatial maps like Gantt charts that are used to display logical relationship among items, and spatial, but nongeographical, maps like star maps that are used by astronomers to present night sky and to locate astronomical objects like stars, galaxies, and constellations. In other words, a map can be defined as symbolic depiction highlighting relationships between elements of that space such as objects, regions, and themes. Some maps are static 2D representations, whereas others are dynamic 3D representations. Moreover, it is not necessary to always have a scale or context for a map like brain maps and genetic maps.

The main objective of this paper is to come up with a generic model of AnyMap, which can be used in any field. As a map has wide applicability, it can change its definition under different contexts; hence, it becomes necessary to define a generic pattern of AnyMap, which could fit all the applications. In order to achieve this goal, the concepts of software stability model (SSM) (Fayad and Altman 2001) will be applied and the pattern so developed will have no influence of application-specific knowledge. Rather, the pattern can be reused for different applications by just hooking the application-specific concrete objects to the pattern

7.1 INTRODUCTION

A map is a symbolic or graphical representation of any real or imaginary objects, regions, or themes in a particular space. They can be 2D or 3D, static or dynamic, logical or spatial, and geographical or nongeographical. The AnyMap pattern introduces this concept of map in a very simple and general manner by using the concept of SSM. This chapter also develops a stable design pattern of the same and discusses the applicability of the pattern in wide applications from different domains. The usage and utilization of map may vary from very simple to very complex, and it may also rely on the usage in many domains for many different purposes. Below are some examples of usage of map and some of its purposes:

General reference. Whenever we need to find some place and we do not know where to look, then we should start with an atlas with an index. There are many sources of maps that can be consulted for general reference, including maps posted in public places. With access to the web and a good search engine, we now have another source of maps to find out *where* places are in our world.

Navigation, control, and route planning. Whether we move on land, at sea, or in the air, we rely heavily on maps to plan our routes and to maintain our path. We also have hiking and biking maps, maps for crawling through caves or orienteering through woods, highway and off-road maps, as well as nautical and aeronautical charts. In addition, there are maps to display rapids in white water and fishing structure in lakes. We can use these maps to plan our routes and then to navigate, when we are on the move.

Communication, persuasion, and propaganda. Many maps are designed and produced to convey a particular image or communicate a particular idea. Because map data must be classified and represented by symbols, in almost all cases the image cannot be very general. Maps that appear in newspapers, accompanying an article, or on TV with a report aid in telling the story. Very often, those maps are not neutral in terms of the message conveyed. Under this category, we might include maps that are used to route traffic in specific directions; maps are employed to get people to register to vote, whereas a number of persons select map projections to convey a specific image; this is particularly true of the Peters projection, which advances the argument that it is time for a new image of the world. And the Australian maps that show the South Pole at the top of the map belong here.

Planning. Because *where* is important, we would use maps to determine where we want to do what. We also turn to maps to determine where a communications tower should be located in terms of reception, visual impact, and zoning and land use restrictions. Urban and regional planning rely heavily on maps for the location of schools and public facilities, for the development of highway, sewer, and water networks, and for the orderly organization of space through zoning and other techniques. We also try to identify areas subject to potential hazards and develop plans for areas containing problems, evacuating those in danger and providing services. Military operations rely heavily on maps for the deployment of troops, for the assessment of enemy positions, and for targeting weapons.

Jurisdiction, ownership, and assessment. Maps are used as legal documents showing the ownership of land and boundaries. Cadastral mapping is that area dealing with the legal systems showing who has rights to property. Land that is subdivided is plotted and is recorded on the maps. Taxes are based on property ownership and assessors rely heavily on maps. In more traditional societies where boundaries have been understood but not documented, efforts are now being made to create maps showing agreed-upon boundaries. These maps are permitting indigenous societies to retain rights to their land against outside forces wanting access to resources.

Understanding spatial relationships. Many maps are made in the process of trying to understand how phenomena are distributed spatially. In some cases, the subject of investigation may be a single variable, and in other cases, a number of variables may be examined in relationship to each other and to other nonspatial variables. The classic example of this is the work of Dr. John Snow in creating the map of the incidence of cholera in London and finding that the patterns led to a public water pump. The development of the concept of plate tectonics was based on a great amount of mapping and map analysis around the world. Police and public officials map data to see if there are patterns in the behavior of crime.

Forecasting and warning. The weathercaster on television is but one component of the use of maps to predict the future of events that play out over the earth's surface and that have the potential for significant damage to systems important to humans. Such forecasting and the dissemination of warnings is done at many scales, ranging from quite localized flash flooding, wildfires, and tornado touchdowns to larger features like hurricane landfalls, severe storms, volcanic eruptions, insect infestations, tsunamis, and sea level rise and high tides. Maps are an important part of the prediction processes and are equally important in forewarning potential victims.

Map compilation/mapping. The making of a map in almost all cases requires the use of maps. Map production is an iterative process, and in that process, a number of maps may be made as we converge on an appropriate design. In many cases, we consult other maps for such things as checking geographic names, confirming boundary changes, or examining land use and topography to better place dots on a map portraying the distribution of dairy cattle in a region.

Decoration, collection, and investment. Maps are collected, sold, and displayed simply because they are maps and many people like the appearance of a map. Historic maps take on value based on their rarity, quality, and area of interest. It is common to see historic map images employed as decoration on clothing, walls, games, and puzzles. And there is the occasional use of map images in advertisements, perhaps as a background.

Storage of information. The topographic maps that are produced by most countries are good examples of this type of map use. These maps are produced to provide a standardized inventory of features that are deemed to be important, for example, boundaries, hydrography, topography, and place names. These types of maps are produced in series, and all maps in the series should be at the same scale and have consistent forms of data capture and representation. The maps are fixed in time, and therefore, the information on the map can be correct only at the time the map was compiled, but much of the information on these types of maps changes slowly, so that maps that are 50 years old may still be useful for the examination of such things as topography and hydrography (see Sidebar 7.3).

Research and analysis. Researchers and biologists use variety of biological maps like genetic map, family tree map, linkage map (see Sidebar 7.1), and chromosomal map for their research and prediction of human traits. The analysis through these maps helps in understanding and developing a cure for a number of genetic diseases. Through these maps only medical practitioners are able to determine diseases at an early stage and cure them.

Geographical information systems. Maps used in these systems are basically called electronic maps or emaps. These maps are used by cartographers at the data gathering survey level. The functionality of these maps has been greatly advanced by new technologies, which simplified the superimposition of spatially located variables onto already existing geographical maps. The superimposition allows local information like rainfall level and wildlife distribution to be integrated onto the same map, which allows more efficient analysis and better decision making. Such superimposition of data on a map led researchers to discover the cause of cholera. Today, these superimposed maps are used by agencies of human kind, wildlife conservatives, and militaries for their work.

Some of the negative impacts of maps are as follows:

1. The ability to use a map depends on the nature of a map. The map should be readable, symbols should be distinguishable and properly defined, and the user should be able to comprehend images.
2. Developing a map takes a lot of time and resources. There are several factors that determine the usefulness of a map. Hence, it becomes necessary to identify all these factors before hand, because if any one factor is ignored, the whole purpose of developing the map is defeated.
3. One aspect of standards and map usability is the scale of the map, but different countries use different scales while developing it. This creates challenge to the user to convert these scales according to their needs and may result in inaccurate results.
4. On a geographical map, all the spatial information like rivers, lakes, and mountains needs to be labeled properly. Over centuries, cartographers have developed the art of placing names on even the densest of maps. This name placement can get mathematically very complex, as the number of labels and map density increases. Therefore, text placement is time consuming and labor intensive.
5. All maps are not accurate. Even the most accurate maps sacrifice a certain amount of accuracy in scale in order to deliver a greater visual usefulness to its users.

For people who practice topic maps, there are always more than one ways of reaching the goal post. The basic structure of topic maps is quite amenable and flexible to such an extent that two different topic map designers faced with the same kind of problem will end up in creating a multitude of different solutions! In fact, there will be a conflicting set of solutions with no single correct patterns among them. If you are developing formal patterns for a given topic map design, you will benefit in almost the same manner as a discerning programmer would benefit from creating design patterns.

7.2 PATTERN DOCUMENTATION

7.2.1 Pattern Name: AnyMap Stable Design Pattern

The basic idea behind choosing the term *AnyMap* is to give this pattern a general form. Generality is the driving force for choosing the term *AnyMap*, as this term applies to all fields with its different types and takes different values yet leads to same meaning. This generality will lead to a stable design pattern for AnyMap by using it as a capability to accomplish mapping as an enduring business theme (EBT) in a variety of applications.

7.2.2 Known As

In general terms, a map is a symbolic or graphical representation of relationships among real or imaginary objects in a particular space. The essential concept of a map can also be compared to the concept of transformation, relationship, metaphor, and binding. It is used to make logical connections between two entities.

Usually, a map is considered to be a geographical map representing some piece of land or water. Nowdays, the definition of map has expanded to a great extent. The map can be geographical or nongeographical, spatial or nonspatial, and 2D or 3D. Earlier, there used to be only static maps, but now due to advancement of technology, dynamic maps can also be built that can interact with users, for example, emaps.

In the domians of physics and mathematics, a map is any mathematical transformation that is applied over and over again in a neat sequence. Sometimes, the term *function* is used instead of map. In medical sciences, the graphical chart used to represent relationship among different components of DNA and chromosomes is also called map-like genetic map or linkage map (see Sidebar 7.1).

Many times, a chart is confused to mean a map. This is quite incorrect, as a chart is a subset of map. Charts are used to represent large amounts of information in tabular, graphical, or function form, so that interpretation is very easy and lucid. For example, Gantt charts are used for projecting or estimating the time needed to complete some work and also to track the progress of the project.

A cartogram, however, is a subset of map in which area is substituted by another mapping variable. Cartogram maps have become very useful these days, as these maps represent all the related information. Though building of such maps is very complex, with the help of technical tools and softwares these maps can be easily created, and they can also be made to interact with the user, like Google maps, which can be zoomed in and out according to the need.

Other similar terms are contour, plot, and so on.

7.2.3 Context

A map is a common tool used for many purposes. Any form of representation of data of one kind by another kind can be thought of as a map.

Geographical map. The most common usage of a map is the geographical map, which graphically represents 3D spatial relationship on a 2D surface and is drawn to scale. It can be used to pinpoint location of a place, a city, or a country or to describe certain features of the earth.

Topographical map. It depicts the contour and elevation of mountains and the depth of oceans on earth, and a geological map, which illustrates geological features of the earth, falls into this category of mapping (see Sidebar 7.3).

Mathematical usage. Another use of a map is commonly found in mathematics and many science disciplines. It has to do with transforming elements in one domain into elements of another domain and is synonymous with the term *function*.

In the field of computers. In computer applications, data mapping and memory mapping are common examples of maps. A site map is widely used in web-based applications, as it provides graphical representation of the various pages and hyperlinks between the pages. It basically gives the layout of the whole site (see Sidebar 7.2).

Biological research. In genetic research, scientists employ the technique of gene mapping to study genetic diseases like genetic maps, linkage maps, and chromosomal maps.

Nongeographical spatial maps. These maps basically represent the sky and are used by astronomers for their study and research, like star maps, maps of planets other than earth, moon map, and solar system maps.

Nonspatial maps. The diagrams or charts that depict the logical relationship among entities also fall under the category of map, like a Gantt chart. Some topological maps, where distance is not important and only the connectivity is significant, also fall under the category of nonspatial maps.

Astronautical maps/charts. These maps are designed to assist in navigation of aircrafts, just like a simple road map for a driver. These maps are used by pilots to determine their position, safe altitude, way to destination, and alternative landing areas in case of emergency.

Floor plan. It is a kind of map used by architects for designing a building structure and its interiors. It is basically a blue print of how the building or the complex will look like from outside and from inside.

Above listed are few contexts of AnyMap. The number of domains and applications, where maps can be used are so vast and widespread; hence, discussing all the contexts in which maps can be used is beyond the scope of this chapter.

The geograpical maps can depict roads, public transports, boundaries of country, state, city, and a particular area and are thus classified as road maps, bus maps, train maps, and so on. All these maps are depicted using the actual area structures. Topographic maps, on the one hand, represent the vast areas of land with the help of contour lines. Image maps, on the other hand, link various parts of images together, so that the user can click on the smaller parts to get further information regarding them.

7.2.4 Problem

The AnyMap design pattern represents the concept of mapping (logically connecting) data in one domain to data in another domain. The source and target data domains of a map can be any domain and could be very different. However, in general, a map is always used for a particular (well-defined) purpose and it determines what the source and target data look like. Therefore, it makes sense to model a generic design pattern that can be used for different purposes involving various types of data domains.

Trying to define a generic model for any kind of map is not that easy. The main difficulty in coming up with such a generic design pattern lies in the fact that the usage of a map and that data used can be quite different. For example, a topographical map that contains contours of mountains and depth of oceans is used for geographical expedition of the earth, whereas a gene map that links genes to specific locations on chromosones is used in researches of genetic diseases. Again, maps generally cannot depict the exact distance due to the large areas they need to depict on the media and hence are drawn according to a specific scale. Small scale results in depiction of finer details. For example, a world map needs a bigger scale than a road map for a city, say San Jose. This consideration of scale is an important aspect in modeling correct patterns. Nowadays, maps are available on various types of media like papers, smart phones, electronic screens (digital), and projections.

Fortunately, by using the SSM to focus on the enduring aspects of the problem, we arrive at a solution for the problem and provide a generic model of AnyMap for any application domain (purpose).

7.2.4.1 Functional Requirements

Functional requirements can be classified as internal requirements and external requirements.

7.2.4.1.1 AnyMap (Visualization) Functional Requirements

7.2.4.1.1.1 Internal Requirements As the name suggests, these requirements are internal to the pattern which means these are the requirements for the proper working of EBT/goal of the pattern and are not visible to the application. These requirements are tightly intervened to the system and changes, if the method used for visualizing the map changes. Some of them are as follows:

1. *Method employed for visualizing a map.* There are various methods through which a map can be visualized, but any method used for creating a map should produce the same result. Thus, all the rules that need to be followed like the signs to be used and the boundary conditions should be defined clearly beforehand and they should be followed by any method used in visualization.
2. *Evidence used to record map.* Depending on the type of map, an appropriate media should be used to store the map. Evidence indicates the existence of map later on, and hence, it is a very important step. Moreover, the format in which a map is stored should also be carefully selected, so that the reader can understand the map easily.
3. *Symbol dictionary.* Usually, on a map, one or many symbols can be used to depict various things. Hence, a standard rule should be followed while selecting symbols during visualization. This will give uniformity to the map, and moreover, it will make the map easier to read. Moreover, a map legend should be created listing all the symbols and their meanings.
4. *Boundary condition.* The boundary of entity or region for which a map is being developed should be kept in mind.

7.2.4.1.1.2 External Requirements

This section highlights the visualize part of the AnyMap stable design pattern as shown in Figure 7.1.

1. *Visualization.* It means analyzing the available data given in any format and creating or plotting any kind of map, be it, say, a geographical map or a chart, from it (Figure 7.1). With the invention of new tools and technology, several tools are present nowadays that aid in the creation of maps. Creating maps also includes labeling, texting, and defining proper scale and symbols and images to depict a variety of things. Hence, map creation is not a single step; instead, it is labor intensive and complex and involves multitude of actions and decisions.
2. *Parties/actors.* AnyActor and AnyParty may request or may generate a visualized form for any entity and/or any region. In the request scenario and in provision scenario, the mechanisms involved may be different and this must be well anticipated and dealt with properly to avoid ambiguity. AnyActor/AnyParty must go through proper mechanisms to request or generate visual form such as any map.
3. *Visualization mechanisms.* Mechanisms should be well defined and implemented to visualize any entity or region. AnyActor/AnyParty must follow the proper, incorporated mechanisms to visualize AnyEntity or AnyRegion. Mechanisms should check AnyCriteria that are required to visualize any entity or any region. AnyCriteria, in turn, validates the selection of the right visualization mechanism to those who holds the criteria true.
4. *Data.* They can be present in any form, but it should represent either some entities or some regions. Data should be complete and well defined and must have the capability to refer to map/s.
5. *Entity.* Entity should be distinct and should have separate existence. It should not be abstract or imaginary. More than one entity can be present in AnyData, but all these entities should be linked and not even one should lie apart.

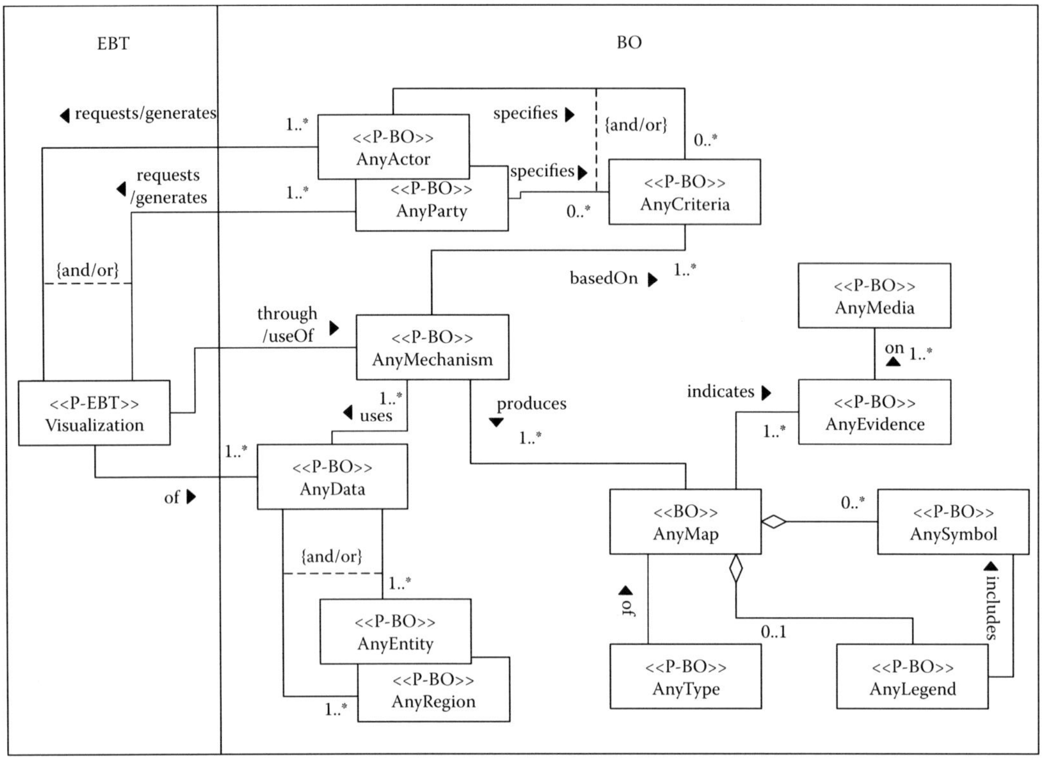

FIGURE 7.1 AnyMap (visualization) stable design pattern class diagram.

6. *Region.* It should have well-defined boundaries and must represent some geographical area. All the characteristics of the region should be explained properly be it physical, human, or functional.
7. *Criteria.* They are set by party while mapping in order to set some standards, like usage of blue color to depict water in all geographical map. These criteria define each and every minute details about how symbols, signs, and images will be used and also define a particular format for labeling. This is necessary in order to achieve uniformity throughout the map and also to avoid confusion among readers.
8. *Type.* Maps are of different types, like geographical maps, linkage maps, and charts, and each type of map has a different purpose. The purpose for which a map is visualized should be kept in mind while visualizing, and the symbols and images used should be in accordance with the purpose.
9. *Legend.* It represents important information on a map, like important buildings, for example, the parliament building; oceans; and structures of common compounds, for example, benzene. Some information are showed as legend and are known to everyone. Hence, these information should be marked properly on a map.
10. *Symbol.* Every map represents information in the form of symbols. These can be some image or keyword, lines, colors, and so forth. The symbols used in a map should be defined properly, so as to avoid any confusion among readers.
11. *Evidence.* Every map has some real existence and some documents or evidences to support its existence. All the evidences of a map should be documented properly and they should also clearly define the purpose of a map.

12. *Media.* Media that are used to visualize maps should be well identified and documented. They can be of different types and usage of each may vary. Media used to present evidences of a map should be appropriate and in accordance with the type of evidence.
13. *Colors/shades/lines.* Different degrees of colors and/or shades are used to distinguish among different properties or area of focuses within the visual form (any map). Different types of lines are used to show different indications, such as focus areas, concentration, and distribution. Colors/shades/lines are considered different entities.

7.2.4.1.2 AnyMap (Navigation) Functional Requirements

7.2.4.1.2.1 Internal Requirements

1. *Type of map.* How navigation is done depends on the type of map at hand; to analyze genetic map, one should have knowledge of DNA sequence and chromosome crossover, but to analyze a road map a party or an actor does not require any specialized knowledge.
2. *Symbol interpretation.* All the symbols and the interpretation rule associated with them should be studied carefully, so as not to make any mistake while navigating a map.
3. *Skills to operate media.* A party or an actor navigating the map should know how to operate the media on which a map is stored. These days, with the advancement of technology, various software applications are used to store different kinds of maps.

7.2.4.1.2.2 External Requirements

1. *Navigation.* It refers to analysis of a map (Figure 7.2) in order to extract information from it. For analyzing, all the symbols, labels, and legends of the map should be studied carefully and all the interpretation rules should be followed. The evidence in support of the map should also be considered.
2. *Parties/actors.* AnyParty or AnyActor may request navigation of AnyMap in order to study AnyEntity or AnyRegion. AnyParty or AnyActor must make use of the evidence available in support of AnyMap and must interpret AnySymbol and AnyLegend correctly. AnyParty or AnyActor must also know how to operate the media on which evidence of AnyMap is present.
3. *Evidence.* Every map has some real existence and documents or evidences to support its existence. All the available evidences of a map should be carefully studied to determine the purpose of the map.
4. *Media.* Media on which evidence of the map should be present should be operated properly, so as to not to destroy the evidence. AnyActor or AnyParty navigating the map should posses the skills to operate the media.
5. *Region.* AnyParty or AnyActor navigating the map should define all the boundaries and characteristics of the region. All the symbols, images, legend, and labels of that region should be analyzed properly.
6. *Type.* Maps are of different types like geographical maps, linkage maps, and charts, and each type of map has a different purpose. The purpose for which a map is navigated should be kept in mind while navigating, and the symbols and images used should be in accordance with the purpose.
7. *Legend.* It is like a dictionary of symbols of a map. A map legend should be used whenever required to avoid confusion while determining the meaning of the symbols.
8. *Symbol.* They form an important part of the map. Hence, while navigating, all the symbols and their meanings should be interpreted correctly. Some symbols are

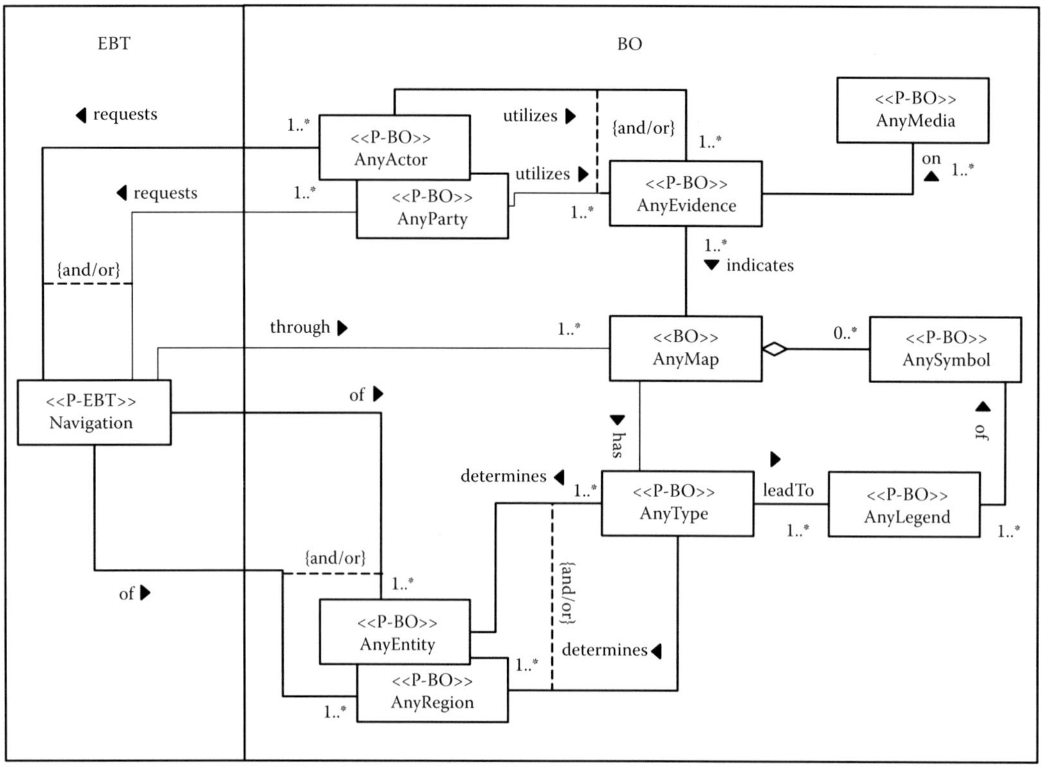

FIGURE 7.2 AnyMap (navigation) stable design pattern class diagram.

very common, but nothing should be taken as granted. Every symbol should be looked up for its meaning. It might happen that the map creator might have created his or her own set of symbols.

9. *Entity.* While navigating, all the entities and their relationships should be analyzed properly. Depending on the purpose, the entities can change their meaning. Hence, the purpose of the navigating map should be clearly understood beforehand.
10. *Colors/shades.* Different degrees of colors and/or shades are used to distinguish among different properties or area of focuses within the visual form (any map). Different types of lines are used to show different indications, such as focus areas, concentration, and distribution. Colors/shades/lines are considered different entities.

7.2.4.1.3 AnyMap (Recording) Functional Requirements

7.2.4.1.3.1 Internal Requirements

1. *Suitable media.* One of the most important requirements is selecting suitable media for recording based on the type of map and target party/actor who will be using the map.
2. *Recording method.* The method of recording is also critical, as the accuracy of the map depends entirely on it.
3. *Dependency on map data and type.* The type of map being recorded also influences the mechanism used for recording, skills needed by party, and even the media selection on which the map resides. All the rules and symbols defined during recording lay heavily on the map type.

7.2.4.1.3.2 External Requirements

1. *Recording*. It refers to recording a map (Figure 7.3), so that it can be made available to any user at later time. The appropriate method and media for recording should be chosen carefully depending on the type of map. The intended users of the map should also be kept in mind.
2. *Parties/actors*. AnyParty or AnyActor may request recording of AnyMap in order to preserve the evidence of existence of that map. AnyParty or AnyActor must consider the type of map being recorded before selecting the type of media.
3. *Criteria*. Depending on the type and amount of data to be recorded, AnyParty/AnyActor should define the criteria for recording. The criteria impose restriction on recording; hence, all the rules should be defined properly and clearly and beforehand.
4. *Data*. They can be present in any form, but it should represent either some entity or a region. They should be complete and well defined and must have the capability to refer to map/s, as they affect the method selected for recording.
5. *Evidence*. Every map has some real existence and some documents or evidence to support its existence. All the evidence of a map should be documented properly and they should also clearly define the purpose of a map.
6. *Entity*. They should be distinct and should have a separate existence. It should not be abstract or imaginary. More than one entity can be present in AnyData, but all these entities should be linked and not even one should lie apart.
7. *Region*. They should have well-defined boundaries and must represent some geographical area. All the characteristics of the region should be explained properly, be it physical, human, or functional.

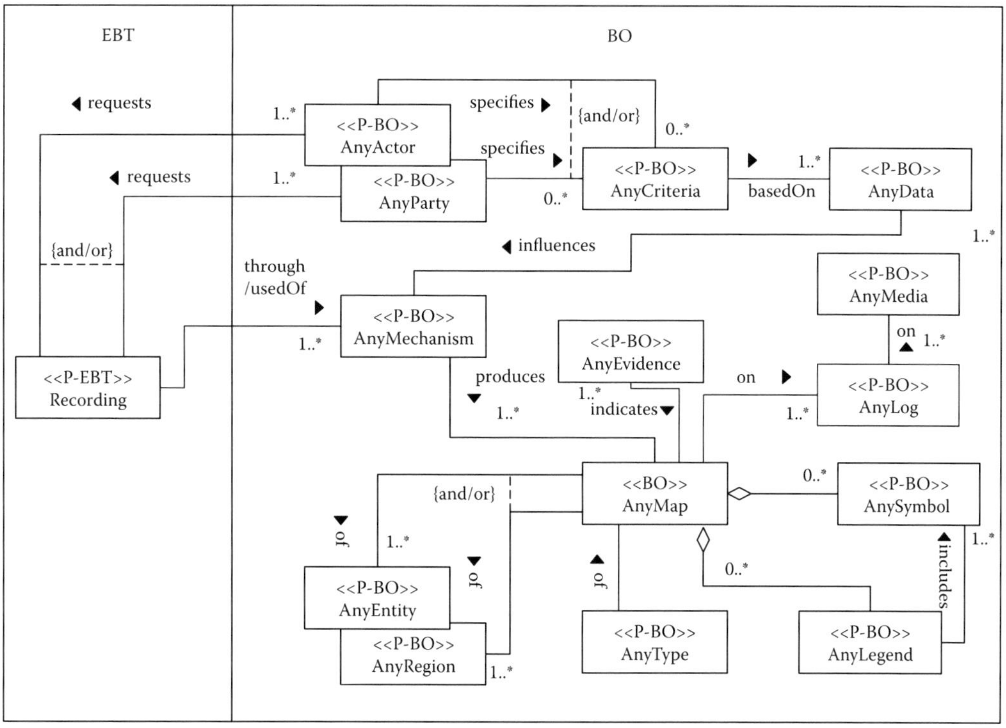

FIGURE 7.3 AnyMap (recording) stable design pattern class diagram.

8. *Symbol*. Every map represents information in the form of symbols. These can be some image or keyword, lines, colors, and so forth. The symbols used in a map should be defined properly, so as to avoid any confusion among readers.
9. *Legend*. It represents important information on a map like important buildings, for example, parliament building; oceans; and structures of common compounds, for example, benzene. Some information are showed as legend and are known to everyone. Hence, these information should be marked properly on a map.
10. *Media*. Media that are used to record maps should be well identified and documented. They can be of different types and the usage of each may vary. Media used to present evidences of a map should be appropriate and in accordance with the type of map.
11. *Mechanism*. They should be well defined and implemented to record the map. AnyActor/AnyParty must follow the proper, incorporated mechanisms to record the map. Mechanisms selected should be appropriate and in accordance with the media used to record the map.
12. *Log*. It represents the format in which map is stored on media. AnyParty should have the knowledge of different formats/logs used for different media.

7.2.4.2 Nonfunctional Requirements

1. *Modeling essentials*.
   a. *Simplicity*. In maps, the technique of *simplicity* is used to achieve the effect of singling out an item or items from their surrounding. Simplicity is one of the underlying map-plotting techniques; a cluttered map distracts the eye and takes away attention from the subject. A simple map can be achieved by getting closer to the subject, which is also one of the main rules of map making. Simplicity is one of the main components of most good maps. The simpler the map, the easier it is for the viewer to comprehend the subject and appreciate it. Cluttered images and backgrounds are less visually pleasing and more likely to cause the subject and lesser objects to confuse each other visually.
   b. *Completeness*. It refers to the presence of all constituent parts with each part fully illustrated. In general, a map is complete if nothing needs to be added to it. It forms an important factor while mapping, and each and every word of the requirement document should be studied carefully.
   c. *Easy to understand or understandability*. It refers to clarity of purpose. This goes further than just a statement of purpose; all of the parts must be clearly illustrated so that it is easily understandable. This is obviously subjective in that the user context must be taken into account: for instance, if the map is to be used by map designer, it is not required to be understandable to the layman. This also includes proper laying of all the interpretation rules and definitions of all signs and symbols used on the map.
   d. *Stability*. If a map is complete, simple, easy to understand, it is most likely to be stable. The map once created should be able to expand and adapt to changing requirements, such that if later on some new features need to be added to the map, then the developer does not have to start from the scratch. Instead, he can use the already existing map and add on the things without any difficulty.
   e. *Testability*. It refers to the disposition to support acceptance criteria and evaluation of performance. Such a characteristic must be built in during the design phase if the map is to be easily testable; a complex design leads to poor testability.

   f. *Visualization*. Maps are visible graphical tools. A picture is worth a 1000 words. This requirement directly relates to understandability and simplicity requirement of the map. The reader of the map should be able to comprehend the things depicted on the map without any external help.
2. *Consistency*. It refers to uniformity in notation, symbology, appearance, and terminology within itself. While designing the map, some standards need to be set in advance so as to avoid confusion during implementation phase. Also, if standards are followed properly, the reader will be able to read the map without any difficulty.
3. *Documentable*. It refers to the capability of being supported by documentary evidence. This is necessary to prove the usefulness of the map.
4. *Portability*. It refers to the ability to be run well and easily on multiple media of a variety of configurations. With the invention of new tools and technology, a number of tools are used while mapping. Hence, it becomes necessary for the developer to ensure that the developed map runs on a number of configurations, for example, Google Maps and Yahoo! Maps.
5. *Structuredness*. It refers to the organization of constituent parts in a definite pattern.

7.2.5 Challenges and Constraints

7.2.5.1 Challenges

- The AnyMap design pattern must not impose any restriction on the kind of data involved in the mapping. In other words, it should not need check on the type of data involved in the mapping. For example, the AnyMap design pattern should apply to geographical data in the same way as it would with computer data or gene data.
- Every map has a purpose and this purpose varies in different applications. While the AnyMap design pattern must generate a map that serves the designated purpose, it must not be dependent on the nature of the purpose.
- The AnyMap design pattern must be usable by different kinds of party. For instance, a person can use a road map for traveling from one place to another, an organization can use a concept map to describe its product marketing strategy, and a software application can use a hash map to store data related by hashed keys.
- Map problems span a fairly wide range of applications and domains, which makes the task of capturing the core concept of a map problem more challenging than it might appear to be.
- Even after extracting the common feature of different types of maps, the difficulty still resides in how these common features can be abstracted in such a way that makes them still valid for all the wide applications where usage of the map exists.
- Maintaining a high level of accuracy in maps is a major challenge.
- Maintaining the simplicity of the map also poses a big challenge. For example, emaps present a number of things at a time, and sometimes, it becomes difficult for the user to comprehend information from the map.
- Deciding on the standards for a map is also difficult. The way maps are labeled and the text language used varies from country to country, and the set of colors used vary by producers, though, the overall image will be fairly simple.
- The conditions and the environment in which a map is used also affect the reader. Developing general model of AnyMap keeping user's environment in mind is not easy.

- With advanced technology, there are various mechanisms through which a map can be visualized. Selecting appropriate mechanism for the type of map to be visualized is not an easy task.
- Selecting an appropriate method for recording depending on the data to be recorded and the type of map is quite challenging.
- While recording, the symbols and legends used to represent a variety of things on the map should be appropriate.

7.2.5.2 Constraints

7.2.5.2.1 Navigation

- Navigation of any map can be requested by or provided to one or more parties or/ and actors.
- Navigation takes place through one or more maps.
- Different maps have different features and some of these features may not apply to other types of maps.
- AnyMechanisms are based on none to more user-defined criteria.
- AnyMap indicates one or more piece of evidence.
- AnyMap and its pieces of evidence are stored and represented on one or more media.
- Navigation can be done for any region and/or any entity that is determined by the type of the requested map.
- The type of the map leads to zero or more legends on the map.
- The type of the map leads to zero or more symbols that exist on the map, which indicates many different things, such as direction and distribution.
- All the symbols defined by the user should be present in at least one legend, so as to avoid confusion while navigation.

7.2.5.2.2 Visualization

- The AnyMap design pattern requires that the user of the map supply the source data, although the pattern does not dictate the format of the source data.
- The format of the source and target (mapped) data is defined by the AnyMappingMechanism, which is unique for each application.
- The AnyMap design pattern does not address how the target (mapped) data are to be interpreted in order to achieve the designated purpose of the map. The interpretation of the mapped data, however, can be aided by AnyInterpretationRule, which is generated as an optional part of the outcome of mapping.
- The AnyMap design pattern is not responsible for scaling or filtering the source data prior to the mapping, but provides a hook for the user to do so via the Any Criteria BO, which represents any additional user requirements.
- Different maps have different features and some of these features may not apply to other types of maps.
- Navigation takes place through one or more maps.
- Navigation of any map can be requested by or provided to one or more parties or/and actors.
- AnyActor can also request for accessibility and/or can define criteria.
- AnyParty/AnyActor can also request for the navigation of any map based on additional user-defined zero or more criteria.
- AnyParty has to follow one or more mechanisms to gain authorized accessibility.

- AnyParty/AnyActor can define none to many criteria for the mechanisms to check for to validate accessibility
- AnyParty has to follow one or more mechanisms to gain authorized accessibility.
- AnyParty/AnyActor can define none to many criteria for the mechanisms to check for to validate accessibility
- AnyMechanisms are based on none to many user-defined criteria.
- Accessibility may be granted or denied to AnyActor/AnyParty, depending upon the mechanism and its criteria.
- AnyMechanism involves at least one or many media types to gain accessibility.
- AnyMedia helps to access one or many entities required for accessibility.
- AnyEntity can take one or many mappings with industrial object (IO) related to the application.
- Accessibility can utilize one or many mechanisms to carry out the task given by the user.
- One or more criteria are needed for validation in order to gain accessibility.

7.2.5.2.3 Recording

- AnyParty/AnyActor should clearly define none or some criteria for recording data into AnyMap.
- AnyCriteria defined by AnyParty/AnyActor should be based on one or more AnyData made available for recording.
- AnyMechanism used for recording should be decided on the basis of one or more AnyData available.
- Recording can be done through one or more AnyMechanism.
- AnyMechanism is capable of producing one or more AnyMap.
- AnyMap can be presented on one or more AnyEvidence.
- AnyMap is stored in one or more AnyLog.
- AnyLog can reside on one or more AnyMedia.
- AnyLegend can be indicated by one or more AnySymbol.
- AnyMap consists of none or any number of AnySymbol and AnyLegend.

7.2.6 Solution Structure and Participants

7.2.6.1 Structures

Figure 7.1 illustrates AnyMap (virtualization) stable design pattern class diagram. Figure 7.2 shows AnyMap (navigation) stable design pattern class diagram. Figure 7.3 represents AnyMap (recording) stable design pattern class diagram.

7.2.6.2 Participants

- *Classes*
  - *AnyMap*. This class represents the map that holds the mapped data and optional rules for interpreting the mapped data.
- *Patterns*
  - *Visualization*. This class represents the goal or EBT of the pattern. AnyMap is visualized by AnyActor or AnyParty on the basis of AnyData available.
  - *Navigation*. This pattern represents the EBT or the goal of AnyMap that represents the purpose served by the map used by a user.

- *Recording.* This class represents the EBT or goal of AnyMap that represents how a map is recorded.
- *AnyParty.* This class represents a group of users, who uses the AnyMap to achieve an AnyPurpose.
- *AnyActor.* This class represents a single user, who can visualize as well as use AnyMap.
- *AnyData.* This class represents the data provided by the user in order to generate AnyMap. This class represents the data in the map translated or mapped from the source data provided by the user. It is part of AnyMap.
- *AnyType.* This class represents the type of map being visualized or navigated or recorded by its users, for example, geographical map, biological map, and chart.
- *AnyMedia.* This class represents the media on which the map is present.
- *AnyMechanism.* This class represents the method specified by the user in order to generate a map. There are several methods and techniques through which a map can be produced and recorded.
- *AnyCriteria.* This class represents optional mapping criteria entered by user to refine quality and scope of a map.
- *AnyEntity.* This class represents the real object, which is visualized and a map is produced.
- *AnyRegion.* This class represents any geographical area for which a map is generated.
- *AnyEvidence.* This class represents the documents that support the existence and purpose of a map.
- *AnyLegend.* This class represents the dictionary of symbols used while producing a map. It contains the complete list of symbols and their definition.
- *AnySymbol.* This class represents the keywords used on a map to depict various things like notations, lines, colors, and shapes.
- *AnyLog.* This class represents the format in which a map is stored on AnyMedia.

7.2.6.3 CRC Cards

Navigation (Navigation) (P-EBT)

Responsibility	**Collaboration**	
	Client	**Server**
To allow navigation of a map	AnyParty	navigates()
	AnyActor	voyage()
	AnyMap	uses()
	AnyEntity	travels()
	AnyRegion	analyze()
		explores()
		determines()

Attributes: route, distance, area, entity, scale, position, range, toolsUsed

Visualization (Visualization) (P-EBT)

Responsibility	Collaboration: Client	Collaboration: Server
To facilitate generation of a map	AnyParty AnyActor AnyMechanism AnyData	produces() generates() visualizes() transforms() converts() creates()

Attributes: sourceData, generatedMap, typeOfMap, dataFormat, symbolsUsed, interpretationRules, mechanismUsed, criteriaFollowed

Recording (Recording) (P-EBT)

Responsibility	Collaboration: Client	Collaboration: Server
To store/record map	AnyParty AnyActor AnyMechanism	records() stores() saves() logs() uses() documents() preserves()

Attributes: sourceData, mediaUsed, typeOfData, dataFormat, methodInvolved, rulesFollowed, skillsNeeded

AnyParty (AnyParty) (P-BO)

Responsibility	Collaboration: Client	Collaboration: Server
To request or generate a visual form, such as any map. Also, requests the navigation of any map	Navigation	requests(), provides(), uses(), defines(), playRole(), group(),
	AnyEvidence	setCriteria(), monitor(), switchRole(), agree(), generate(), navigates(), utilizes(), requires(),
	Or Visualization AnyCriteria Or Recording AnyCriteria	

Attributes: partyName, type, members, location, areaOfExpertise, workHours, accessFor, activity

AnyActor (AnyActor) (P-BO)

Responsibility	Collaboration	
	Client	**Server**
To request or generate a visual form, such as any map. Also, requests the navigation of any map	Navigation	gainsAccess(), uses(), defines(), navigate()
	AnyEvidence	participate(), interact(), organize(), request(), explore(), utilizes(), requests()
	Or Visualization AnyCriteria Or Recording AnyCriteria	

Attributes: name, rank, typeOfAuthority, accessFor, members, category

AnyMechanism (AnyMechanism) (P-BO)

Responsibility	Collaboration	
	Client	**Server**
To incorporate means/methods for generating a visual form	Visualization AnyCriteria AnyData AnyMap Or Recording AnyData AnyMap	used(), hasBase(), makesUseOf(), execute(), activate(), pause(), attach(), status(), generate(), utilizes(), archives(), classifies()

Attributes: isStated, isValid, methodUsed, skillsRequired, rulesFollowed, accuracy, toolsUsed, output, intermediateState

AnyMap (AnyMap) (BO)

Responsibility	Collaboration	
	Client	**Server**
To represent data in a visual form	Navigation	returnMap(), illustrate()
	AnyEvidence	focusOn(), includeSymbols(), mayIncludeLegend(), showEvidences(), hasType(), aidsNavigation(),

(Continued)

AnyMap (AnyMap) (BO)

Responsibility	Collaboration	
	Client	**Server**
	AnyType	consistsOf()
	AnySymbol or	
	AnyLegend	
	AnySymbol	
	AnyType	
	AnyMechanism	
	AnyEvidence or	
	AnyMechanism	
	AnyEntity, AnyRegion, AnyEvidence, AnyLog, AnySymbol, AnyLegend	

Attributes: mapName, size, relatedTo, focusOn, evidences, meetCriteria, scale, colorsUsed, shadeUsed, type, purpose

AnyData (AnyData) (P-BO)

Responsibility	Collaboration	
	Client	**Server**
To provide source data of a map	AnyEntity, AnyRegion, AnyMechanism, Visualization	returnData()
	Or	returnType()
	AnyMechanism	returnLegalRange()
	AnyCriteria	belongsTo()
		formCollection()
		represents()
		hasValue()

Attribute: type, isReliableData, dataSource, id, name, property, model, application, format, belongTo, domain, context

AnyEntity (AnyEntity) (P-BO)

Responsibility	Collaboration	
	Client	**Server**
To define entities that can be accessed	Navigation	utilizes(), defines(), type(), update(),
	AnyType	new(), performFunction(), status(), determines()
	Or	
	AnyData (in Visualization)	
	Or	
	AnyMap (Recording)	

Attributes: nameOfEntity, typeOfEntity, useOf, usedFor, descriptionOfEntity, status, position, states

AnyRegion (AnyRegion) (P-BO)

Responsibility	Collaboration	
	Client	**Server**
Represents the form of data to be visualized	Navigation	exists(), represents(), hasBoundary(), determineMapType(), actAsData()
	AnyType or AnyData or AnyMap	

Attributes: location, boundary, physicalCharacteristic, functionalCharacteristic, name, type, size, area

AnyType (AnyType) (P-BO)

Responsibility	Collaboration	
	Client	**Server**
Defines the type of map in question	AnyEntity, AnyRegion, AnyMap, AnyLegend (Navigation) or	classifies(), categorizes(), distinguishes(), sorts(), makesClass(), organizes(), separates()
	AnyMap(Visualization) or AnyParty, AnyMechanism (Recording)	

Attributes: name, status, number, basis, ruleForClassification, parameterUsed, factor, inspirationForClassification

AnyCriteria (AnyCriteria) (P-BO)

Responsibility	Collaboration	
	Client	**Server**
To define visualization criteria for the map	AnyParty	validates(), (TRUE, only when validation passes), providesBase(), verify(), apply(), prioritize(), exhibit(), imposes(), limits(), influences(), standardizes()
	AnyActor AnyMechanism Or AnyData AnyParty AnyActor	

Attributes: nameOfCriteria, numberOf, checkedBy, implementedBy, leadsTo, condition, property, priority

AnyEvidence (AnyEvidence) (P-BO)

Responsibility	Collaboration	
	Client	**Server**
To support map presence and purpose	AnyParty, AnyActor, AnyMap, AnyMedia (Navigation) or AnyMedia, AnyMap (Visualization) Or AnyMap(Recording)	supports(), provesExistence(), documents(), showsValidity(), definesPurpose(), indicates()

Attribute: evidenceld, name, type, status, validity, approvedBy, format, purpose

AnyMedia (AnyMedia) (P-BO)

Responsibility	Collaboration	
	Client	**Server**
The technique through which evidence of a map is stored	AnyEvidence (Visualization) or AnyLog (Recording)	stores(), displays(), actAsMedium(), works(), executes(), operates(), showsEvidence()

Attribute: mediaId, type, name, technologyUsed, rulesToOperate, skillsNeeded

AnyLegend (AnyLegend) (P-BO)

Responsibility	Collaboration	
	Client	**Server**
To define symbols used on a map	AnyType, AnySymbol (Navigation) or AnySymbol, AnyMap (Visualization) or AnyMap, AnySymbol (Recording)	defines(), explains(), lists(), identifies(), compiles(), clarifies(), discloses()

Attribute: name, languageUsed, format, numberOfSymbols, id, type, length, symbolsDefined

AnySymbol (AnySymbol) (P-BO)

Responsibility	Collaboration	
	Client	**Server**
To represent real objects on a map	AnyLegend, AnyMap or AnyMap, AnyLegend (Recording)	represents(), shortens(), indicates(), marks(), simplifies()

Attributes: id, name, type, format, colorUsed, value, importance

7.2.7 Consequences

The AnyMap design pattern satisfies its objective and provides a base pattern that is adaptable to applications in different domains.

Understandability. The AnyMap design pattern presents the enduring concept of mapping in an easily understandable fashion. It accomplishes this through the Mapping EBT, and by using a basic AnyMappingMechansim, it produces AnyMap based on a user-supplied AnyMapType.

Adaptability. The AnyMap design pattern is generic enough to be applicable in multiple domains that require mapping. This is illustrated through the use of a generic AnyMappingMechansim controlled by AnyMapType. It can be further infered from the two examples given under the applicability section.

Stability. The concept of mapping is described in generic terms, without using any domain-specific IOs. Applications in different domains can use this concept by substituting IOs (industrial objects) specific to the application. Examples of domain-specific IOs may be different mapping mechansims, different mapping criteria, and different data type involved in the mapping.

Extensibility. The pattern can be extended by plugging in the application-specific context classes such as instances of AnyParty, AnyMappingMechanism, and AnyInterpretationRule. Thus, the system provides a high level of extensibility to suit applications in various domains.

The good thing with the AnyMap design pattern is that AnyParty can specify the criteria and get the map according to their liking. But the bad thing is that AnyParty is responsible to interpret the mapped data. This can be a big problem, because AnyParty can have different interpetations of the map depending on their understanding.

7.2.8 Applicability with Illustrated Examples

7.2.8.1 Case Study 1: Navigation—Google Road Map for Planning Driving Routes

Google road maps are widely used for navigation. They are used in day-to-day life for traveling from one place to another, for finding a particular address, for searching near well-known buildings and intersections, and so on. This application serves as an example, where the generic model of navigation developed above for AnyMap is used for finding shortest route from one point to another. The traveler can easily interpret the routing directions using the map legend available along with google map. Besides showing the route, the map shows the distance between the two points and how much will be needed to cover the distance with various options of traveling like by car and by public transport.

Use Case: Navigate through Google Road Map

Actors	Roles
AnyParty	Traveler

Classes	Type	Attributes	Operations
Navigation	EBT	route distance scale entity	explores() navigates()
AnyParty	BO	Name workHour activity member	requests() navigates() utilizes()
AnyMap	BO	name purpose evidence scale	hasType() aidsNavigation()
AnyEvidence	BO	Id name type status	indicates() documents() provesExistence()
AnyMedia	BO	name type Id technologyUsed	showsEvidence() displays()
AnyType	BO	status number basis factor	leadsTo() classifies()
AnySymbol	BO	name type format value	represents()
AnyLegend	BO	name type languageUsed format	definesSymbol()
AnyRegion	BO	location boundary Area Size	determines()
AnyEntity	BO	name Use description position	determines()
GoogleEarth	IO	technologyUsed version systemRequirement	displaysMap()
Traveler	IO	name Id status	navigates() specifies()

(Continued)

Classes	Type	Attributes	Operations
		qualification	
GoogleMapService	IO	parameter	accepts()
		algorithmUsed	generates()
			selects()
GoogleMap	IO	Line	showsRoute()
		symbol	directs()
		color	
Address	IO	location	providesLocation()
		Area	
		country	
MapLegend	IO	symbol	defines()
		format	
		language	
ShortestRoute	IO	distance	actsAsOutput()
		Area	showsDirection()
		path	

Use Case Description

1. Navigation is requested by AnyParty for exploration and Traveler who wants to navigate the map for driving direction inherits from AnyParty.
   How to do navigation? What type of map is used by traveler? How the map will show driving direction?
2. Traveler specifies address for direction on a map.
   How does Traveler specifies address? What is the format of address? Is the format of address fixed?
3. GoogleMapService accepts address and then generates GoogleMap showing the route.
   How the address acts as input for GoogleMapService? How GoogleMap is generated? What is the technology used?
4. The ShortestRoute selected by GoogleMapService acts as output for Traveler.
   Which algorithm is used by MapService to select shortest path? How does the algorithm work?
5. AnyParty utilizes AnyEvidence and AnyEvidence is present on AnyMedia like GoogleEarth.
   How the evidence is utilized? What is the purpose of evidence? What kind of media is used?
6. AnyEvidence indicates AnyMap, which has many AnyType.
   How the map is classified into various types? How evidence indicates map?
7. AnyType leads to AnyLegend and AnyLegend defines a list of AnySymbol, which represents various things on a map.
   How are type of map and legend related to each other? What is the format of legend? How are symbols defined by legend?
8. AnyMap aids in navigation.

How does map helps in navigation? Who does navigation? What are the techniques used?

9. Navigation is done for AnyRegion or AnyEntity and they both determine AnyType.
   What is the definiton of entity? How region and entity decide the type of map?
10. AnyEvidence proves existence of AnyMap and documents it.
    What is the format of the document? How is the existence and purpose of a map proved?

Aternatives:

1. The address specified by traveler is not correct.
2. GoogleMapService also shows alternative routes which are not the shortest ones.

Class Diagram—A class diagram is shown in Figure 7.4.
Class Diagram Description

1. Navigation is requested by AnyActor or AnyParty.
2. Traveler inherits from AnyParty.
3. Traveler specifies address, which is a part of AnyRegion.
4. Address is accepted by GoogleMapService and it generates GoogleMap.

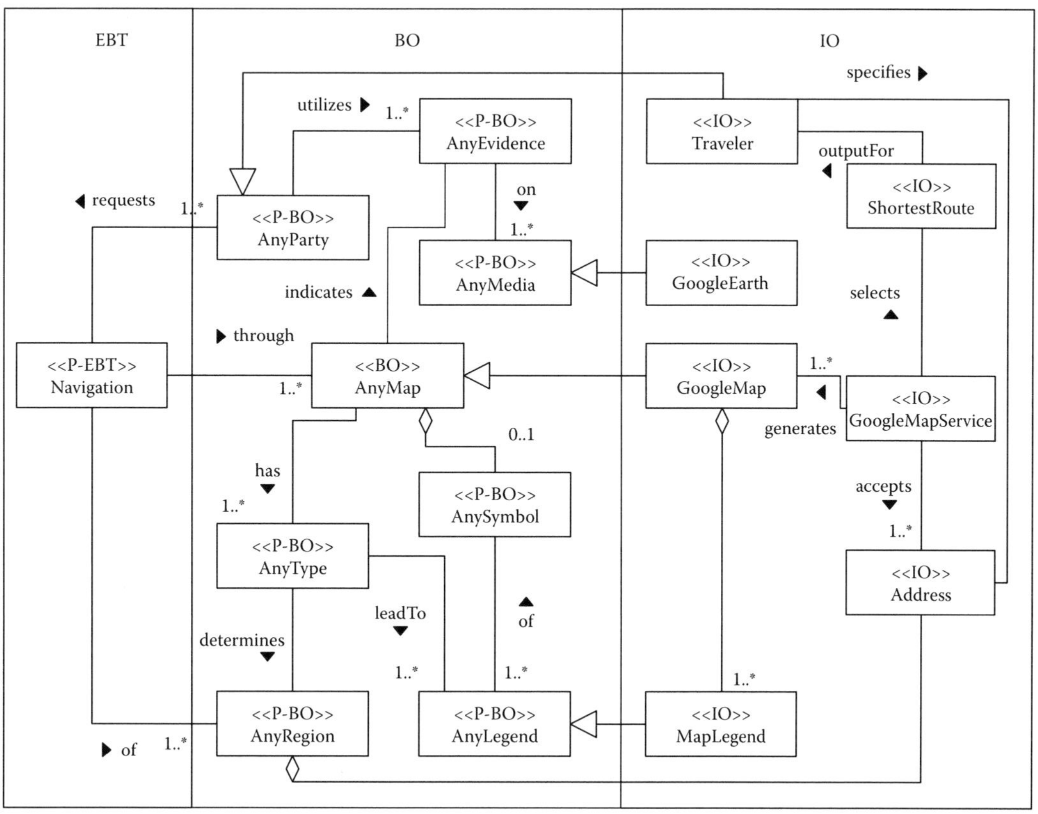

FIGURE 7.4 Class diagram for navigation through Google Road Map.

5. GoogleMapService selects ShortestRoute that acts as output for traveler.
6. AnyParty utilizes AnyEvidence on AnyMedia.
7. GoogleEarth inherits from AnyMedia.
8. AnyEvidence indicates AnyMap, which has many AnyType.
9. GoogleMap inherits from AnyMap and is on GoogleEarth.
10. AnySymbol is a part of AnyMap.
11. AnyType leads to AnyLegend, which consists of AnySymbol.
12. MapLegend inherits from AnyLegend and is a part of GoogleMap.
13. Navigation is done through AnyMap.
14. Navigation is of either AnyEntity or AnyRegion and they both determine AnyType.

Sequence Diagram—A sequence diagram is shown in Figure 7.5.
Sequence Diagram Description

1. Navigation is requested by AnyParty(Traveler).
2. AnyParty(Traveler) specifies AnyRegion(Address).
3. AnyRegion(Address) is accepted by GoogleMapService.
4. GoogleMapService selects ShortestRoute.
5. ShortestRoute acts as an output for AnyParty(Traveler).
6. AnyParty utilizes AnyEvidence.
7. AnyEvidence indicates AnyMap.
8. AnyMap has AnyType.
9. AnyType leads to AnyLegend(MapLegend).
10. AnyLegend(MapLegend) is a part of AnyMap(GoogleMap).
11. AnyMap helps in navigation.

7.2.8.2 Case Study 2: Mathematical Mapping

Consider the problem where a person uses a certain mathematical function for the purpose of computation. For each value X in the domain of function F, a value in the range represented by F(X) is the expected mapping. In this case, AnyParty will be extended with one IO that will represent the mathematician. For each computation, the mathematican will provide a value in the *domain* of the function and the function will return a value of the *range* of the function. In some cases, a boundary condition may need to be applied to the function. Such a boundary condition is implemented as an extended object of AnyCriteria.

Use Case: Map Mathematical Data

Actors	Roles
AnyParty	Mathematician

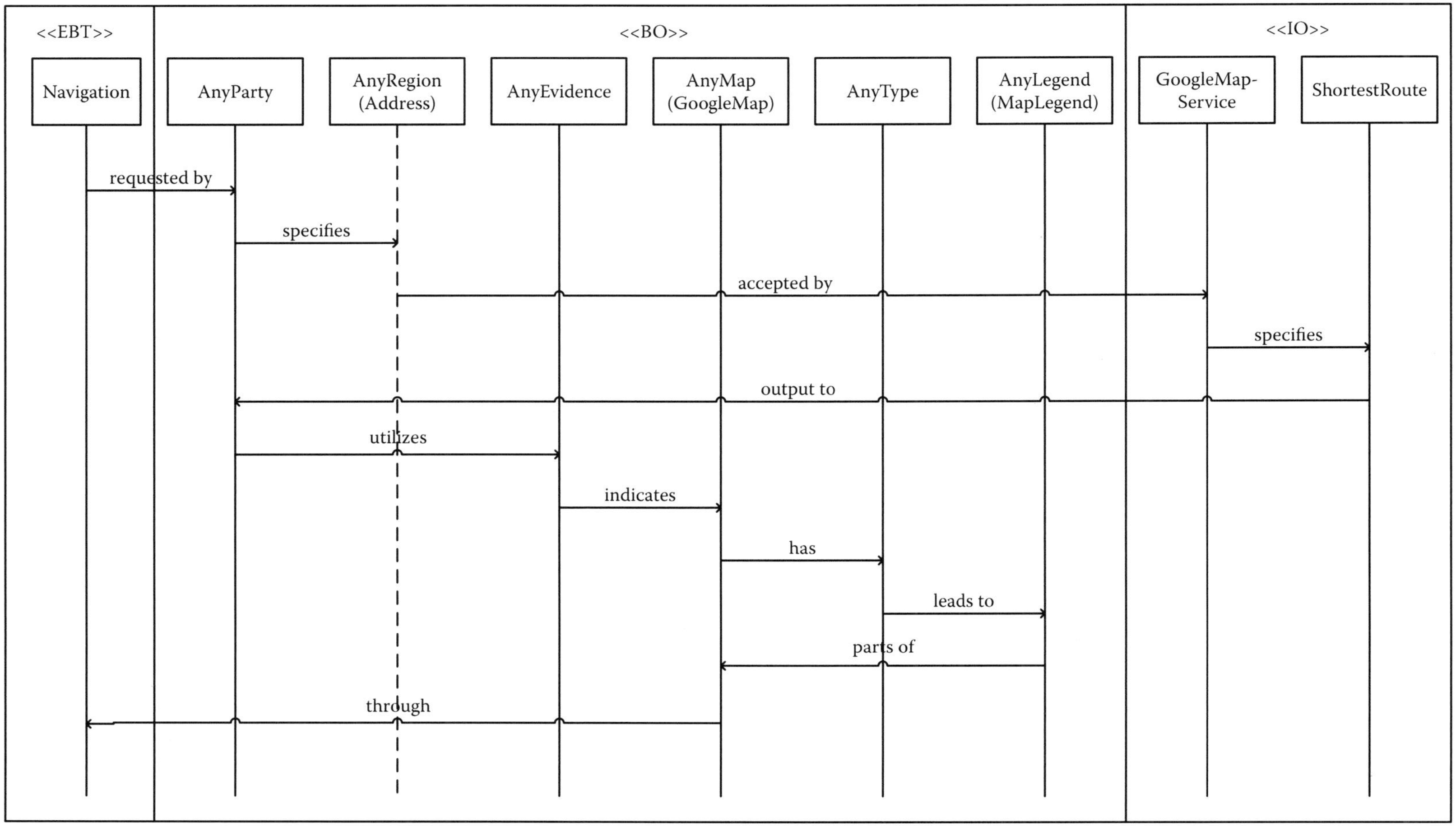

FIGURE 7.5 Sequence diagram for navigation through Google Road Map.

Classes	Type	Attributes	Operations
Visualization	EBT	sourceData generatedMap mapType	visualizes() produces()
AnyParty	BO	Name workHour activity member	requests() specifies() defines()
AnyCriteria	BO	name implemented condition	limits() influences()
AnyMap	BO	name purpose evidence scale	hasType() consistsOf()
AnyEvidence	BO	id name type status	indicates() documents()
AnyMedia	BO	name type id technologyUsed	showsEvidence() displays()
AnyMechanism	BO	methodUsed accuracy output	uses() generates()
AnySymbol	BO	name type format value	represents()
AnyLegend	BO	name type languageUsed format	definesSymbol()
AnyEntity	BO	name use description	exists()
AnyType	BO	status number basis factor	classifies() categorizes()
Thesis	IO	name topic researchArea	provesExistence() recordsResearch()
Mathematician	IO	name id researchArea	requests() defines()

(*Continued*)

Classes	Type	Attributes	Operations
MathematicalFormula	IO	parameter range	devises()
MathematicalMap	IO	purpose function entity	typeOf() presents()
BoundaryCondition	IO	range entryValue exitValue	limits() impacts()
Function	IO	expression boundaryValue limit	isMap() containsElement()
Relation	IO	element value mapping	inheritsFromEntity() relates()
DomainElement	IO	numberOfElement typeOfElement value	partOfFunction()
CodomainElement	IO	numberOfElement typeOfElement value	partOfFunction()

Use Case Description

1. Visualization is requested by AnyParty and Mathematician inherits from AnyParty, who is researching and wants to develop a new mathematical function.
   What is visualization? What kind of research mathematician is doing? What is the area of research? What knowledge is required for developing a function?
2. Mathematician first needs to define the AnyCriteria(BoundaryCondition) for the function.
   How to set BoundaryCondition? What is its purpose? How will it help in developing function?
3. AnyCriteria(BoundaryCondition) defined by Mathematician limits AnyMechanism (MathematicalFormula), which is used in devising AnyMap(Function).
   On what basis boundary condition is set? How mechanism for developing map/function is restricted? What is the mechanism used in developing function?
4. AnyMap(Function) is documented on AnyEvidence and this evidence can be present on AnyMedia(Thesis).
   What is the purpose of documentation? What type of media is used for storing function? Who prepares thesis? Thesis is approved by whom?
5. Thesis is prepared by Mathematician as a proof of their research.
   What is the format of thesis?
6. AnyParty specifies AnyCriteria and AnyCriteria influences AnyMechanism.
   What type of criteria is specified by party? How does criteria infleunces mechanism?

7. AnyMechanism makes use of AnyData which is of the type AnyEntity.
   What kind of entity is used in visualization? What should be the format of input data?
8. Relation inherits from AnyEntity and it relates DomainElement and CodomainElement, which are a part of Function.
   What are domain element and codomain element? How are they related? How does function represents them? What are their values?
9. AnyMap has many AnyType and one such type is MathematicalMap, which is used in the field of mathematics to represent any function or graph.
   How many types of map can exist? Can any function be represented as map? What other types of map are used in the field of mathematics?
10. AnyMap consists of AnySymbol and AnyLegend. AnyLegend acts like a dictionary of symbols and symbols are used to represent shorthand notations used on map.
    What are the uses of AnySymbol and AnyLegend? What is the format of AnyLegend? How many symbols are used on a map?
11. AnyMap is generated by AnyMechanism and that mechanism aids in visualization.
    What mechanism is used? What are the steps in mechanism? How is visualization done through this mechanism?
12. AnyData acts as input in visualization and through these data AnyMap is generated.
    What is the input to visualization? How is map generated through these data?

Alternatives:

1. AnyMechanism(MathematicalFormula) is faulty and thus AnyMap(Function) so returned is inaccurate.
2. AnyParty(Mathematician) might miss some AnyCriteria(BoundaryCondition) when specifying them.

Class Diagram—A class diagram is shown in Figure 7.6.
Class Diagram Description

1. Visualization is requested by AnyParty(Mathematician).
2. Mathematician defines AnyCriteria(BoundaryCondition).
3. AnyCriteria(BoundaryCondition) limits AnyMechanism(MathematicalFormula).
4. AnyMechanism(MathematicalFormula) implements AnyMap(Function).
5. AnyMap is indicated by AnyEvidence.
6. AnyEvidence on AnyMedia(Thesis).
7. AnyMedia(Thesis) is prepared by AnyParty(Mathematician).
8. AnyParty specifies AnyCriteria.
9. AnyCriteria influences AnyMechanism.
10. AnyMechanism uses AnyData.
11. AnyEntity(Relation) inherits from AnyData.
12. AnyEntity(Relation) relates DomainElement and CodomainElement.
13. DomainElement and CodomainElement are a part of AnyMap(Function).
14. AnyMap has AnyType and MathematicalMap is a type of AnyMap.
15. Function is a part of MathematicalMap.
16. AnyMap consists of AnySymbol and AnyLegend and AnyLegend includes AnySymbol.

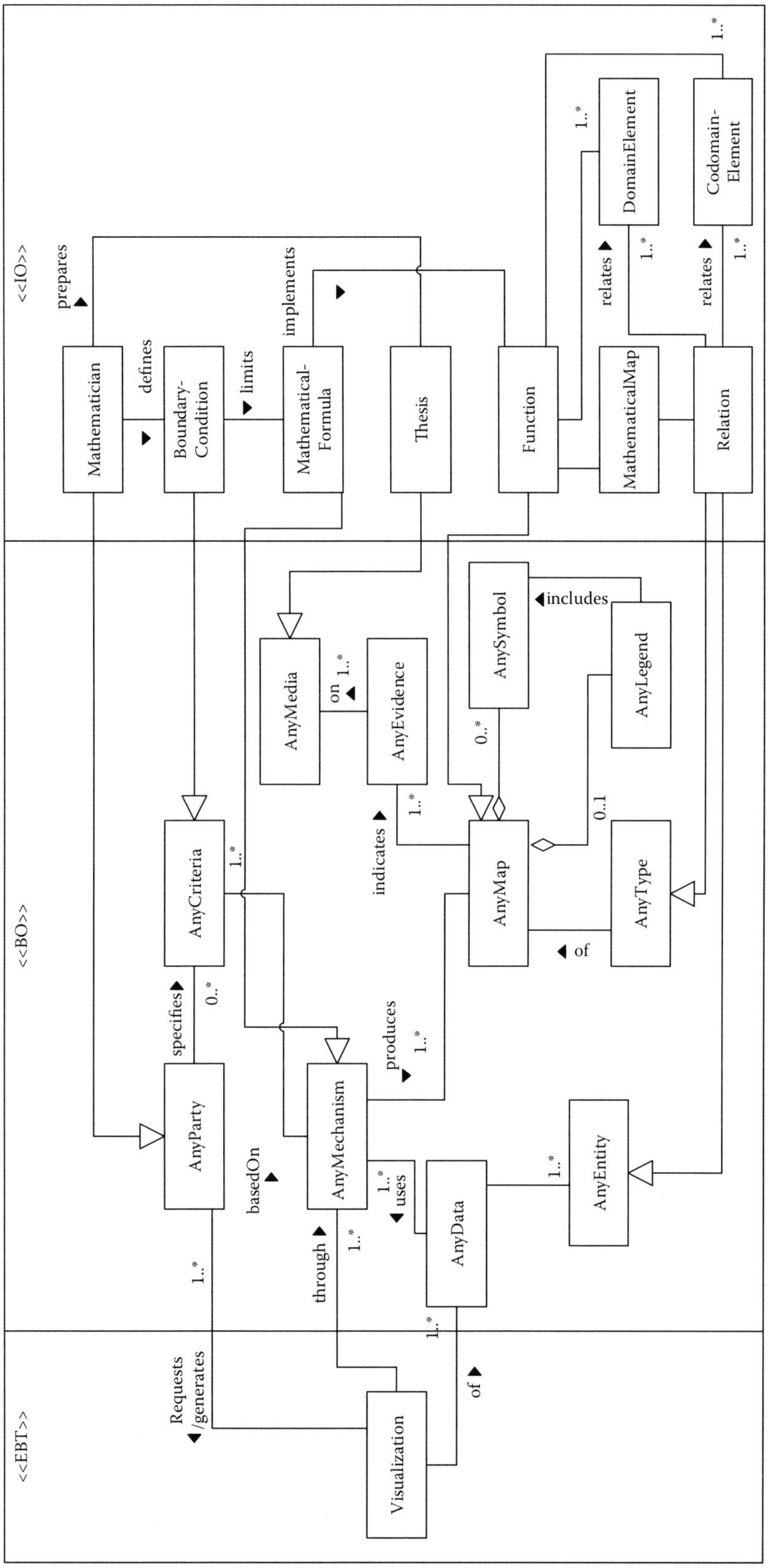

FIGURE 7.6 Class diagram for mathematical mapping.

17. AnyMap is produced by AnyMechanism.
18. AnyMechanism aids in visualization.
19. Visualization is of AnyData and AnyEntity inherits from AnyData.

Sequence Diagram—A sequence diagram is shown in Figure 7.7.
Sequence Diagram Description

1. Visualization is requested by AnyParty(Mathematician).
2. Mathematician defines AnyCriteria(BoundaryCondition).
3. AnyCriteria(BoundaryCondition) limits AnyMechanism(MathematicalFormula).
4. AnyMechanism(MathematicalFormula) implements AnyMap(Function).
5. AnyMap is indicated by AnyEvidence.
6. AnyEvidence on AnyMedia(Thesis).
7. AnyMedia(Thesis) is prepared by AnyParty(Mathematician).
8. AnyParty specifies AnyCriteria.
9. AnyCriteria influences AnyMechanism.
10. AnyMechanism uses AnyData/AnyEntity(Relation).
11. AnyEntity(Relation) relates DomainElement.
12. DomainElement is a part of AnyMap(Function).
13. AnyMap is produced by AnyMechanism.
14. AnyMechanism aids in visualization.

7.2.9 Related Patterns and Measurability

7.2.9.1 Traditional Model versus SSM

- A traditional model, as shown in Figure 7.8, is based on tangible objects, that is, IOs, which are physical objects and are unstable. However, a stable model relies on three concepts—EBT, BO, and IO, which are nontangible. These nontangible objects make the stability model very stable. The EBTs represent elements that remain stable internally and externally. The BOs are objects that are internally adaptable, but externally stable, whereas IOs are the external interfaces of the system.
- In traditional modeling, because we only design as much as is needed for a specific application in question and do not think of its applicability in other domain, the application that results is very specific to the application problem in question. In stability modeling, we use general enduring concepts, and hence, the resulting pattern can be used for building numerous applications. In short, the resulting pattern can serve as a building block for diverse application domains.
- The cost and maintenance of the traditional model is very labor intensive, costly, and time consuming, because of its unstable nature. However, the stable model takes very less time to develop and it is very easy and less labor intensive to maintain.
- The traditional model is neither adaptable to the changing need of the requirement nor extendable, whereas the stable model can be used in a wide variety of applications, by just hooking the application IOs to the general pattern.
- Challenges and constraints of an application are easier to determine in the stable model as compared to the traditional model and all the challenges and constraints defined in the stable model are applicable for all applications in any domain.

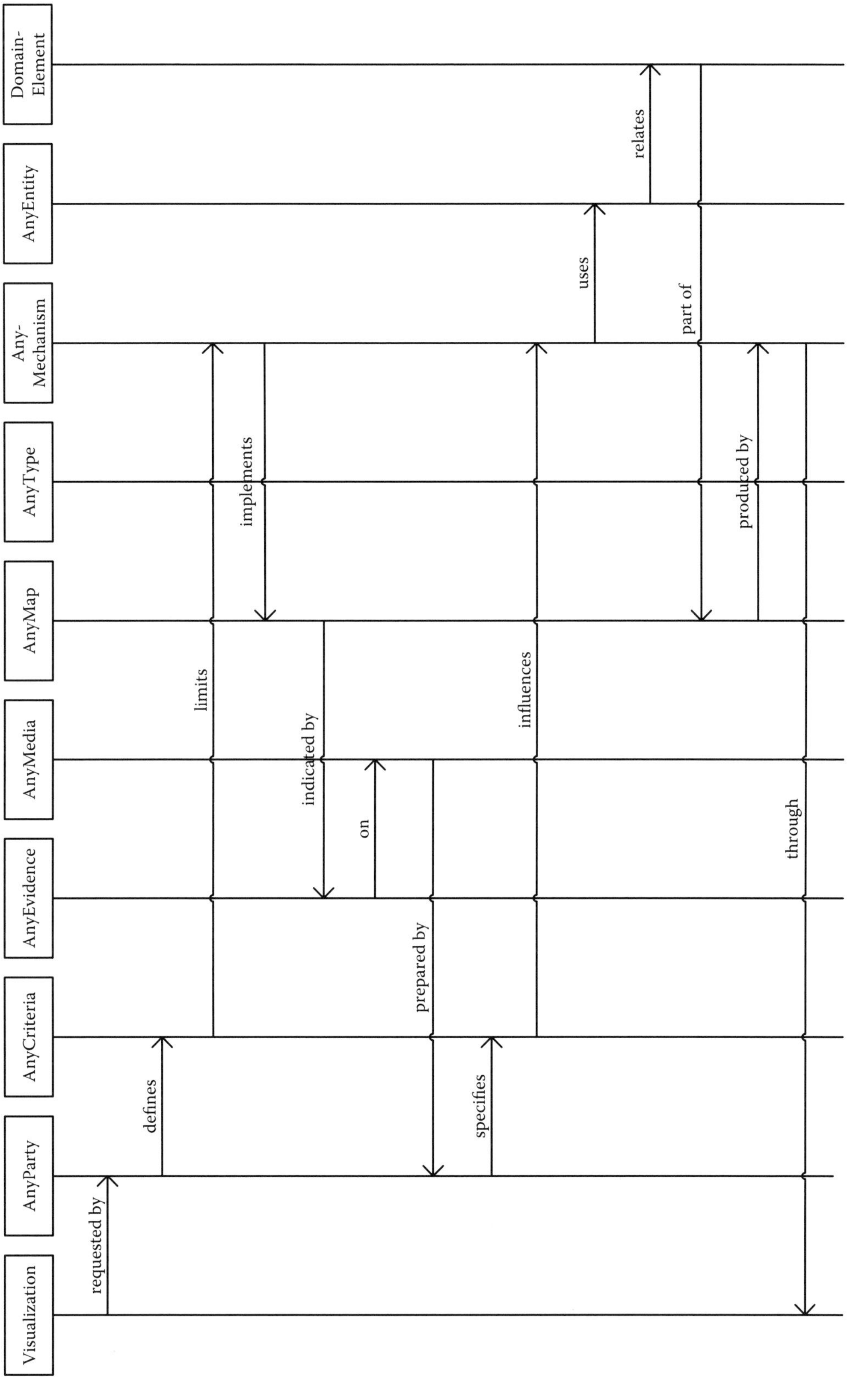

FIGURE 7.7 Sequence diagram for mathematical mapping.

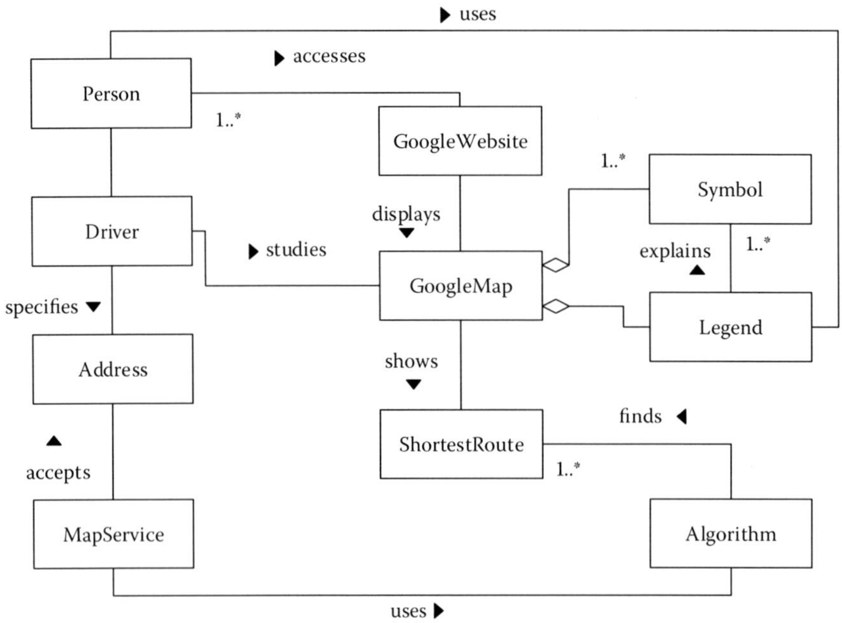

FIGURE 7.8 Traditional model of a map.

- The interdependency among the classes in the traditional model is very high such that a small change in one class disturbs the whole model. But this is not the case in the stable model.
- The number of classes in the traditional model is more to represent any application and they are very application specific, but in the stable model the number of classes is limited, which makes the stable model easier to understand and implement.
- It is very difficult to define multiplicity of relationship in the traditional model, whereas in the stable model, the multiplicity constraint is more obvious and easy to determine.

7.2.9.2 Measurability

7.2.9.2.1 Quantitative Measure

Factors on which quantitative measures can be applied are as follows:

1. *Quantity aspect of EBTs, BOs, and IOs.* The more the number of patterns, the more the lines of code that will result, while developing the system. In addition, as lines of code increase, error propagation rate will also increase and it will be difficult to maintain accuracy in pattern development. Quantitative aspects show that EBTs, BOs, and IOs should be selected in such a way that they should cover all the necessary patterns required in modeling, and yet they should be developed using a manageable number of lines of code, which will result in lesser error propagation.
2. *Number of classes.* The second aspect of quantitative metrics is that the stability model has lesser number of classes with focus on explicit as well as implicit factors, as compared to the traditional model. The stability model relies on the concept of EBTs, BOs, and pluggable IOs. As a result, the base pattern remains stable and has the capability of representing a large number of applications by just hooking the

appropriate IOs with the base pattern. This reduces the number of classes required to represent an application by a drastic amount.
3. *Cost estimation.* Determining and developing estimation or measurement metrics is far easier and less time consuming as compared to that in the traditional model, because we know the base pattern of a stable model well in advance.
4. *Coupling among classes.* Coupling represents how tightly the classes bind together and depend on each other. In the traditional model, coupling among classes is very high. As a result, even a small change or modification to any class in the traditional model ripples through and affects the entire model, whereas in a stability model, change in one class does not affect the whole model and remains restricted to that particular BO.
5. *Constraints.* They represent the multiplicity of the class and are very easy to define in the stability model as compared to the traditional model.

7.2.9.2.2 Qualitative Measure

Qualitative measure of a pattern relates to usability, stability, scalability, and maintainability of the pattern. If you can use pattern for a number of applications without any significant changes in the system, then the pattern will have a number of qualitative qualities. Moreover, the pattern should be reusable in a wide variety of application. Besides these features, the maintenance cost of the pattern should also be very low. Stable model approach to develop patterns supports all these features, whereas traditional model does not. Patterns developed by using traditional model are quite specific to many applications and thus cannot be used repeatedly. Moreover, they are meant for only one specific application. For a new application with the same base, an entirely new pattern has to be developed which incurs a lot of cost and resources. However, stability model is opposite to the traditional model. One single pattern only supports a variety of different application with the same goal.

7.2.10 Modeling Issues, Criteria, and Constraints

7.2.10.1 Abstraction

In stable model, classes are classified into three layers: EBT, BO, and IO. EBT represents the goal of the pattern (the basic or core of the pattern and its purpose). BOs are tangible objects, but externally stable, and they can adapt to any applications. This BO layer encapsulates the pattern behavior from the application and thus, results in reusability of the pattern in wide range of applications. The third layer is IO and is specific to the application, which actually represents the application and is hooked to the second layer of BO. Thus, the core of the pattern lies in EBT and BO, and thus, it becomes necessary to discuss at length on the selection of EBT and BO, so that the pattern can fit to any application in any domain without major changes.

Map cannot be the ultimate goal of the pattern. Hence, in this pattern, map is taken as BO and AnyMap design pattern is developed.

The next step is to find the ultimate goal or EBT for this design pattern. There can be a number of goals associated with map like mapping, navigating, recording, and analyzing. After much discussion, we came with three goals of map—navigation, recording, and visualization. Visualization means to form a visual picture of something. Mapping is not an appropriate EBT here, as map also relates to graphs, charts, and functions besides geographical maps. The other goal was taken as navigating, which means studying or analyzing the map. Recording was also taken as one of the EBT, because without recording a map the other two goals are useless. We could not find one goal which can fit these three goals of map, so we decided on taking three EBTs instead of one in order to cover every aspect of map application.

Next would be the selection of BO for both the EBTs.

- *Visualization*
  - AnyActor is not the only one generating the map, as a group of people might work together to create a map like mathematical map and biological map. Hence, AnyParty was also chosen.
  - AnyActor/AnyParty, depending on the type of map they are developing, defines the rules or criteria of map interpretation, which also sets the rules for standardizing the map symbol. Another BO that becomes important for this pattern is AnyCriteria.
  - Now, the next class or BO to think is AnyMechanism or AnyMethod through which measurement is performed.
  - AnyData in this pattern represents something that needs to be mapped. It can be any location or any real object. To represent a location, AnyRegion BO is used and the object is represented by AnyEntity.
  - Depending on whether map is being developed for AnyRegion or AnyEntity, the type of map is selected. Hence, AnyType is also taken as one BO.
  - There should be something to document the map and should prove its existence. Hence, AnyEvidence and AnyMedia are also taken as BOs.
- *Navigation*
  - Again, navigation can be done by either a single person or a group of people who can study the map. Thus, AnyActor and AnyParty both are considered.
  - Navigation can be done with any type of map and can be done for anything. Hence, AnyType, AnyRegion, and AnyEntity are taken as BOs.
  - AnyActor/AnyParty navigating should have map on something. So, AnyEvidence is taken as BO that displays map and shows its existence. Moreover, the map evidence will be present on some media; hence, AnyMedia is also selected as one of the BO.
- *Recording*
  - Recording can be done by anybody. Hence, both AnyActor and AnyParty were chosen as BOs of the pattern.
  - Some rules have to be followed while recording. Hence, AnyParty or AnyActor should be defined before starting the process. As a consequence, AnyCriteria was also chosen.
  - AnyCriteria is based on the type of data and the amount of data to be recorded. So, AnyData was also considered.
  - Recording has to be done on some media. Thus, AnyMedia forms an important part of the pattern.
  - Different media uses different formats for storage. To represent these formats, AnyLog BO was taken.
  - There has to be some properly defined method via which AnyMap can be stored on AnyMedia. Thus, AnyMechanism was also chosen as one of the BOs.
  - AnyEvidence was also taken as it presents a proof for the existence of AnyMap.

- AnyLegend and AnySymbol also form a part of the pattern as AnyMap uses them to symbolize various things.
- AnyMap is of AnyRegion or AnyLegend. Hence, they also form a part of the pattern.
- In the end, using all these BOs, a pattern was developed, and this pattern was complete by itself and was able to support all the applications from any domain.

7.2.11 Design and Implementation Issues

In the design phase, the BOs and EBTs so decided are taken and a pattern is formulated by using them. This phase involves the tedious task of deciding on the attributes and operations for each EBT and BO. Once the attributes and operations are finalized, the constraints associated with each one of them are listed in order. Then, the relationship among BOs and EBTs is defined and a stable pattern is designed. The challenges and constraints associated with the pattern as a whole are also taken into consideration. After the design phase is over, the next phase is implementation phase. In this phase, the pattern is applied to any desired application. For this, the IOs are first defined based on the context of the application and then hooks are created between pattern and IOs of the application. Thus, in implementation phase, the pattern is developed for the application by simply hooking the IOs of the application to the BOs of the pattern. One way of developing the hooks is via interface (see Figure 7.9).

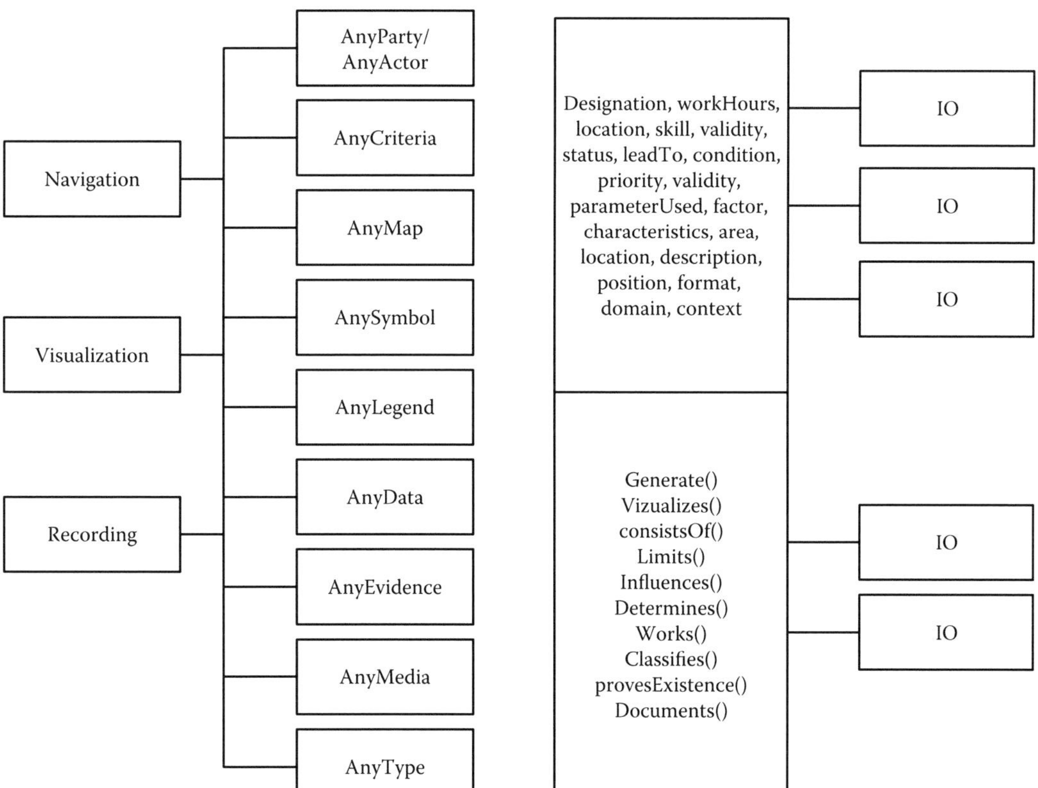

FIGURE 7.9 Hooking BOs to IOs.

Interface is a function that would list all operations of BOs in a combination required to connect BOs to IOs. Thus, BOs will connect to IOs via interface. It will also increase functionality. All the links, which are used to connect to IOs, will be included in interface.

7.2.12 Formalization

Formalization in XML (AnyMap [Recording]):

```
<pattern>
<title>
"recording"
</title>
            <sort>
            <title>
                  "Recording"
            </title>
            <sort>
                  <title>
                  "Recording"
                  </title>
                  <sort>
                              <title>
                              "name"
                        </title>
                              <type>
                              String
                              </type>
                              <universe>
                              {…}
                              </universe>
                  </sort>
                  <sort>
                        <title>
                        "type"
                        </title>
                        <type>
                              String
                        </type>
                              <universe>
                        {…}
                              </universe>
                  </sort>
                  <sort>
                        <title>
                        "description"
                        </title>
                        <type>
                              String
                        </type>
                              <universe>
```

```
      {…}
            </universe>
</sort>
<function>
      <title>
            "allowReusability"
      </title>
      <type>
      Constant: allowReusability
      </type>
</function>
<function>
            <title>
            "providePerseverance"
            </title>
            <type>
            Constant: providePerseverance
            </type>
            </function>
      <function>
            <title>
            "definePurpose"
            </title>
            <type>
            Constant: definePurpose
            </type>
            </function>
</sort>
<sort>
      <title>
      "AnyParty"
      </title>
      <sort>
                  <title>
                  "name"
            </title>
                  <type>
                  String
                  </type>
                  <universe>
                  {…}
                  </universe>
      </sort>
      <sort>
            <title>
            "type"
            </title>
            <type>
                  String
            </type>
      </sort>
```

```
            <sort>
                  <title>
                  "description"
                  </title>
                  <type>
                        String
                  </type>
                        <universe>
                  {…}
                        </universe>
            </sort>
            <function>
                  <title>
                              "accessData"
                        </title>
                        <type>
                              Constant : accessData
                        </type>
            </function>
            <function>
                  <title>
                              "performRecording"
                        </title>
                        <type>
                  Constant : performRecording
                        </type>
            </function>
            <function>
                        <title>
                        "useMedia"
                        </title>
                        <type>
                        Constant : useMedia
                        </type>
                  </function>
      </sort>
      <sort>
            <title>
            "AnyType"
            </title>
            <sort>
                  <title>
                  "name"
            </title>
                  <type>
                  String
                  </type>
                  <universe>
                  { …}
                  </universe>
            </sort>
```

```
            <sort>
                  <title>
                  "type"
                  </title>
                  <type>
                        String
                  </type>
            </sort>
            <sort>
                  <title>
                  "description"
                  </title>
                  <type>
                        String
                  </type>
                        <universe>
                  {…}
                        </universe>
            </sort>
            <function>
                  <title>
                        "classifies"
                  </title>
                  <type>
                  Constant : classifies
                  </type>
            </function>
      <function>
            <title>
                        "organize"
                  </title>
                  <type>
                  Constant : organize
                  </type>
      </function>
      <function>
            <title>
                        "stateStandard"
                  </title>
                  <type>
                  Constant : stateStandard
                  </type>
      </function>
      </sort>
      <sort>
      <title>
      "AnyMechanism"
      </title>
            <sort>
                  <title>
                  "name"
```

```
      </title>
            <type>
            String
            </type>
            <universe>
            {…}
            </universe>
</sort>
<sort>
      <title>
      "type"
      </title>
      <type>
            String
      </type>
      <universe>
      {…}
      </universe>
</sort>
<sort>
      <title>
      "description"
      </title>
      <type>
            String
      </type>
            <universe>
      {…}
            </universe>
</sort>
<function>
      <title>
            "specifyOperation"
      </title>
      <type>
      Constant : specifyOperation
      </type>
</function>
<function>
      <title>
            "records"
      </title>
      <type>
      Constant : records
      </type>
</function>
<function>
      <title>
                  "utilizeMedia"
            </title>
            <type>
```

```
                    Constant : utilizeMedia
                    </type>
          </function>
    </sort>
          <sort>
          <title>
          "AnyEntity"
          </title>
          <sort>
                    <title>
                    "name"
               </title>
                    <type>
                    String
                    </type>
                    <universe>
                    { …}
                    </universe>
          </sort>
          <sort>
               <title>
               "type"
               </title>
               <type>
                    String
               </type>
               <universe>
                    {…}
               </universe>
          </sort>
          <sort>
               <title>
               "description"
               </title>
               <type>
                    String
               </type>
                    <universe>
               {…}
                    </universe>
          </sort>
          <function>
               <title>
                         "distinguishEntity"
                    </title>
                    <type>
                    Constant : distinguishEntity
                    </type>
          </function>
          <function>
               <title>
```

```
                        "occupySpace"
                  </title>
                  <type>
                  Constant : occupySpace
                  </type>
      </function>
      <function>
            <title>
                        "specifyEntity"
                  </title>
                  <type>
                  Constant : specifyEntity
                  </type>
      </function>
</sort>
</sort>
      <sort>
      <title>
      "AnyLog"
      </title>
      <sort>
                  <title>
                  "name"
            </title>
                  <type>
                  String
                  </type>
                  <universe>
                  { …}
                  </universe>
      </sort>
      <sort>
            <title>
            "type"
            </title>
            <type>
                  String
            </type>
            <universe>
                  {…}
            </universe>
      </sort>
      <sort>
            <title>
            "description"
            </title>
            <type>
                  String
            </type>
                  <universe>
            {…}
```

```
                    </universe>
        </sort>
        <function>
              <title>
                          "residesOn"
                    </title>
                    <type>
                    Constant : residesOn
                    </type>
        </function>
        <function>
              <title>
                          "givesSpace"
                    </title>
                    <type>
                    Constant : givesSpace
                    </type>
        </function>
        <function>
              <title>
                          "showCapacity"
                    </title>
                    <type>
                    Constant : showCapacity
                    </type>
        </function>
  </sort>
  </sort>
        <sort>
        <title>
        "AnyMedia"
        </title>
        <sort>
                    <title>
                    "name"
              </title>
                    <type>
                    String
                    </type>
                    <universe>
                    { …}
                    </universe>
        </sort>
        <sort>
              <title>
              "type"
              </title>
              <type>
                    String
              </type>
              <universe>
```

```
                    {…}
              </universe>
        </sort>
        <sort>
              <title>
              "description"
              </title>
              <type>
                    String
              </type>
                    <universe>
              {…}
                    </universe>
        </sort>
        <function>
              <title>
                          "create"
                    </title>
                    <type>
                    Constant : create
                    </type>
        </function>
        <function>
              <title>
                          "open"
                    </title>
                    <type>
                    Constant : open
                    </type>
        </function>
        <function>
              <title>
                          "save"
                    </title>
                    <type>
                    Constant : save
                    </type>
        </function>
  </sort>
  </sort>
</pattern>
```

7.2.13 Testability

If AnyMap design pattern can be used, as it is without changing the core design and by only plugging IOs for infinite number of applications, then AnyMap pattern can be said to be testable. In applicability section, two widely different applications are illustrated and they do not require changing the core design of the pattern. Using the scenarios listed in this chapter, many such scenarios can be deduced and proved that AnyMap pattern is indeed testable.

The above-presented recording pattern is generally applicable to all the domains, and it is designed in such a way that it should be applicable in any type of scenario. If the designed pattern can be applied to all the application, then it is said to be testable. If the derived model is applied with hooking some IOs to related BOs pertaining to that particular scenario, then the pattern is called stable.

There are certain conditions that must be satisfied in order to accomplish successful recording. If these requirements and conditions are not maintained, then recording pattern may fail in end. One of the most basic requirements is to choose proper recording medium depending upon what entity to be recorded. Second is proper security measures should be taken care to prevent the loss of recorded media in case of any natural disaster, theft, and other intrusion-related malicious activities. For example, if a company is recording all the details of their employees in a computer database and then to any other digital media like CD and DVD. In that case, if the system crashes due to virus or disturbance while recording, the company may lose the data, and here in this case, the recording pattern will fail.

Some scenarios in which the above patterns will not give correct result and will fail are as follows:

1. *Visualization.* This pattern fails when the criterion defined by party or actor is not properly followed while implementing mechanism to generate a map. This mistake can mess up with the various symbols used for notation purposes and it might happen to symbols mean that the same or two different things on a map are denoted by a single symbol.
2. *Navigation.* While navigating, the party or actor analyzing the map should carefully read all the interpretation rules and purpose of the map properly. Moreover, the person should also make proper use of legend as any negligence while deciphering the map can lead to inappropriate results.
3. *Recording.* Recording of a map can go wrong when AnyParty/AnyActor does not follow proper mechanism and all the rules laid down for the media chosen.

7.2.14 Business Issues

This section covers business rules, their structural elements, and properties:

- A map can be used by individual, group, organization, software, or a concept.
- A map connects one kind of data to another kind of data.
- Data mapped from or to can be of any form. It can be information, physical objects, or intangible concepts.
- The mechanism employed to perform the visualization can be any method.
- There should be some data ready for visualization.
- Data used for visualization should have real existence and should not be imaginary.
- All the rules influencing the mechanism used in visualization should be defined properly.
- The symbols used in map should be defined in legend properly.
- Any exceptions with the mechanism involved in visualization should be laid down clearly beforehand.

- The map so created should be documented properly using the appropriate media.
- While navigating, all the symbols should be deciphered correctly.
- The choice of media for recording should be done keeping the type of map and intended users in mind.
- The format in which map is stored should be in accordance with the media type.
- The mechanism used for recording a map should be chosen according to the type of the map.

Business rules control and manage the behavior of the system. They impose constraints on the system and tell the system *what* it should do. Business rules are atomic in nature and thus they cannot be broken down into smaller pieces without causing a loss of information. They must be defined prior to defining requirements of the system.

Elements of business rules are as follows.

1. **Business Items**
   This element corresponds to different classes forming the pattern. Stable pattern consists of classes at various levels: EBT, BO, IO, and they have different functions and responsibilities at each level. Some of the business rules defined for business items are as follows:
   - Each class should be capable of at least one function.
   - Classes should be able to work independently.
   - IO classes should interact with BO classes only.
   - EBT classes should interact with BO classes.
   - Classes should be able to reflect the specificity of application.
   - EBT class should represent the ultimate goal of the pattern.
   - BO classes when combined should be able to represent a pattern depicting the meaning of EBT/s involved.
2. **Properties**
   Properties in business language corresponds to attributes and operations of classes in stable language. Business rules related to properties are as follows:
   - The operations defined for the class should be unique and generic, such that they can be used to represent any application.
   - The class should be able to carry out the responsibility assigned to it.
   - The attributes of the class must cover all the distinct aspects of the class.
   - The operations defined for the class should be such that the class is able to perform them independently, as well as in cooperation with other classes.
3. **Relationships**
   It presents the interdependency among classes and in what manner one class relates to the other class. Business rules defined at this level are as follows:
   - One relation can connect only two classes.
   - Every class should relate to another class through some relation. No classes in a pattern can standalone.
   - Relation can be simple relation connecting two classes, that is, association, or it can be *a kind of* or *a part of* relation.
   - Every associative relation has some multiplicity. The default is one to one.
   - Every association relation has some name that represents the type of connection between the two classes.

4. Facts

These represent business or common terms that can occur in the form of EBT or BO. Some of them are as follows:

1. *Navigation*. It is the process of reading, analyzing, and controlling.
2. *Visualization*. It is the process of forming visual pictures.
3. *Recording*. It is the process of documenting.
4. *Legend*. It is the dictionary of symbols used on map.
5. *Boundary*. This term represents any geographical boundary or functional boundary.
6. *Notation*. It represents symbols like color, lines, and images used on map.

5. Constraints

These represent the restrictions imposed on pattern. Some of the constraints are as follows:

1. One or more criteria can influence AnyMechanism.
2. AnyMap should be represented on AnyEvidence.
3. Visualization/navigation can be done by one or many actor/party.
4. AnyMap should have at least one AnyType.
5. AnyMap can be of AnyEntity/AnyRegion.
6. Visualization can be done through one or more AnyMechanism.

Based on the above elements, some generic business rules for the pattern are given as follows:

1. AnyData used for visualization can be AnyEntity or AnyRegion.
2. AnyRegion used for visualization should have well-defined boundary.
3. AnyEntity should have real existence and should not be imaginary.
4. AnyParty/AnyActor involved in visualization should define all the criteria properly.
5. AnyMechanism used in visualization should follow AnyCriteria defined.
6. AnyMap should be indicated on AnyEvidence to prove its existence.
7. While navigating, AnyParty/AnyActor should use AnyLegend properly to understand AnySymbol.
8. AnyType of AnyMap decides whether map is for AnyRegion or AnyEntity.
9. AnyMap should be available for navigation.
10. AnyCriteria should be validated for visualization.
11. AnyMap is stored on AnyMedia in the form of AnyLog.
12. AnyParty/AnyActor requires AnyData of map to decide AnyMechanism and AnyMedia for recording.
13. Quality of AnyLog depends on the capabilities of AnyMechanism.
14. AnyLog is a faithful reproduction of AnyType.
15. All recording mechanisms should be defined.

7.2.15 Known Usage

The AnyMap pattern can be used in many applications in our daily life. Road maps are the most common form of AnyMap pattern. But almost every application that tries to relate one set of things with another set of things can use the AnyMap pattern. The dictionary that we use often to look up the meaning of a word is a form of map. The directory in a shopping mall is another form of map. These are the more *physical* or tangible kind of AnyMap patterns. At the more abstract level, the functions used by mathematicians and physicals are applications of AnyMap. Computer applications find many usage of this pattern in data mapping, internet domain name servers, and site mapping. The possibilities are many and endless. With global positioning system (GPS) installed in cars, personal digital assistants (PDAs), and mobile phones being in vogue, AnyMap design pattern provides ample usage for easy implementation. Satellite maps are used by military to get an idea of enemy's terrain and positions. Site maps are useful to first-time users to access required information correctly (see Sidebar 7.2).

Number of scenarios where this pattern can be used are as follows:

- This pattern can be used as geographical map, for teaching in the class, for finding routes, and for pinpointing any location, in GPS.
- This pattern can be used as biological map, to make a map of chromosomal crossover like in genetic map and linkage map (see Sidebar 7.1).
- Every function is a type of map. Hence, this pattern can also be used for representing any mathematical function.
- This pattern can also be used by astronauts for their research, as they can use this pattern to develop a map of constellations, galaxies, and stars.
- Through this pattern, site maps can be developed which aid in developing websites.

7.2.16 Tips and Heuristics

1. In designing the AnyMap pattern, we try to not just look at the most obvious map pattern. We attempt to research and try to exhaust all possible forms of a map and can try to extract commonalities among all the maps that we could find, so that our pattern will be as generic and extensible as possible.
2. We first tried to identify the EBT of the pattern, which serve to anchor our design goals.
3. In looking for the BOs that provide capabilities for the identified EBT, we will select objects/concepts that are more stable, which cannot change easily over time.
4. We would also make sure that our pattern can be used by any party, human, or nonhuman (i.e., software) alike and the pattern can be scaled to different types of data involved.
5. EBT must represent the goal of the pattern.
6. Intuition and experience is required in order to find correct EBT for the pattern.
7. BOs provide capabilities to achieve the goal of the pattern. Identification of BO requires spending some time in thinking and coming up with correct BOs.
8. BOs provide hooks to which specific IOs can be plugged and getting varied applications in diverse domains. This reduces the cost by encouraging reusability.
9. Designing class diagram using EBTs and BOs serves as the basis for coming up with correct sequence diagram. Drawing sequence diagram using class diagram is much easier and accurate.
10. Writing a clear and thorough description for class diagram and sequence diagram helps in understanding the concept behind the specific design pattern.

11. Describing patterns is a hard job and requires careful and calibrated work.
12. Metamodel is totally different than stable model and it is a traditional model.
13. Pattern design must be generic, so that it can be applied to applications spread across various domains.

SUMMARY

AnyMap is modeled by using the SSM by identifying the corresponding EBTs and BOs. This model can be used for different domains and IOs can be extended according to the application. The model represents the core knowledge of the pattern in different applications and is presented as EBTs and BOs. The model is explained with two specific applications that perform well based on this model.

Though building a stable design pattern for AnyMap that can be reused and reapplied across diverse domain is always difficult and requires thorough understanding of the problem, it is worth the effort and time. Modeling AnyMap pattern by using SSM can result in reusable, extensible, and stable pattern.

The correct identification of EBTs and BOs for AnyMap is the most challenging task and requires some prior experience. Once EBTs and BOs are correctly identified, next main challenge is to determine the relationship between EBTs and BOs, so that AnyMap pattern can hold true in any context of usage for data. Once this is done, depending on the application, the IOs are attached to the hooks so provided by the BOs. Thus, using AnyMap pattern as a basis, infinite number of applications can be built by just plugging in the application-specific IOs to the pattern. This results in reduced cost, effort, and stable solution. Hence, AnyMap design pattern is very useful.

OPEN RESEARCH ISSUES

How to test the AnyMap design pattern effectively is one of the open issues that are left to the user to probe further. Some pointers have been provided under the testability section, but they need further research.

One open issue is to come up with one ultimate goal for AnyMap design pattern, which will contain goals like visualization, navigation, and recording.

List all the possible pitfalls of this pattern.

REVIEW QUESTIONS

1. Explain what do you mean by the term *map*?
2. What is the usage of the AnyMap stable design pattern?
3. Can the term *map* be used in any other context than what you thought of?
4. Can AnyMap design pattern be used interchangeably with AnyChart design pattern? Explain your answer.
5. Which pattern can AnyMap be used interchangeably? Justify.
6. What problem does the AnyMap design pattern solve?
7. In what context is the AnyMap design pattern (Recording) being applied?
8. In what context is the AnyMap design pattern (Navigation) being applied?
9. In what context is the AnyMap design pattern (Visualization) being applied?
10. Name a few scenarios for the application of AnyMap design pattern (Recording).

11. Name a few scenarios for the application of AnyMap design pattern (Navigation).
12. Name a few scenarios for the application of AnyMap design pattern (Visualization).
13. What are the challenges faced in implementing the AnyMap design pattern (Recording)?
14. What are the challenges faced in implementing the AnyMap design pattern (Navigation)?
15. What are the challenges faced in implementing the AnyMap design pattern (Visualization)?
16. What are the constraints faced in implementing the AnyMap design pattern (Recording)?
17. What are the constraints faced in implementing the AnyMap design pattern (Navigation)?
18. What are the constraints faced in implementing the AnyMap design pattern (Visualization)?
19. Discuss briefly the functional requirements of AnyMap design pattern (Recording).
20. Discuss briefly the nonfunctional requirements of AnyMap design pattern (Recording).
21. Discuss briefly the functional requirements of AnyMap design pattern (Navigation).
22. Discuss briefly the nonfunctional requirements of AnyMap design pattern (Navigation).
23. Discuss briefly the functional requirements of AnyMap design pattern (Visualization).
24. Discuss briefly the nonfunctional requirements of AnyMap design pattern (Visualization).
25. Explain AnyMap pattern model (Recording) with the help of class diagram and CRC cards.
26. Explain AnyMap pattern model (Navigation) with the help of class diagram and CRC cards.
27. Explain AnyMap pattern model (Visualization) with the help of class diagram and CRC cards.
28. What are the design and implementation issues for the given AnyMap design pattern (Recording)?
29. What are the design and implementation issues for the given AnyMap design pattern (Navigation)?
30. What are the design and implementation issues for the given AnyMap design pattern (Visualization)?
31. Provide some patterns related to the AnyMap design pattern (Recording).
32. Provide some patterns related to the AnyMap design pattern (Navigation).
33. Provide some patterns related to the AnyMap design pattern (Visualization).
34. Explain usage of AnyMap design pattern (Recording) with two examples other than the ones provided in this chapter.
35. Explain usage of AnyMap design pattern (Navigation) with two examples other than the ones provided in this chapter.
36. Explain usage of AnyMap design pattern (Visualization) with two examples other than the ones provided in this chapter.
37. How does traditional model differ from the stability model? Explain using the AnyMap design pattern model (Recording).

38. How does traditional model differ from the stability model? Explain using the AnyMap design pattern model (Navigation).
39. How does traditional model differ from the stability model? Explain using the AnyMap design pattern model (Visualization).
40. Enlist some of the business issues encountered for the AnyMap design pattern (Recording).
41. Explain procedure for testing the AnyMap design pattern.
42. Discuss some of the real-time usages of AnyMap design pattern (Recording).
43. Discuss some of the real-time usages of AnyMap design pattern (Navigation).
44. Discuss some of the real-time usages of AnyMap design pattern (Visualization).
45. What are the lessons learned by you from this pattern.
46. List some of the domains in which AnyMap design pattern (Recording) can be applied.
47. List some of the domains in which AnyMap design pattern (Navigation) can be applied.
48. List some of the domains in which AnyMap design pattern (Visualization) can be applied.
49. What is the trade-off of using this pattern?
50. List some advantages of using AnyMap design pattern (Recording) in real applications.
51. List some advantages of using AnyMap design pattern (Navigation) in real applications.
52. List some advantages of using AnyMap design pattern (Visualization) in real applications.
53. Can you think of any scenarios where AnyMap design pattern will fail? Explain each scenario briefly.
54. Describe how the developed AnyMap design pattern would be stable over time.
55. List some of the testing patterns that can be used to test AnyMap design pattern.
56. Can you think of any other goal which is not covered by AnyMap design pattern?
57. Briefly explain how AnyMap design pattern (Recording) supports its objective.
58. Briefly explain how AnyMap design pattern (Navigation) supports its objective.
59. Briefly explain how AnyMap design pattern (Visualization) supports its objective.
60. Examine the functional requirements of all the patterns involved—Are there any missing requirements? Discuss them.
61. Examine the nonfunctional requirements of AnyMap design pattern—Are there any missing requirements? Discuss them.
62. Try to list few more business rules for the pattern.
63. Try to develop test cases for the application developed in 29 above.

EXERCISES

1. Think of few scenarios, where AnyMap stable design pattern (Recording) is applicable and come up with corresponding class diagram, use case, and sequence diagram as shown in the solution and applicability sections for each of the scenarios.
2. Think of few scenarios, where AnyMap stable design pattern (Navigation) is applicable and come up with corresponding class diagram, use case, and

sequence diagram as shown in the solution and applicability sections for each of the scenarios.

3. Think of few scenarios, where AnyMap stable design pattern (Visualization) is applicable and come up with corresponding class diagram, use case, and sequence diagram as shown in the solution and applicability sections for each of the scenarios.

Research and Development

New product design and development is more than often a crucial factor in the survival of a company. In an industry that is fast changing, firms must continually revise their design and range of products. This is necessary due to continuous technology change and development, as well as other competitors and the changing preference of customers. A system driven by marketing is one that puts the customer needs first and only produces goods that are known to sell. Market research is carried out, which establishes what is needed. If the development is technology driven, then it is a matter of selling what it is possible to make. The product range is developed, so that production processes are as efficient as possible and the products are technically superior, hence possessing a natural advantage in the market place (Wong and Tong 2012; Koen et al. 2007).

Utilize the AnyMap pattern as an application of a research and development:

a. Draw a class diagram based on the AnyMap pattern to show the application of research and development.
b. Document a detailed and significant use case as shown in case study 1.
c. Create a sequence diagram of the created use case of b.

Indexing and Dictionary Search

Nowadays, dictionary forms an important part of life. Dictionary not only means word dictionary, it exists in various forms like encyclopedia, atlas, thesaurus, and WordNet. Every domain requires dictionary, and hence, the format of dictionary also varies according to the domain and its application. According to Nielsen 2008a, a dictionary may be regarded as a lexicographical product that is characterized by three significant features: (1) it has been prepared for one or more functions; (2) it contains data that have been selected for the purpose of fulfilling those functions; and (3) its lexicographic structures link and establish relationships between the data so that they can meet the needs of users and fulfill the functions of the dictionary. In real sense, a dictionary is another term for map. Instead of showing all the information is a tree format or visual presentation, the things are listed out and indexed (Büttcher et al. 2010; Spink et al. 2001).

Utilize the AnyMap pattern as an application of indexing and dictionary search:

a. Draw a class diagram based on the AnyMap pattern to show the application of dictionary search.
b. Document a detailed and significant use case as shown in case study 1.
c. Create a sequence diagram of the created use case of b.

Site Map Development

A site map is a graphically vivid representation or display of the site plan or the architecture of a given website (see Sidebar 7.2). It could be a simple document used as a sophisticated planning tool for website design and construction. It could also be a simple web page providing a comprehensive list of all the pages in a given website, well organized in a hierarchical pattern. Site maps can help a site visitors by saving the time needed to navigate to a particular web page. Site maps help web masters to optimize the web pages by using a search engine optimization technology (Morville and Rosenfeld 2006; Brandman et al. 2000).

Utilize the AnyMap pattern as an application of site map development:

a. Draw a class diagram based on the AnyMap pattern to show the application of development of site map.
b. Document a detailed and significant use case as shown in case study 1.
c. Create a sequence diagram of the created use case of b.

PROJECTS

1. *Topographic map* (Harvey 1980; Kraak and Ormeling 1996). A topographic map is a type of map characterized by large-scale detail and quantitative representation of relief, usually using contour lines in modern mapping, but historically using a variety of methods. Traditional definitions require a topographic map to show both natural and man-made features (see Sidebar 7.3).
2. *Thematic map* (Slocum et al. 2009). A thematic map displays spatial pattern of a theme or series of attributes. Thematic maps emphasize spatial variation of one or a small number of geographic distributions. These distributions may be physical phenomena, such as climate, or human characteristics, such as population density and health issues. These types of maps are sometimes referred to as graphic essays that portray spatial variations and interrelationships of geographical distributions. Location, of course, is also important to provide a reference base of where selected phenomena are occurring. Petchenik (1979) described the difference as *in place, about space.* Although general reference maps show where something is in space, thematic maps tell a story about that place.
3. *Geologic map.* A geologic map or geological map is a special-purpose map made to show geological features.
4. *Dymaxion map* (Fuller 1943). The Dymaxion map or Fuller map is a projection of a world map onto the surface of a polyhedron, which can then be unfolded to a net in many different ways and flattened to form a 2D map that retains most of the relative proportional integrity of the globe map.
5. *Mercator projection* (Monmonier 2004). The Mercator projection is a cylindrical map projection presented by the Flemish geographer and cartographer Gerardus Mercator, in 1569. It became the standard map projection for nautical purposes, because of its ability to represent lines of constant course, known as rhumb lines or loxodromes, as straight segments. Although the linear scale is constant in all directions around any point, thus preserving the angles and the shapes of small objects (which makes the projection conformal), the Mercator projection distorts the size

and shape of large objects, as the scale increases from the equator to the poles, where it becomes infinite.

6. *Linkage Map.* A linkage map (Griffiths et al. 1993) is a genetic map of a species or experimental population that shows the position of its genes or gene markers relative to each other in terms of recombination frequency, rather than as physical distance along each chromosome. Every living organism has countless number of genes in them which control a number of traits like skin color, eye color, and blood type, and these genes are present all over the chromosomes. A chromosome is an organized structure of protein and DNA that forms an essential part of all the living cells.
   a. Name two to three ultimate goals of each of the above maps.
   b. List all the functional requirements and nonfunctional requirements of each of the ultimate goals.
   c. List 10 challenges for the two or three ultimate goals combined for each area.
   d. Name 10 different applications for each of the goals
   e. Name five different applications for the two or three ultimate goals combined.
   f. Select a significant use case per application and describe each one of them with test cases.
   g. Map each of the use cases in point f into a sequence diagram.

SIDEBAR 7.1 Genetic or Linkage Map

Genetic mapping (also popularly known as linkage mapping) is an exciting topic of considerable scholarly interest. It is unique and special in its approach, and it provides a researcher a series of new and revealing set of information and details on genetic disposition of man. One example of genetic mapping is the precious clue that tells more about chromosomes, and the one that contains the gene, and the exact location where it lies within the chromosome. Genetic or linkage maps are quite beneficial in finding out those single genes responsible for rare genetic disorders and diseases. Genetic maps are quite handy in providing invaluable clues on the nature of diseases and their properties. A genetic map is not a physical map or gene map (Griffiths et al. 1993), that is, the descriptive representation of the structure of a single gene.

REFERENCE

Griffiths, A. J. F., J. H. Miller, D. T. Suzuki, R. C. Lewontin, and W. M. Gelbart. Chapter 5. *An Introduction to Genetic Analysis*, 5th Edition, 1993.

SIDEBAR 7.2 Site Map

A site map (Morville and Rosenfeld 2006) is a graphically vivid representation or display of the site plan or the architecture of a given website. It could be a simple document used as a sophisticated planning tool for website design and construction. It could also be a simple web page providing a comprehensive list of all pages in a given website, and all these pages are well organized in a hierarchical pattern. Site maps can help a site visitors by saving the time needed to navigate to a particular web page. Site maps help web masters to optimize the web pages by using a search engine optimization technology.

REFERENCE

Morville, P. and Rosenfeld, L. Information Architecture for the World Wide Web: Designing Large-scale Web Sites. O'Reilly & Associates, Inc. Sebastopol, CA, 3rd edition, December 2006.

SIDEBAR 7.3 Topographic Map

A topographical or contour map (Courant, Robbins, and Stewart 1996; Harvey 1980; Kraak and Ormeling 1996) is a special map that provides voluminous details and quantitative facts of land relief, by usually using contour lines in modern mapping, but historically using a variety of methods. In modern mapping, researchers used a contour line pattern to create a topographical map. Classical definition for topographical map means a detailed and accurate graphic representation of cultural and natural features on the ground (The Centre for Topographic Information). The study or discipline of topography, while interested in relief, is actually a much broader field, which takes into account all natural and fabricated features of terrain (Kraak and Ormeling 1996).

A *contour map* or *topographic map* is a map illustrated with contour lines, for example, a topographic map, which thus, shows valleys and hills and the steepness of slopes (Tracy 1907). The *contour interval* of a contour map is the difference in elevation between successive contour lines (Tracy 1907). A *contour line* (also *levelset*, *isopleth*, *isoline*, *isogram*, or *isarithm*) of a function of two variables is a curve along which the function has a constant value (Courant, Robbins, and Stewart 1996). In cartography, a contour line (often just called a *contour*) joins points of equal elevation (height) above a given level, such as mean sea level (Courant, Robbins, and Stewart 1996).

REFERENCES

Courant, R., H. Robbins, and I. Stewart. *What Is Mathematics? An Elementary Approach to Ideas and Methods*. New York, NY: Oxford University Press, p. 344, 1996.

Harvey, P. D. A. *The History of Topographical Maps: Symbols, Pictures and Surveys*, Thames and Hudson, 1980.

Kraak, M.-J., and F. Ormeling. *Cartography: Visualization of Spatial Data*. London, UK: Longman, p. 44, 1996.

Tracy, J. C. *Plane Surveying: A Text-Book and Pocket Manual*. New York, NY: Wiley, p. 337, 1907.

8 AnyContext Stable Design Pattern

Language is not merely a set of unrelated sounds, clauses, rules, and meanings; it is a total coherent system of these integrating with each other, and with behavior, context, universe of discourse, and observer perspective.

Kenneth L. Pike

Context is a topic of immense interest in different fields and domains, such as computer science, knowledge engineering, and software engineering. It is a topic that would need an entire book; however, that is not our goal in this context. The goal is, instead, the simplest form of understanding, for example, stable software patterns, and determining why it is so important and critical in the development of a knowledge map.

For example, assume we have you with your friends religiously watching a football match. This is giving a weird and confusing look to a friend just because he or she made a simple comment on space shuttles, and how NASA started collecting relevant information on Mars, during a football game! If so, you are not alone. Knowing the context in which you are directing your comments will determine whether these comments will be accepted, understood, or even cherished.

Imagine that you are in a very important requirement-review meeting with a very important customer. You and your team are trying to arrive at a consensus on the requirements for the customer's new web-based biometric system. In this instance, it is possible that you would be wasting your and other people's time, and even losing the customer in the process. During the analysis of the requirements you may start talking about the customer's requirements and your hiking experiences in North Dakota. The latter will simply be out of context here, because of the fact that the hiking topic is irrelevant to the web-based biometric system.

However, if the customer's new project relates to the development of new *all-terrain* hiking shoes, then the idea of bringing up your hiking experiences in North Dakota in the meeting may form an excellent input, especially, because you would be providing input as a customer and an analyst.

Can context awareness be the important key to unlimited success? Context as a logical fence that will help you to determine what belongs inside the fence and what does not. The remainder of this chapter will navigate you through this concept's process and how it would be represented as a stable software pattern.

8.1 INTRODUCTION

The word *context* is a very common jargon that is used in different fields, for example, computer science, software engineering, and linguistics. According to Dey (2001), this word means "any information that can be used to characterize the situation of an entity.

An entity is a person, place, or object that is considered relevant to the interaction between the user and an application, including both the user and the application." Metaphorically speaking, this definition alludes to a logical fence that encloses relevant information about the domain in question and ignores irrelevant ones.

The impact of having an in-context problem perusal in specific areas of software engineering, for example, software modeling, software development, and software management, is explicitly exhibited in the quality, cost, life span, and level of maintenance of a final software product. The less accurate the software requirements being modeled, the harder it will be to achieve a high-quality solution, the higher the chances to be canceled, and the more the efforts needed to arrive at the solution.

In fact, having an in-context problem perusal will benefit software practitioners (e.g., analysts, architects, developers, and testers), because it will give them a mental construct that will guide them toward the development of a cost-effective and high-quality software product. In addition to this, it will also reduce the clash of ideas and opinions among software practitioners, because everybody has the same frame of reference of the problem under discourse. However, the means to accomplish an in-context problem perusal has not been fully realized until now. For this sole reason, we will carry out an exhaustive study on the concept of context, with the sole purpose of centralizing a common understanding of its structure and importance via stable software patterns.

8.2 PATTERN DOCUMENTATION

8.2.1 Pattern Name: AnyContext Stable Design Pattern

Context is essential in communicating correct information. The objective of being in context is to deliver relevant information to stakeholders based on the environment and current interactions. It helps us in passing the message across the board in a clever and effective manner. The objective of a pattern is to generalize the idea of context so that one can use it as a base for initiating future interactions.

The AnyContext design pattern abstracts encapsulation of any entity based on certain conditions and thereby achieves a general pattern that is usable across any type of application. This pattern is required to model the core knowledge of context without tying the pattern on a specific application or domain; hence, developers are choosing the name AnyContext.

8.2.2 Known As

Developers misunderstand and misconceive context to be the same as domain, background, or situation. Although context is the set of circumstances or facts that surround a particular event, occurrence, background, and situation refer to locality, and hence, it is not possible to use it in place of context. Similarly, a domain refers to a field of action, thought, and influence. Thus, you cannot use it in line with the context. Superficially, only circumstance comes close to the meaning of context. However, you cannot use *circumstance* interchangeably with *context*, as both have different connotations.

8.2.3 Context

Context is a term that is widely used in various fields, such as computer science, linguistics, and negotiations. The term is used regardless of its area of application and is the main goal of context to enclose any type of relevant information that is possible to use by characterizing any type of entity such as a person, system, and class, based on varied conditions or circumstances. Consider for instance a canonical example of context that is found in programming languages, such as the scopes of the methods represented by curly brackets (i.e., methodName() {….}).

Everything within these curly brackets defines a context. It defines the variables, statement, or operations. The curly brackets encapsulate information that belongs to the method itself. You can find another instance of context in clearance systems. In clearance systems, a person is assigned a number that determines access to the boundary of determined information. Another example is medical expert systems, where knowledge from a domain expert that was stored and processed in a database will be used to identify the context of the system.

8.2.4 Problem

The realization of context as a reusable asset for different solutions is a complex and mostly a vague idea. This activity can lead to misinterpretations, constant debates, and recurrent cost/effort/time while addressing it. In this instance, context is usually addressed in different ways that can share a minimal aspect of it or nothing. There are a number of characteristics and behavior for the implementation. For instance, on the one side, we have the encapsulated context pattern by Allan Kelly, and on the other side, we have the context object pattern by Arvind Krishna and Michael Style. Both patterns implementations are different in structure and their implementation mode. They both refer to a concept, though in different ways, which is the same regardless of an area of application. The main reason for this is a poor and incomplete analysis of what the structure of a context is, and in the end, when requirements change. They will be totally different as in their structure and behavior. So, how can we capture and share an enduring structure and behavior of a context that could span several applications, if its structure and behavior remain the same? In this case, the requirements must satisfy the following aspects:

1. *Wide recurrence.* AnyContext is required in all systems that belong to any domains (banking, web applications, medicine, engineering, etc.).
2. *Unlimited scope.* Existing context patterns are limited to certain contexts. Consequently, it is fairly hard to adapt these patterns to handle context in all domains.
3. *Generality.* The AnyContext pattern should be general enough to form a base for developing any context in any application.
4. *Wide applicability.* A pattern that represents a base for modeling any context should have an appropriate and correct level of flexibility, so that the developer can apply the pattern to the desired application.
5. *Specificity.* Utilizing the AnyContext pattern must show, at the same time, the specificity to the domain. This point seems to contradict point 4, which it does not. What

we are saying that any context is used to represent a wide applicability to any domain, and shows specificity of a particular domain and its applications as well.

8.2.5 Challenges and Constraints

8.2.5.1 Challenges

- To provide a generic definition of context based on business and software perspectives as a whole unit.
- To provide the means and methods for a proper encapsulation of any entity's relevant information.
- To cope with different types of conditions or circumstances that may qualify an entity in a systematic manner.
- To allow a context to include the definition of any type of actor, rather than relying of specific actors' roles.
- To decouple the definition of any type of entity from its actual implementation in specific environments.
- To come up with the right abstractions that identify what a context is, how it is structured, and how it will behave.

8.2.5.2 Constraints

- AnyActor must understand the conditions or circumstances that qualify the entity of interest, before determining this entity's context of application.
- AnyContext's definition must wrap its rationale (encapsulation) within its structure, so that it could take advantage of the rationale's functionality, without the necessity of exposing its rationale to the outside world.
- The circumstance or conditions that may qualify an entity must serve as significant inputs in the encapsulation activity.
- One or many contexts share the same unique goal, which is encapsulation.
- One or many contexts must be determined by an actor.
- A context can hold at least one condition that drives the encapsulation of information.
- The conditions that are aggregated in a context are established by at least one actor.
- One or many conditions qualify one entity of interest.
- Entities can be composed by one or many entities.

8.2.6 Solution: Pattern Structure and Participants

The solution is divided into pattern's structure and participants.

- *Pattern structure*
  The structure of this pattern is provided using a class diagram.
- *Participants*
  The participants of the pattern are classified into two types: *classes* and *patterns*. Classes are individual objects that construct something that appears in any

traditional class diagram. Patterns encapsulate a collection of classes and the associations that stem from these classes.

- *Classes*
  - *AnyContext.* It represents the enclosed scope of the relevant information about any type of entity.
- *Patterns*
  - *AnyActor.* It represents the actor that determines the context based on certain circumstances or conditions.
  - *AnyEntity.* It identifies the entity to be qualified and represented within context.
  - *AnyCondition.* It represents the varied and certain/uncertain events, situations, or conditions qualifying a particular entity.
  - *Encapsulation.* It represents the localization of any relevant information about any type of entity.

As shown in Figure 8.1, the class diagram consists of five classes that embody the canonical structure and behavior of the concept context. The Encapsulation class represents the ultimate goal of the AnyContext class. Any activity related to the enclosing of any entity in a determined context will be addressed by the Encapsulation class. The class responsible for providing stimuli to the AnyContext class to enclose entities is the class AnyActor. The AnyCondition class is responsible for qualifying entities in a selected context (AnyContext).

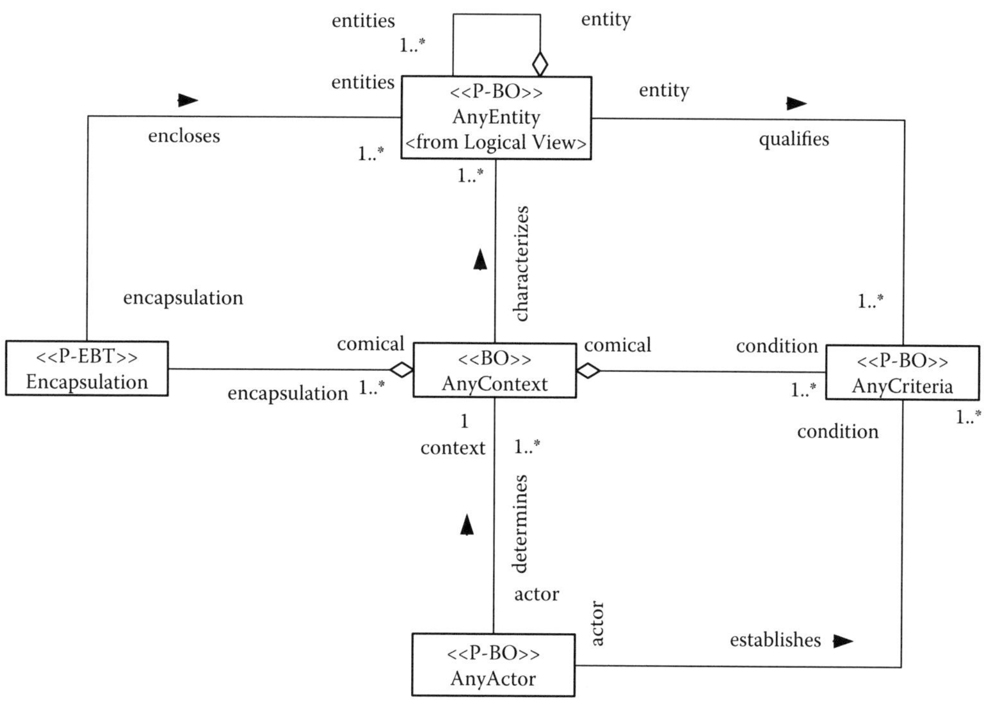

FIGURE 8.1 The structure of the AnyContext stable design pattern.

8.2.6.1 CRC Cards

AnyContext (Context) (Type: BO)

Responsibility	Collaboration	
	Client	**Server**
Represents the canonical structure and behavior of the context concept	Encapsulation AnyActor AnyCondition AnyEntity	determines(), characterize(), accomplish(), enlist(), release(), lookupConditions()

Attributes: container: Map, goal: Encapsulation, conditions: List, domain: String, actor: AnyActor

Encapsulation (Encapsulation) (Type: EBT)

Responsibility	Collaboration	
	Client	**Server**
Localizes the relevant information of a set of enclosed entities	AnyContext AnyEntity	Init(), encapsulates(), disperse(), retrieve(), isProtected()

Attributes: contextTracker: Map, entities: List, isProtected: Boolean

AnyEntity (Entity) (Type: BO)

Responsibility	Collaboration	
	Client	**Server**
Represents the distinct types of entities that can be characterized by a determined context	Encapsulation AnyContext AnyCondition AnyEntity	add() remove() peek() presents() define() acknowledge() belongs()

Attributes: entity: AnyEntity, entities: List, conditions: List, context: AnyContext

AnyCondition (Condition) (Type: BO)

Responsibility	Collaboration	
	Client	**Server**
Declares the set of circumstances that helps the characterization of a set of entities	AnyContext AnyEntity AnyActor	qualify() apply() evidence() validates()

Attributes: entity: AnyEntity, characteristic: int, behavior: int, minQualifier: int, maxQualifier: int, basicQualifier: String[], modifier: int

AnyActor (Actor) (Type: BO)

Responsibility	Collaboration	
	Client	**Server**
Captures the distinct roles of an actor dealing with a context	AnyCondition AnyContext	acts() starts() stops() acknowledges() establishes() disregards() specify() draw() becomes() register() unregister()

Attributes: roles: Map, contexts: List, listeners: List, activityLog: List

8.2.6.2 Consequences

- Because contexts are ruled by a different set of conditions, determined by different sorts of actors and entities, and deployed over different environments, we need to use different sets of new patterns to turn a pattern more application specific. Examples of these patterns are the AnyLog stable design pattern, in the case, for example, we need to record the activities of an actor. Another example is found in data mining, component testing, and so on. In this case, we need to integrate the AnyMedia stable design pattern into the current solution.
- The AnyContext design pattern is generic enough to serve as a building block for applications in diverse domains. Using the AnyContext design pattern for AnyEntity will require that the conditions are corrected, evaluated, and analyzed. In addition, AnyActor must have enough flexibility to allow us to choose the right conditions. However, this does not mean that the pattern is incomplete, as this is the nature of patterns—they need to be used with other components.
- The good thing with the AnyContext design pattern is that the pattern has been derived with stability alone in mind. It has captured the enduring knowledge of the business and its capabilities and will stand the test of time. But the disadvantage with it is that it might result in incorrect results when the entity is used out of context.
- The AnyContext pattern has the following benefits:
  - *Flexibility.* A very good thing about the AnyContext pattern is that it is very flexible, per the conditions stated by AnyActor. As the conditions change, the context of the entity can be easily altered.
  - *Reusability.* The AnyContext pattern is a very stable pattern. It can be reused in many different scenarios spread across many different fields.

8.2.6.3 Applicability with Illustrated Examples

The following two scenarios provide two possible uses for this pattern. For the sake of simplicity, we did not include the complete pattern's model.

8.2.6.3.1 Case Study 1: Context Sensitivity Customization in Component-Based Systems

The main idea is to create a system that will ensure proper context sensitivity customization of component-based applications. This customization will be realized by the systematic propagation of calling and deployment contexts, along with the provided contextual properties, within the component-based systems (CBS).

Class Diagram (see Figure 8.2)

Use Case Id: 1.0
Use Case Title: Customize of Context Sensitivity in CBS Scenario

Actors	Roles
AnyActor	MiddlewareApp

Classes	Type	Attributes	Operations
Encapsulation	EBT	contextTracker: Map	encapsulate()
AnyActor	BO	name role	acknowledges() establishes() disregards()
AnyEntity	BO	entities:List conditions: List	peek() presents() acknowledge()
AnyContext	BO	container: Map goal: Encapsulation conditions: List domain: String actor: AnyActor	characterize() accomplish()
AnyCondition	BO	entity:AnyEntity, basicQualifier:String[], modifier: int	qualify() apply() evidence()
Component	IO	id description	define()
MiddlewareApp	IO	type identificationNo	acknowledge()
Collaboration	IO	characteristic qualifier	validate()
EnvironmentScope	IO	boundary	specifyCondition()

Use Case Description

1. AnyActor (MiddlewareApp) determines AnyContext (EnvironmentScope). How does AnyParty (MiddlewareApp) determine AnyContext (EnvironmentScope)?
2. This AnyContext (EnvironmentScope) is used to characterize AnyEntity (Component). On what basis is AnyEntity (Component) characterized by AnyContext (EnvironmentScope)?
3. AnyEntity (Component) must qualify AnyCondition (Collaboration). How is AnyCondition (collaboration) qualified by AnyEntity (Component)?

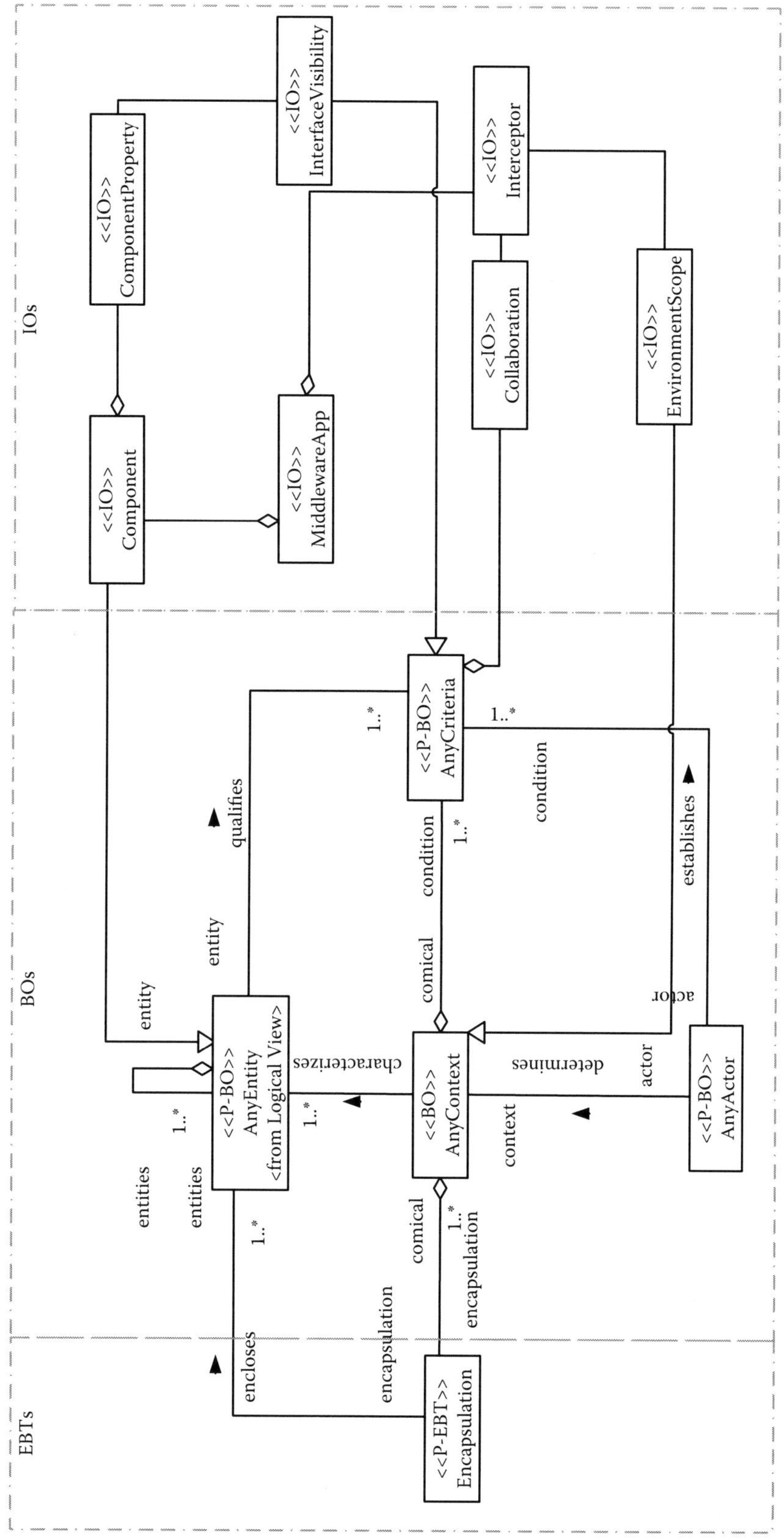

FIGURE 8.2 The context sensitivity customization in CBS scenario.

4. In addition AnyActor (MiddlewareApp) must establish the AnyCondition (Collaboration) so established previously. How is the synchronization between two cases of establishing AnyCondition (Collaboration)?
5. Thus in this way AnyEntity (Component) establishes encapsulation (EBT) of AnyContext (EnvironmentScope). Is this establishment of encapsulation reliable and useful?

Alternatives:

1. AnyContext (EnvironmentScope) determined by AnyActor (MiddlewareApp) is out of context.

8.2.6.3.2 Case Study 2: Context-based Storage System in Mobile Environments

The proposed systems will facilitate the access and management to stored information from users across mobile environments based on the users involved situation or state. For example, the system will facilitate relevant information to the user according to the users' rights or permissions with respect to accessing information, the environments from where the user is accessing the information, the available bandwidth, and so on.

Class Diagram (see Figure 8.3)

Use Case Id: 1.0
Use Case Title: Customize of Context Sensitivity in CBS Scenario

Actors	Roles
AnyActor	Browser

Classes	Type	Attributes	Operations
Encapsulation	EBT	contextTracker: Map	encapsulate()
AnyActor	BO	name role	acknowledges() establishes() disregards()
AnyEntity	BO	entities:List conditions: List	peek() presents() acknowledge()
AnyContext	BO	container: Map goal: Encapsulation conditions: List domain: String actor: AnyActor	characterize() accomplish()
AnyCondition	BO	entity: AnyEntity, basicQualifier:String[], modifier: int	qualify() apply() evidence()
Document	IO	id description	define()
Browser	IO	type identification capability	renderContent()
AbstractCondition	IO	characteristic qualifier	validate()
SecureAccess	IO	security	provideSecureEnvironment()

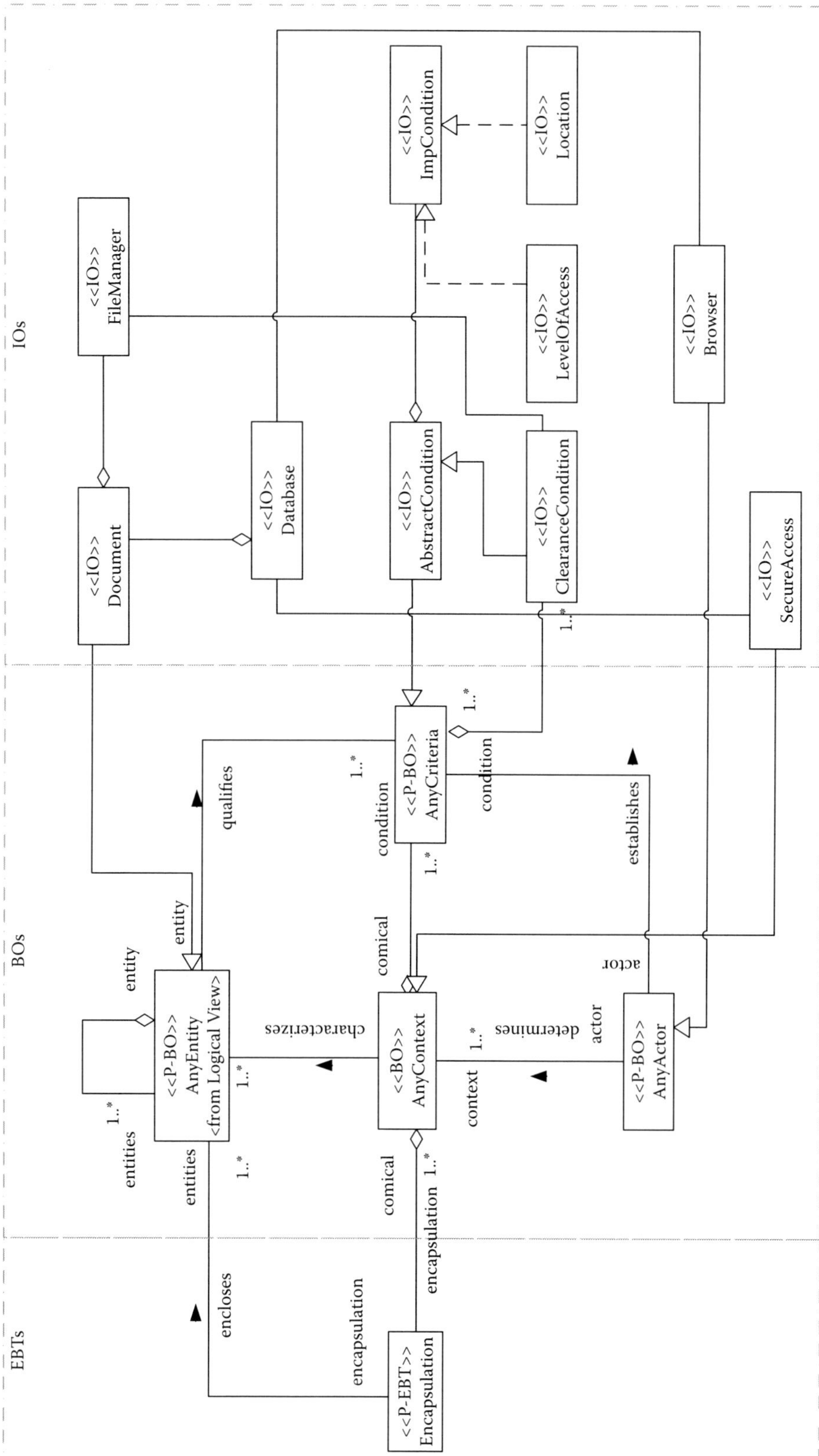

FIGURE 8.3 The context-based storage system scenario.

Use Case Description

1. AnyActor (Browser) determines AnyContext (SecureAccess). How does AnyParty (Browser) determine AnyContext (SecureAccess)?
2. This AnyContext (SecureAccess) is used to characterize AnyEntity (Document). On what basis is AnyEntity (Document) characterized by AnyContext (SecureAccess)?
3. AnyEntity (SecureAccess) must qualify AnyCondition (AbstractCondition). How is AnyCondition (AbstractCondition) qualified by AnyEntity (SecureAccess)?
4. In addition, AnyActor (Browser) must establish the AnyCondition (AbstractCondition) so established previously. How is the synchronization between two cases of establishing AnyCondition (AbstractCondition)?
5. Thus in this way, AnyEntity (SecureAccess) establishes encapsulation (EBT) of AnyContext (SecureAccess). Is this establishment of encapsulation reliable and useful?

Alternatives:

1. AnyContext (SecureAccess) determined by AnyActor (Browser) is out of context.

SUMMARY

A pattern is an enduring solution to a problem in context and is modeled based on the stability model. As a result, the pattern is scalable and reusable over any domain that uses AnyContext and is stable over time too. The advantages mentioned can be inferred from the two examples discussed under applicability section. The main problem is group of forces acting on the system. The definition of problems in context and the constraints on the problem must be realized. There are many challenges to the solution of the problem and therefore the definition of problem and conditions must be known for design of pattern. The pattern can be designed by using unified modeling language (UML) modeling tools like use case diagram, class diagram, and sequence diagram. There are various matrixes that can be used for the design of patterns such as encapsulation, abstraction, and openness. The implementation of patterns can be carried out by using programming languages like Java, C++, and PHP hypertext preprocessor (PHP). The testing of the implemented pattern can be done by conducting various tests.

OPEN RESEARCH ISSUES

The AnyContext design pattern is difficult to understand, as it covers the rather intricate subject of context. Thus, it is necessary to be clear in what respect the term *context* is being analyzed and modeled. Because this explanation is not in the scope of the book, it is left as an open-ended discussion for readers.

REVIEW QUESTIONS

1. What do you mean by the term *context*?
2. Is context an analysis pattern or design pattern? Explain.
3. Is the following statement true or false? Explain your answer: Context is business object.

4. Explain the stability model in your own words.
5. What are the problems with the AnyContext design pattern?
6. List the constraints and challenges with the AnyContext design pattern?
7. With what other patterns can the AnyContext pattern be used? Explain.
8. List some design and implementation issues with regard to AnyContext design pattern.
9. Describe the AnyContext design pattern using CRC cards, class diagram, and sequence diagram.
10. Choose any two domains where AnyContext can be used and show how the above-mentioned design pattern fits in. Provide the class diagram, use case, as well as sequence diagram for each illustration.
11. How will you test the AnyContext pattern?
12. Describe some of the business rules of the AnyContext pattern.
13. Give some known usages of the AnyContext pattern.
14. What are the tips and heuristics that one must apply to the AnyContext design pattern?
15. Give the context where AnyContext pattern can be applied.
16. State some business issues for the AnyContext pattern.
17. How does the AnyContext pattern measure against the traditional context pattern? Explain using the stability and metamodel.
18. Are the following statements true or false?
    a. AnyContext is a BO in the negotiation analysis pattern.
    b. AnyContext is a BO in the research analysis pattern.
    c. AnyContext is a BO in the modeling analysis pattern.
    d. AnyContext is a BO in the analysis analysis pattern.
    e. AnyContext is a BO in the personalization analysis pattern.
    f. AnyContext is a BO in the dynamism analysis pattern.
    g. AnyContext is a BO in the searching analysis pattern.
    h. AnyContext is a BO in the classification analysis pattern.
    i. AnyContext is a BO in the visualization analysis pattern.
19. Discuss briefly the functional requirements of AnyContext.

EXERCISES

1. Name and design a few EBTs where AnyContext will be part of its BOs.
2. Briefly describe a few scenarios of AnyContext.

PROJECTS

AnyContext stable design pattern is included in many EBTs' and BOs' structures. Consider

1. The following EBTs
    a. Analysis
    b. Self-confidence
    c. Negativity
    d. Sensitivity
    e. Delegation

2. The following BOs
   a. Assumption
   b. Brainstorming
   c. Subject
   d. Dialog
   e. Responsibility

Answer the following questions for each of the EBTs and BOs [Hint: Use Appendix A: Pattern Documentation–Detailed Template]:

1. List all the functional requirements of each of the concepts
2. List all the nonfunctional requirements of each of the concepts
3. List five challenges for each of the concepts
4. List ten constraints for each of the concepts
5. Draw a class diagram of each of the concepts
6. Describe briefly three applications for each of the concepts
7. For each described application in 6, do the following:
   a. Attach IOs to the drawing class diagram in 5
   b. Provide a significant use case
   c. Map the use case into a sequence diagram

Section IV

Knowledge Maps, Development, and Deployment

Section IV discusses knowledge maps as road maps for building systems of patterns; development scenarios of the knowledge maps; and deployment, verification and validation, and quality factors of the knowledge maps. Section IV also contains three chapters and seven sidebars.

Chapter 9 is titled "Knowledge Maps: System of Patterns," and it defines the knowledge maps representation structures, quality factors, and properties; compares knowledge maps versus traditional pattern languages; and shows samples of architectural patterns. This chapter concludes with a summary and open research issues. This chapter also covers some review questions, exercises, and projects.

Chapter 10 is titled "Development Scenarios: Setting the Stage," and it discusses the implementation issues of the knowledge maps, such as type versus class within type-oriented paradigm, design type specification template, the uses of contracts in design types' specification, the hook specification template, and others. The chapter also discusses the aspect-oriented modeling. This chapter concludes with a summary and open research issues. This chapter also provides review questions, exercises, and projects.

Chapter 11 is titled "Deployment, Verification and Validation, and Quality Factors," and it discusses the conceptual and practical points of views of deployment of knowledge maps, quality factors, and verification and validation process. This chapter concludes with summary and open research issues. This chapter also provides review questions, exercises, and projects.

Sidebar 9.1 is titled "Domain Analysis," and it provides different definitions of domain analysis.

Sidebar 9.2 is titled "What is the System of Patterns?" and it provides different definitions of systems of patterns.

Sidebar 9.3 is titled "Knowledge Map of the CRC Cards," where the template is filled with a sample of CRC cards knowledge map (Fayad's knowledge map template as shown in Appendix B).

Sidebar 9.4 is titled "Knowledge Map of the Software Project Management," where the template is filled with a sample of software project management knowledge map (Fayad's knowledge map template as shown in Appendix B).

9 Knowledge Maps
System of Patterns

A good plan is like a road map: it shows the final destination and usually the best way to get there.

H. Stanley Judd
1978

Knowledge, in literary terms, is defined as the expertise and skills acquired or gained by a person by means of prior experience accumulated, by education, or as a result of the theoretical, subject-wise, or practical understanding of a particular domain. Knowledge could also mean acquiring expertise of what is already known in a particular field of study or understanding the *facts* about a subject, or gaining awareness or familiarity about a fact or situation because of previous experience.

The management of available knowledge comprises of a number of practices used or employed in an institution to identify, detect, create, represent, distribute, and enable adoption of a number of insights and experiences. These could comprise knowledge, either contained in individuals or embedded in institutional processes and practices. A knowledge map is an enduring document or useful feature that can help people to create or set up an elaborate road map for understanding the available pool of knowledge and its sharing and reuse in many different applications and systems.

9.1 INTRODUCTION

In our belief and understanding, *knowledge* is the mirror of experience gained or accumulated by practice and/or study of a particular discipline. Once an individual acquires this knowledge, he or she can use it as a driving force to prevent experienced pitfalls, bottlenecks, and sloppy processes to reoccur and to create new or streamline previously addressed environments in the search of cost-effective and reusable solutions. Therefore, the question now is how we can represent such knowledge in a straightforward and coherent manner, so that individuals can use it repeatedly to solve recurrent problems.

Existing software patterns and pattern languages emerged as valid answers to the aforementioned question. Developers created them to leverage best practices and lessons learned in the form of standard solutions and architectural styles to solve a set of recurrent problems (Schmidt, Fayad, and Johnson 1996). Software patterns are standard and established solutions, whereas pattern languages are architectural styles that encompass a set of interrelated patterns (Appleton 1997). Let us consider software patterns as a particular recipe for a complete meal and pattern languages as the cookbook containing a set of useful recipes.

Now that we have provided simple analogies to explain what software patterns and pattern languages are, we can further explore what their disadvantages or inefficiencies are, especially in terms of context capturing, organization, and synthesis. Since their appearance, about a decade ago, software patterns and pattern languages have been thought of as a great potential for reducing and improving the software development life cycle (Buschmann 1996; Coplien 1996; Devedzic 1999; Fernandez 1998; Fowler 1997; Gamma et al. 1995; Oestereich 1999; Schmidt, Fayad, and Johnson 1996). In fact, they have been successfully applied to the development of large- and small-scale solutions (Buschmann 1996; Fernandez 1998; Fowler 1997). However, the completion of a particular solution is not always a strong indicator of the existence of efficient and systematic processes. Moreover, even though there have been successful stories of use of software patterns and pattern languages, their applications have not been as coherent or straightforward as they should be.

Most software patterns and patterns languages are created after an exhaustive software development practice—sometimes with very good experience and sometimes with extremely poor practices (Gamma et al. 1995; Laplante and Neill 2006). Throughout these practices, developers are able to find themselves applying the same fine-grained and often continuously repeated solutions to solve *specific* problems. In the end, these fine-grained solutions were turned into patterns that would solve problems of the same sort. However, when it comes to the creation of large-scale solutions from weaving a set of individual patterns, practitioners realize that it is neither an easy nor a systematic task. Consequently, the reduction and ease of the software development life cycle, via patterns, is still a dream difficult to accomplish.

Knowing that the idea or conception of creating and weaving such solutions still constitutes a distant dream, we are motivated to explore or probe the area and current literature and come up with an approach that will make this elusive dream come true. This approach is the knowledge maps approach. This approach facilitates a focused road map for knowledge sharing and knowledge reuse. This road map will guide us throughout the process of applying and weaving software patterns to build systems. To understand this road map, we need to describe some of the properties of knowledge maps, such as knowledge partitioning and intersection, measurement, reusability, stability, infinite software architecture generation, return of investment (ROI), remote knowledge, context awareness, and generality. These properties will be described in the subsequent sections of this chapter.

9.2 REPRESENTATION OF KNOWLEDGE MAPS: STRUCTURE, QUALITY FACTORS, AND PROPERTIES

Another way to define knowledge maps is as a topology of software patterns. First, they are a topology of software patterns, because they represent a logical arrangement of patterns according to the patterns' rationale, scope, and nature. Second, knowledge maps, as a topology, facilitate a flexible and coherent environment, where practitioners can systematically plan, select, and weave patterns to build systems.

The building of systems is possible by two means. The first one is that part of the creation of knowledge maps includes the definition of a road map for knowledge representation and reuse. The definition of this knowledge representation will require an understanding of what knowledge to represent in a particular discipline (to assure a more generic nature we prefer using *discipline* instead of *application*); how to extract and create this stream of knowledge; and how to efficiently and effectively represent it. One will also need to understand how to implement its chosen schema; how to adapt the knowledge when facing new problems; how to analyze the knowledge; and, finally, how to exploit this knowledge to create accurate solutions (Cercone and McCalla 1987). The second one is that the knowledge maps facilitate a systematic and a straightforward approach to generate, filter, develop, and deploy the pertinent features of a domain via the software stability concepts approach.

In representing the knowledge maps as the groundwork for any discipline and as the answer to problems, such as pattern composition and traceability, we must illustrate how knowledge maps are structured.

9.2.1 Structure of the Knowledge Maps

The structure of knowledge maps consists of five major stages that correspond to the software stability models (SSMs) (Mahdy and Fayad 2002). These stages form the representation scheme that guides knowledge extraction, generation, and synthesis. The knowledge maps show, by means of use of stable patterns, the overall processes to solve a set of problems within the discipline in question, such as performance evaluation and software testing. Knowledge maps will ensure that both managers and the technical staff easily understand these processes. For example, in online bargain systems, concepts such as negotiation, trading, and customer satisfaction are well known, by both managers and technical staff. For this reason, these concepts must be taken into consideration when building these types of systems. Knowledge maps, then, allow managers and technical staff to use these concepts as part of their bargain systems' solution. This is possible by exploring the context, where these activities happened, who is executing them, the media in which they are being executed, which mechanism will be used to execute them, as well as the expected result.

The stages of knowledge maps emphasize the elements pertinent and relevant to the software development life cycle, as well as these elements' collaboration and interactions in accordance with customer requirements. Analysis concerns are also achieved by design concerns (capabilities), knowledge concerns (combine analysis and design concerns) forming stable patterns that lead to development concerns of unlimited applications using industrial objects (IOs), and deployment concerns show the best quality factors of the knowledge concerns. The course of this collaboration is shown on a priority basis. It starts from the element most crucial for the subject matter to exist, and then it proceeds to the ones with lower priority. Being a low-priority element does not mean that this element is not important within a patterns' topology. On the contrary, the level of priority gives you the order of how each of the elements that form the subject matter's rationale will be handled or selected, by either managers or a technical staff member (most likely by the technical staff).

The elements' collaboration is either permanent or temporary. It is permanent when these elements, along with their associations, are part of a conceptual solution (e.g., a stable analysis pattern [SAP]). It is temporary when an already defined conceptual solution needs to be extended by adding an extra aspect (e.g., goals connected to subgoals or capabilities connected to other capabilities). This temporary collaboration happens at an analysis patterns level, a design pattern level, or an architectural level.

To ensure consistency on how the knowledge maps are represented, we will incorporate a set of intuitive symbols, along with their specific notations, that will simply the use and deployment of the knowledge maps. To comply with UML (Unified Modeling Language) notation, we will also provide how each symbol is represented using UML standards (Oestereich 1999).

9.2.1.1 The Notation Used in Knowledge Map Structures

The first perspective of the knowledge maps targets a specific segment of professionals—the nontechnical people (e.g., managers). However, this does not mean that technical people cannot use it. On the contrary, they can use it to gather an expanded global idea of the subject of interest.

The main categories of the knowledge maps are represented by 20-sided polygons (icosagon). and consists of (1) analysis concerns or goals, (2) design concerns or capabilities (the synergy between goals and capabilities), (3) development concerns, and (4) deployment concerns or quality factors (see Figure 9.1).

The elements in the main categories are represented by hexagons. A gray line drawn between the element in question and its category represents the nature of these elements. The nature of these elements is also perceived by how the elements' names are numbered. For example, the elements from the goals category have their names numbered 1. Figure 9.2 shows the entire numbering conventions for each one of the elements in the knowledge maps. Regarding their naming conventions, they follow a *camel casing* style (e.g., ClassName).

The associations between the pertinent elements (sources) of one category and their subsequent elements of another category (destinations) are denoted by the thickness of the line according to the subsequent elements nature. For instance, from goals to capabilities, the association is fine gray line (the same as capability). See Figure 9.3 for more details.

One important fact is that the gray line between each category indicates the transition between categories. This transition implies the set of heuristics and evaluation indicators discussed throughout this book, such as the ones used to find goals, capabilities, development scenarios, and quality factors.

Have you noticed that in Figure 9.3 there are associations drawn neither from capabilities to the development scenario elements nor from capabilities to the quality factors? Instead, all these drawn associations depart from the hexagon representing the SArchPs. This happens because SArchPs enclose all the goals (at least two), along with their pertinent capabilities, as a whole and independent base that can be reused elsewhere, regardless of the forthcoming requirements.

9.2.1.2 Knowledge Map Template

The following knowledge map template (see Appendix B) is used throughout this book to fully document and chronicle knowledge maps. A full knowledge map template consists of the following sections:

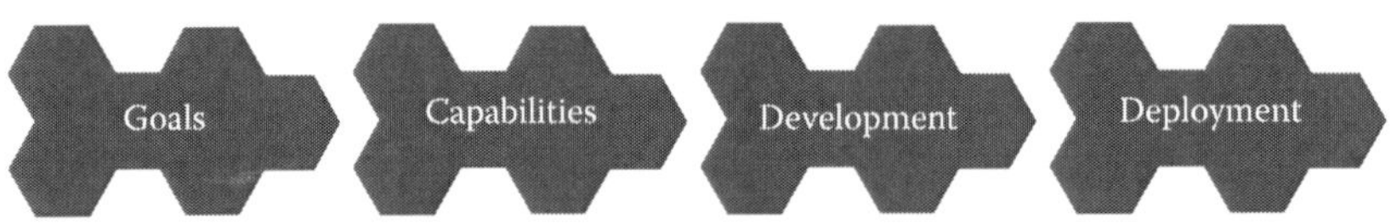

FIGURE 9.1 The main categories of knowledge maps.

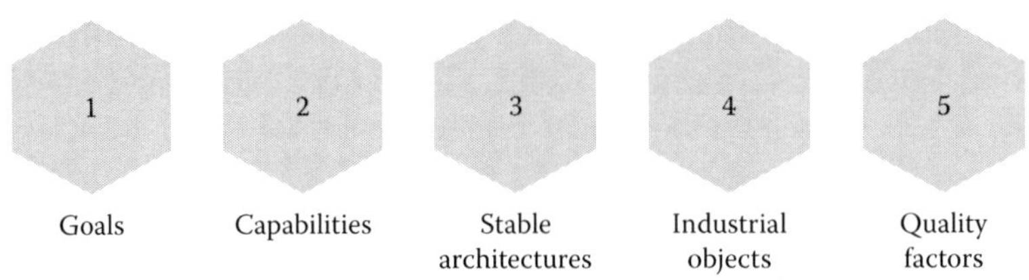

FIGURE 9.2 The elements of knowledge map categories.

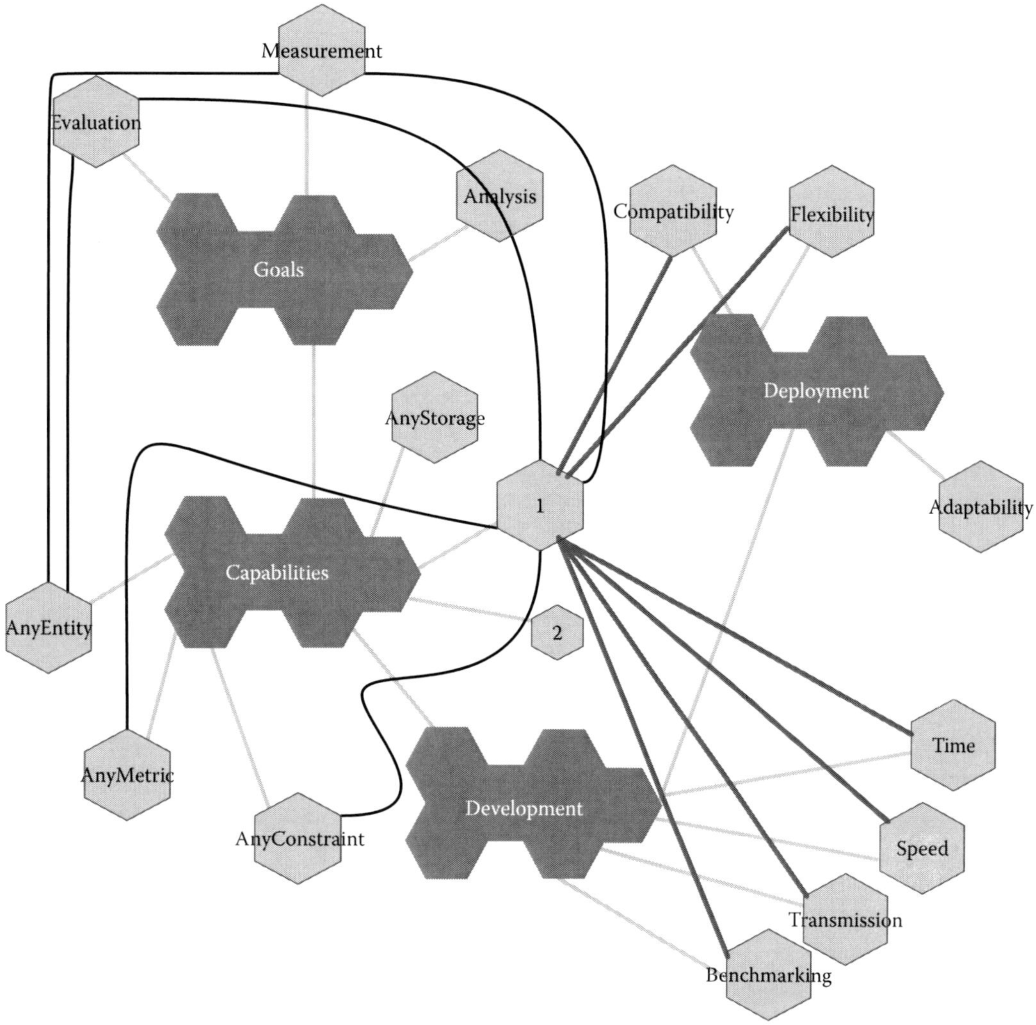

FIGURE 9.3 The associations between the elements of knowledge map categories.

1. *Knowledge map name*. What is the knowledge map called? The knowledge map name is a noun or noun phrase and it follows the UML specifications of a class name. It also can be called stable pattern language. Consider finding out the knowledge map of e-learning as a domain so that the knowledge map name is e-learning knowledge map, knowledge map of e-learning, e-learning stable patterns language, or patterns languages of e-learning.
2. *Known as*. What are the other names for the knowledge map?
3. *Knowledge map domain/subject/topic description*. What is the description of the domain, subject, or topic?
4. *Enduring business themes (EBTs)/goals*. What are the ultimate EBTs or goals of this knowledge map? Find the ultimate goals of the domain, subject, or topic by asking the following questions: What is the domain, subject, topic, or concept for? What problems does this domain, subject, topic, or concept solve? What are the goals of the domain, subject, topic, and concept?

5. *Business objects (BOs)/properties.* What are the properties/BOs of each of the goals of the knowledge map? What are the BOs that can accomplish each of the ultimate goals of the knowledge map?
6. *Knowledge map (core knowledge).* It maps each of the goals into its BOs.

For samples of knowledge maps, refer to Sidebars 9.3 and 9.4, and Chapter 13.

9.2.1.3 Structure of Knowledge Maps and Its Relationship with UML

The second perspective, UML perspective (Oestereich 1999), of knowledge maps targets another specific segment of professionals—the technical people (e.g., analysts, designers, and programmers). Managers, with strong technical skills, can certainly understand this perspective, so they can guide their crew members throughout the software development life cycle.

The individual elements that form the structure of the knowledge maps have the following meanings within the UML's realm:

- *Goals.* They represent the specification classes that deal with the problem space and emphasize on the rationales and objectives of any given domain. They are represented as classes, as in any other traditional class diagram. They follow a *camel casing* style as their primary naming convention. Classes of this sort are marked using the two types of stereotypes (see Figure 9.4 for more details):
  - <<EBT>> indicates that classes of this sort are atomic EBTs.
  - "<<Pattern-EBT>>" or "<<P-EBT>>" indicates that these classes have a second level of abstraction that encloses other classes and, in some cases, other patterns.
- *Capabilities.* Like goals, capabilities are represented as classes, as in any other traditional class diagram. They also follow the same naming convention as goals. Classes of this sort are marked using either the stereotype <<BO>> or <<Pattern-BO>>, or <<P-BO>>. Their main purpose is to achieve the goals of a subject matter. See Figure 9.5.

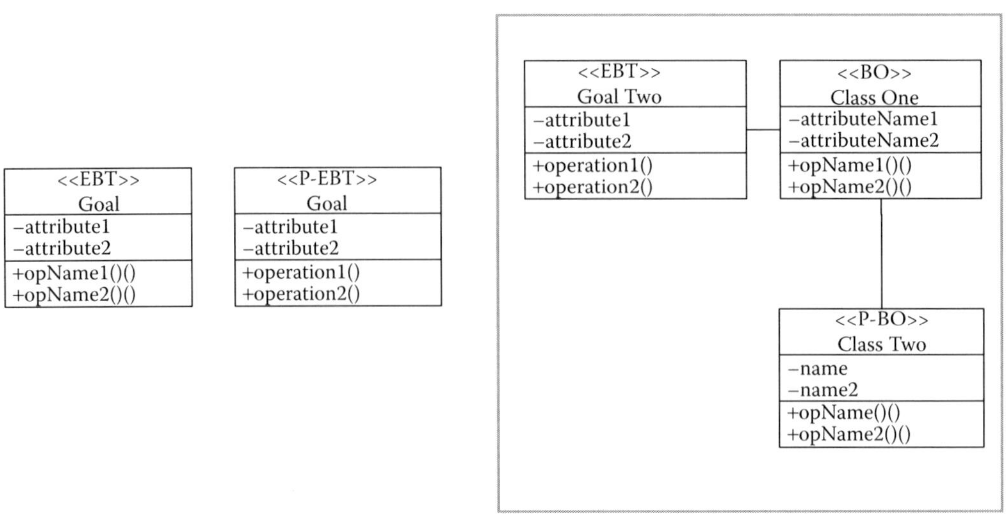

FIGURE 9.4 The UML representation of goal.

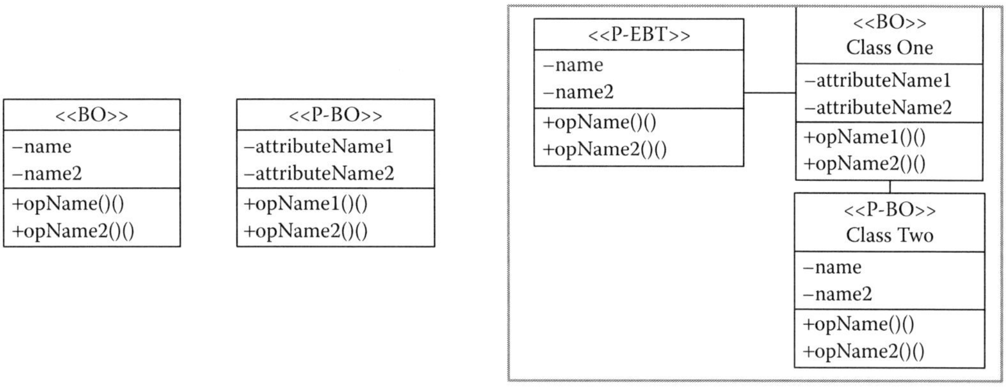

FIGURE 9.5 The UML representation of capabilities.

  - <<BO>> indicates that classes of this sort are atomic BOs.
  - <<Pattern-BO>> or <<P-BO>> indicates that these classes have a second level of abstraction.
- *Development scenarios.* These enclose the specific contexts to where our stable patterns will be applied. These contexts are represented by a set of classes, along with their interrelations and constraints. Classes of this sort are marked with the stereotype <<IO>> to convey their volatile nature (see Chapter 6 for details).
  - <<IO>> indicates that these classes are objects that proportionally change according requirements.
  - These classes are attached to the core formed by our stable patterns, by means of using classes with a determined hooking code (e.g., hooks). Classes of this sort are marked with the stereotype <<Hook>>.
  - <<Hook>> indicates that these classes are extension points of the stable patterns included in the knowledge maps.
- *Quality factors.* They, within knowledge maps, are considered as the deployment *goals* of your subject matter. Therefore, they will follow the same representation and same stereotype and naming conventions as the main goals of your subject matter.
- *Associations.* They, in UML, are represented by lines drawn between one or more classes. At the end of each association, multiplicities, role names, navigability, and qualified attributes (if necessary) are indicated. Likewise, they need to have a name; this name is usually set in *italics*. Because our purpose is not to give a full description of associations' notation, we recommend that the reader access Oestereich (1999) for more specific details.
- *A three-layer representation.* The SSMs (Mahdy and Fayad 2002) are the visual representation of the software stability concepts approach. They facilitate an explicit separation of concerns of subject matter by clearly dividing it into three visual layers: EBT layer (goals), BO layer (capabilities), and IO layer (context). Figure 9.6 shows what we mean by a three-layer representation (Fayad 2002a, 2002b, 2002c; Fayad and Altman 2001; Fayad, Ranganath, and Pinto 2003).

How the knowledge maps are structured and organized depends solely on the compliance of certain quality factors. Some of these quality factors were extracted from the enterprise

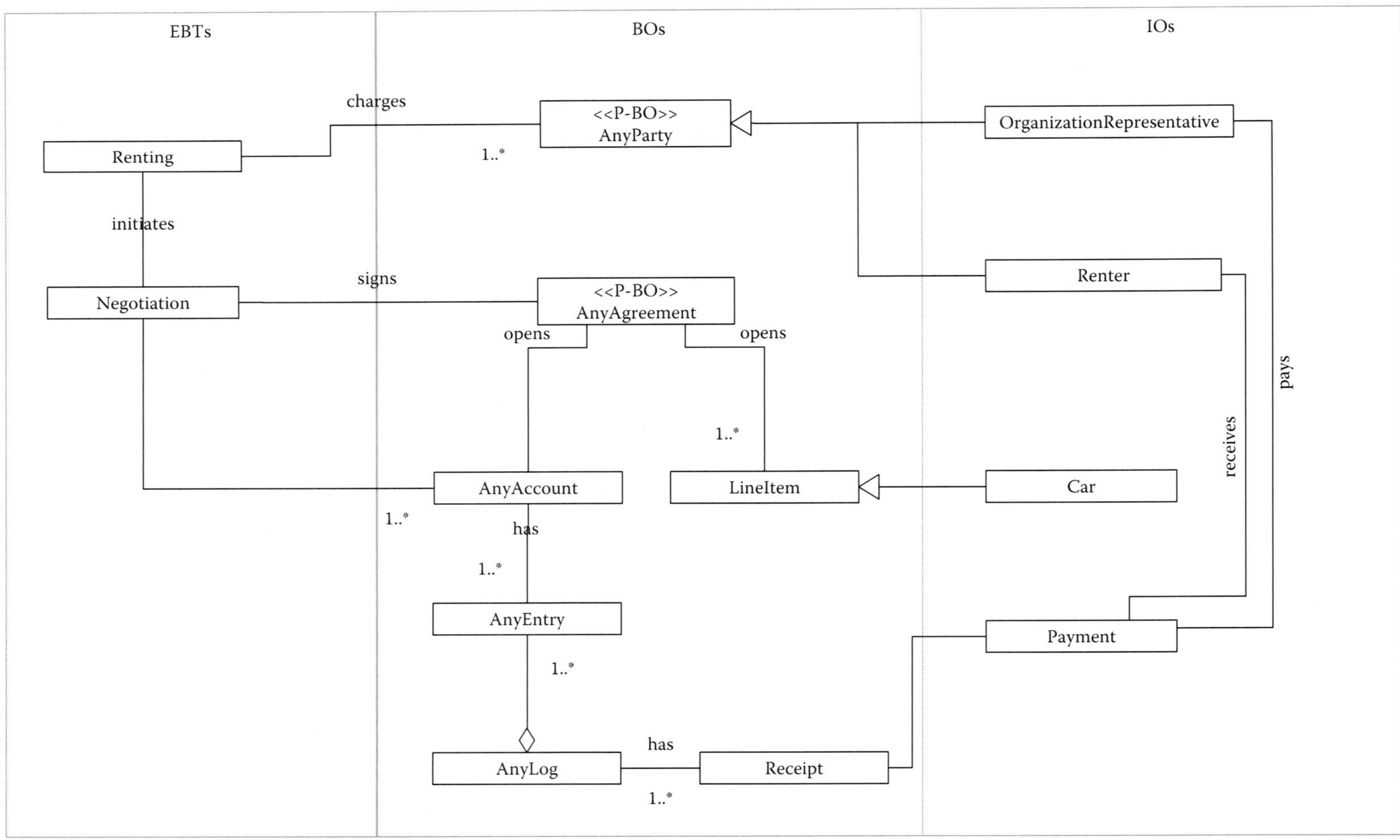

FIGURE 9.6 The three-layer representation of knowledge maps.

software architectures literature (Fayad, Hamu, and Brugali 2000) and applied to the deployment of knowledge maps. We will illustrate these quality factors in the following sections.

9.2.1.4 Quality Factors in Knowledge Maps

Knowledge maps can guide individuals to plan, design, and solve a set of recurrent problems within a given context. What we mean by *within context* is the ability to focus on the actual problem, to acknowledge and agree on the set of constraints, pitfalls, trade-offs, and to obtain tidbits of advice pertinent to the problem, while ignoring irrelevant and useless details. Such qualities are not conceived out of sheer magic; rather, they are conceived as the product of compliance with a set of quality factors. These quality factors are as follows:

1. *Stability.* Knowledge maps must be defined at a knowledge level by using the two primary artifacts of the software stability concepts approach: EBT and BO. Due to the enduring and reusable nature of both artifacts, they allow the knowledge maps to serve as the base for creating stable solutions capable of facing ever-changing problems without any struggle (see Figure 9.7) (Fayad 2002a, 2002b; Fayad and Altman 2001; Fayad, Ranganath, and Pinto 2003; Mahdy and Fayad 2002).
2. *Scalability.* This concentrates on the ability of the knowledge maps, along with their enclosed elements, to adapt to evolving needs and insights without unnecessary effort (Fayad, Hamza, and Sanchez 2005), by means of either enabling or disabling software patterns from and to the knowledge map and extending or reducing the peripheral of their knowledge maps without any complication or collapse (Fayad, Hamza, and Sanchez 2005). Figure 9.8 summarizes what we mean by scalability.
3. *Traceability by rationale.* This quality concentrates on the ability of the enclosed patterns or elements of the knowledge maps to be successfully traced back to their original goal or usage rationale (Hamza 2002; Hamza and Fayad 2003) after their implementation.
4. *Generative nature.* This quality represents both the ability of the knowledge maps to reproduce and originate complete solutions based upon a set of intertwined patterns and the patterns' connective rationale. These are the guidelines for selecting patterns and their interconnection.
5. *Adequacy.* Knowledge maps can satisfy the requirements of their intended and established purposes, along with their consequences (Fayad, Hamu, and Brugali 2000). In other words, the knowledge maps offer a *goodness of fit*, regardless of the context of application. This quality extends from the following:
   a. Visualizing and monitoring each of the elements in the knowledge maps

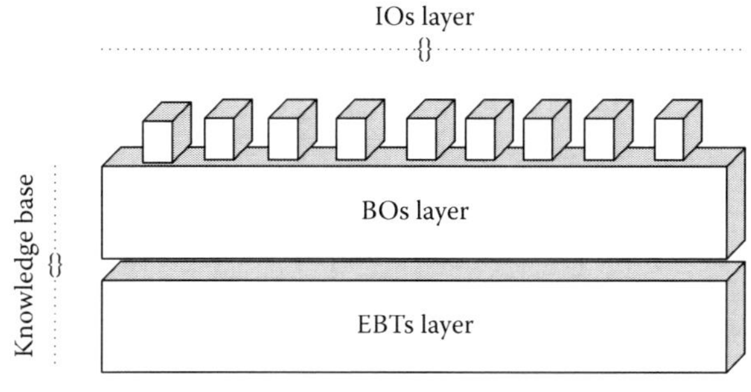

FIGURE 9.7 Stratification of the software stability concepts.

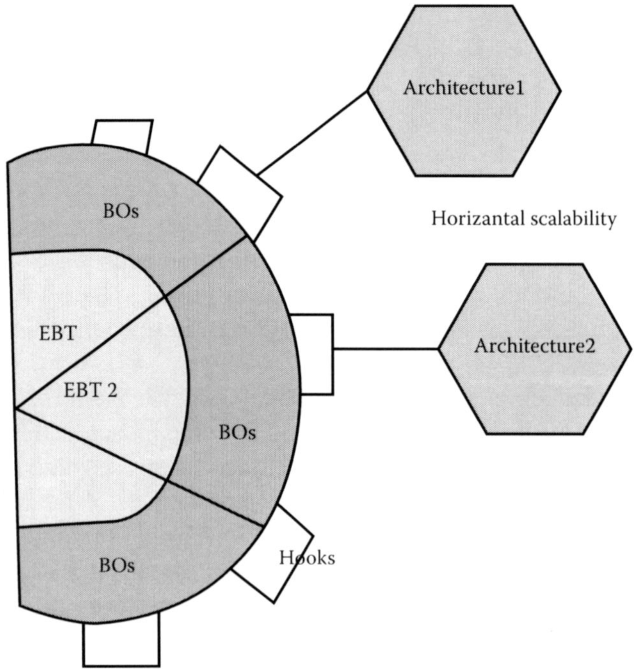

FIGURE 9.8 The achievement of the scalability quality.

   b. Efficient modeling and documentation means
   c. The ability of attaining and representing objective knowledge and the rationale of what is known
   d. Searching and recognizing capabilities
6. *Verification and validation (V&V) ability.* This quality concentrates on providing the means for verification and validation of selected paths (synergies of goals and capabilities) and decides whether the selected path is the most suitable selection or not.

Depending on the fulfillment of the aforementioned quality factors, the properties of the knowledge maps will enable the process of building cost-effective systems.

9.2.1.5 Properties of Knowledge Maps

The proper execution of the knowledge maps' quality factors in conjunction with a proper knowledge representation (structure of the knowledge maps) will make obvious of more than 24 properties of the knowledge maps and the benefits that stem or arise from them. These properties will streamline the process of building systems from patterns. They are partitioning, intersection, measurement, reusability, infinite software architectures generation, faster ROI, direct and remote knowledge, context awareness, and generality. The following are some of the properties of knowledge maps:

1. *Partitioning property of the knowledge maps.* This property is driven by a common object-oriented technology's mantra, *divide and conquer.* In other words, a particular domain is partitioned as follows.
   a. First, we break a domain of interest into small sets of understanding or subdomains. Put in this way, instead of dealing with a general domain that contains a

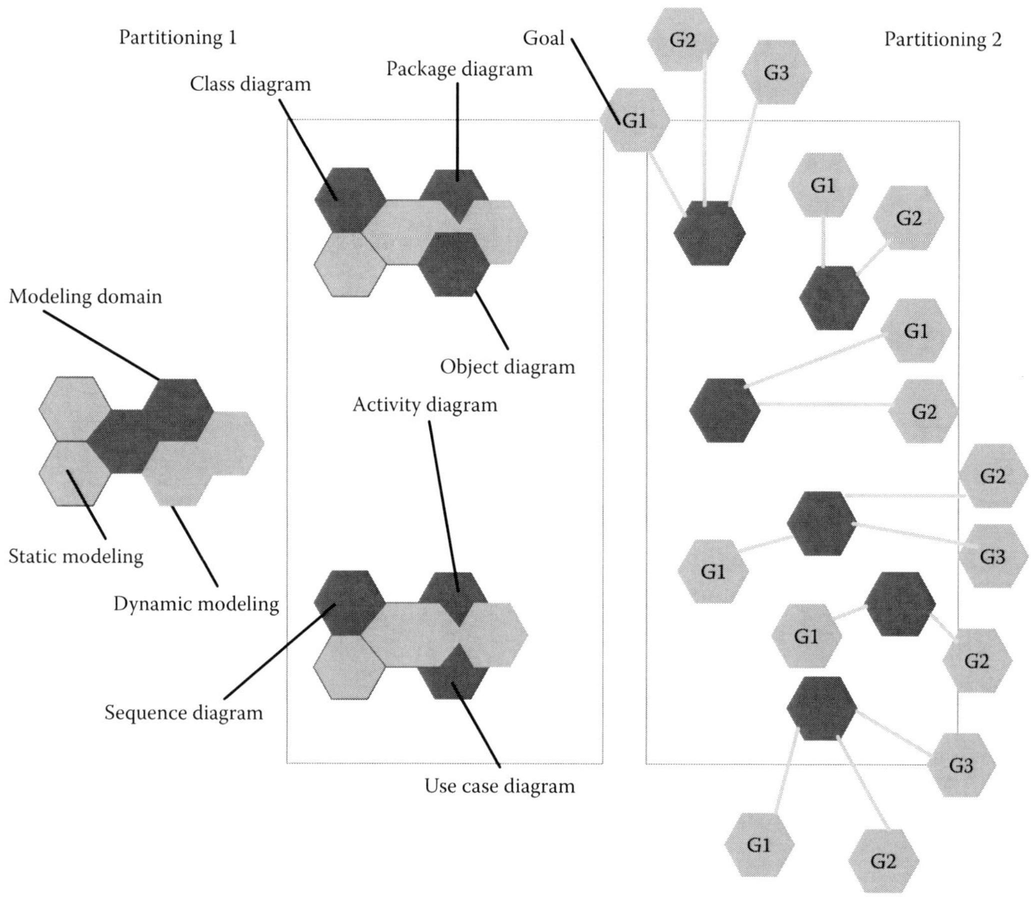

FIGURE 9.9 The partitioning of the modeling domain.

dizzy array of information, you can break this domain into fine-grained sets of understanding, where you will be able to manage them with ease. Figure 9.9 shows an example of domain partitioning using modeling as the domain of interest.

b. Second, we allocate each one of the concerns that form a particular domain or subdomain based on their nature or purpose in the patterns' topology: EBT, BO, or IO. Figure 9.10 shows this particular property of the knowledge maps (see Chapters 3 and 4).

2. *Intersection property of the knowledge maps.* Through the sharing of concerns determined among two or more patterns' topologies and their proper partitioning, we will be able to enhance the associated domain understanding and extension (e.g., remote and associate knowledge) (see Figure 9.11).
3. *Measurement property of the knowledge maps.* The number of goals and capabilities enclosed in a pattern's topology drives software measurement. It is used as an indicator for measuring the cost, time, and effort that must be allocated to produce a particular software outcome. For example, imagine a customer approaches you and requests your services to develop a collaborative environment solution. He or she will want to know how long it will take you to develop such a solution, how many classes will be necessary to complete the solution, how many developers

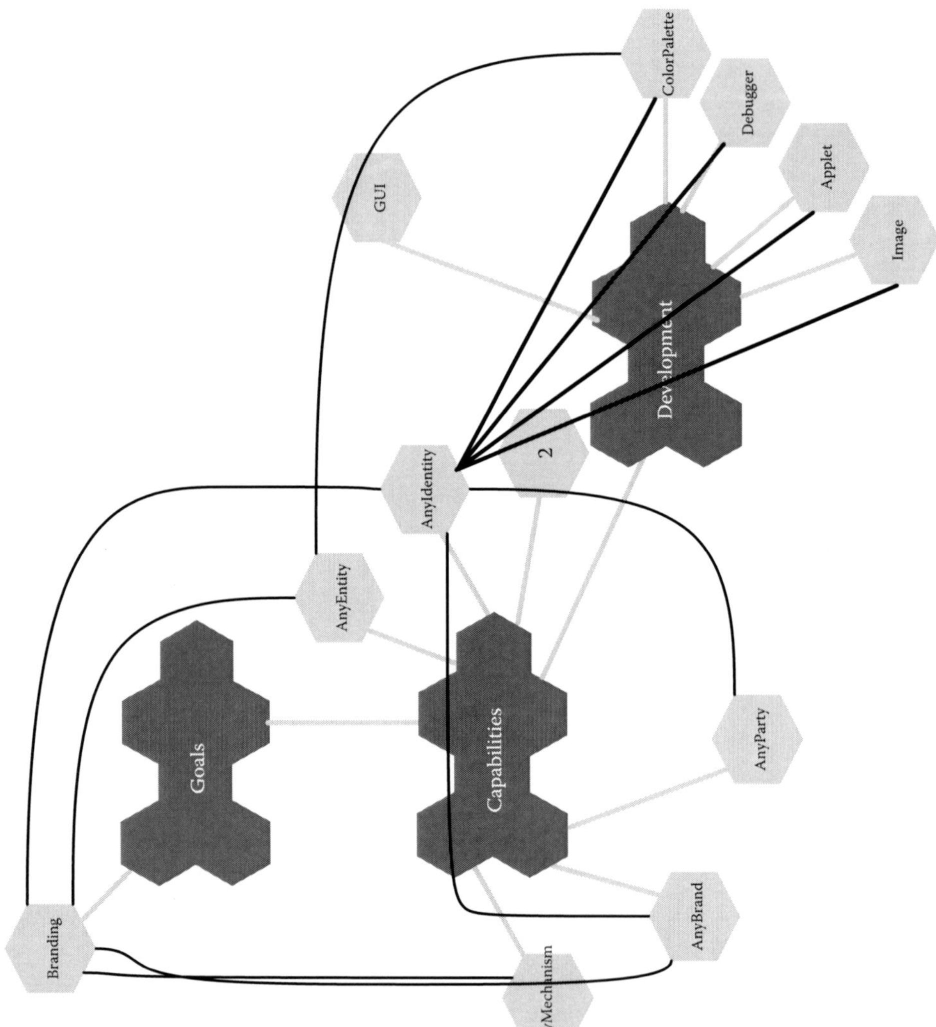

FIGURE 9.10 Separating concerns in the branding concept. (From Sanchez, H. A. "Laying the Foundations for Branding as a Stable Analysis Pattern." Paper presented at the 19th European Conference on Object-Oriented Programming, Glasgow, UK, July 25–29, 2005.)

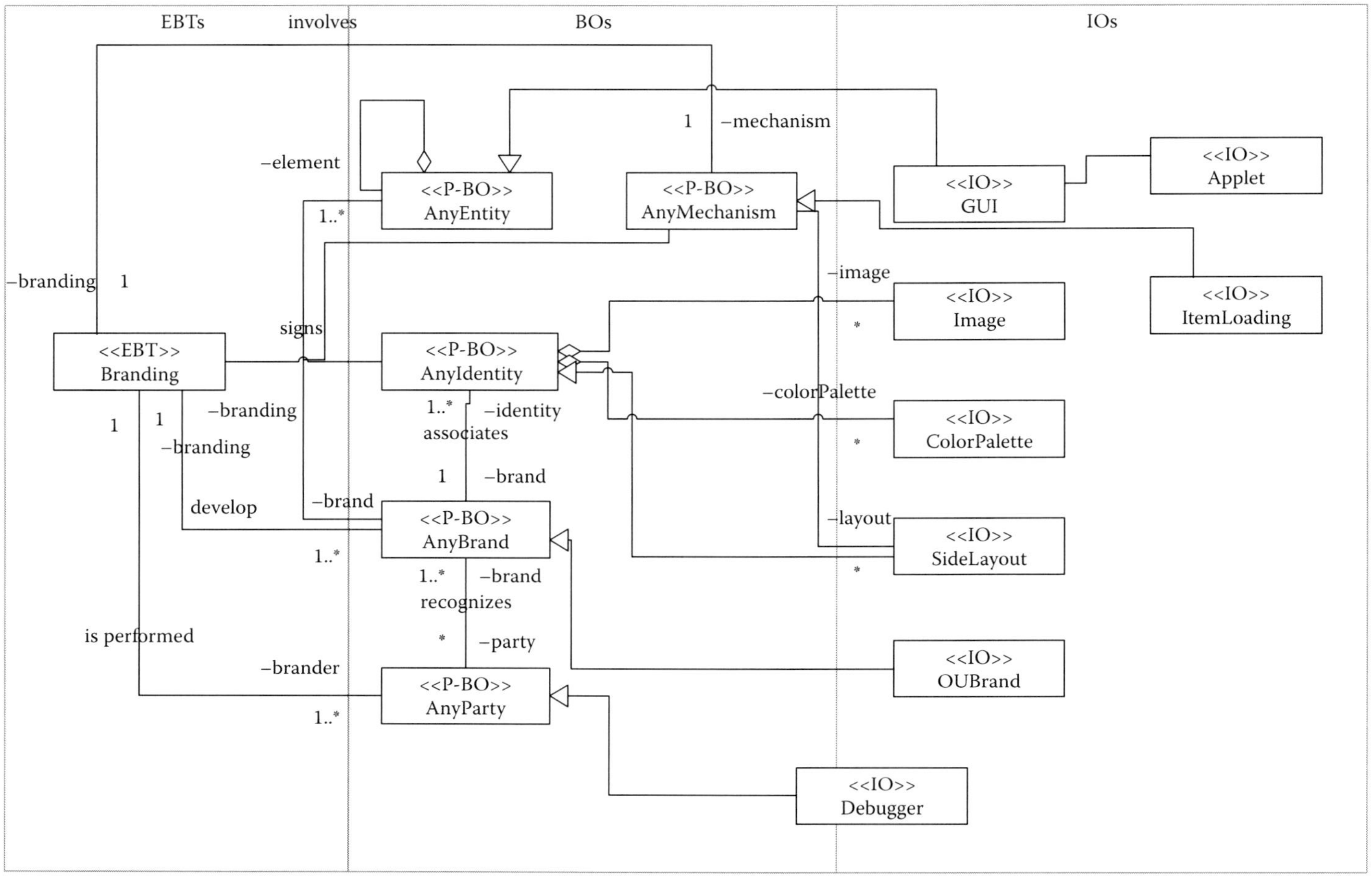

FIGURE 9.10 (Continued) Separating concerns in the branding concept. (From Sanchez, H. A. "Laying the Foundations for Branding as a Stable Analysis Pattern." Paper presented at the 19th European Conference on Object-Oriented Programming, Glasgow, UK, July 25–29, 2005.)

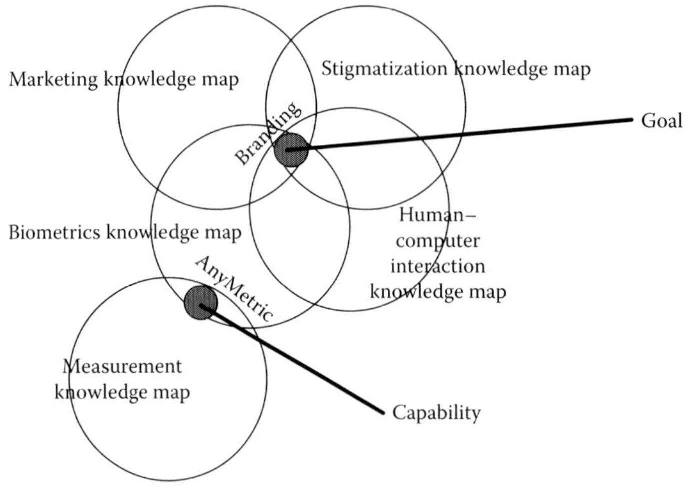

FIGURE 9.11 Illustration of Intersection property of the knowledge maps.

will be required, and how much effort will be required from them to develop such a solution. These questions will be answered by means of knowing the number of goals, along with the capabilities of these goals that the solution encloses.

4. *Generality and reusability property of the knowledge maps*. This property deals with the degree to which the enclosed software patterns, or the knowledge map itself, can be reused elsewhere. This also provides the necessary logistics to create on-demand stable architectures. Knowledge maps ensure that all their software patterns are at a proper level of abstraction and generality. The proper levels of abstraction and generality of these patterns will allow us to handle the same problems regardless of their context of application. By doing so, we will be able to create new environments or solve problems using reusable solutions; rather than reinventing the wheel every time, we face new problems of the same sort. Figures 9.12 and 9.13 show this property using the SAP named *branding* (Sanchez 2005) in two contexts: biometrics and marketing.
   a. As you can see, the branding SAP is similar for the two contexts, due to the generic and reusable nature inherited from the software stability concepts approach (Sanchez 2005). The only thing that changes here is the application context. These contexts will be attached to the core by using hooks (Fayad, Hamza, and Sanchez 2005).
5. *Infinite stable software architectures*. Due to the exploitation of useful patterns synergies from available software patterns, we are able to generate a vast number of stable software architectures. Therefore, we can accelerate software architecture production and work flow pattern selection, and simultaneously, we can provide an intuitive process execution to address reusability, scalability, adaptability, and so on.
6. *Faster ROI*. Due to the stable and reusable nature of the knowledge maps, software architectures will be easier and cheaper to construct, with a significant reduction in cost and innovation, which means a faster ROI.
7. *Context awareness and assessment of the knowledge maps*. Domain understanding and assessment sit side by side in the conceptualization of the knowledge maps. By using the set of heuristics and questions described in Appendix A and Chapters 1

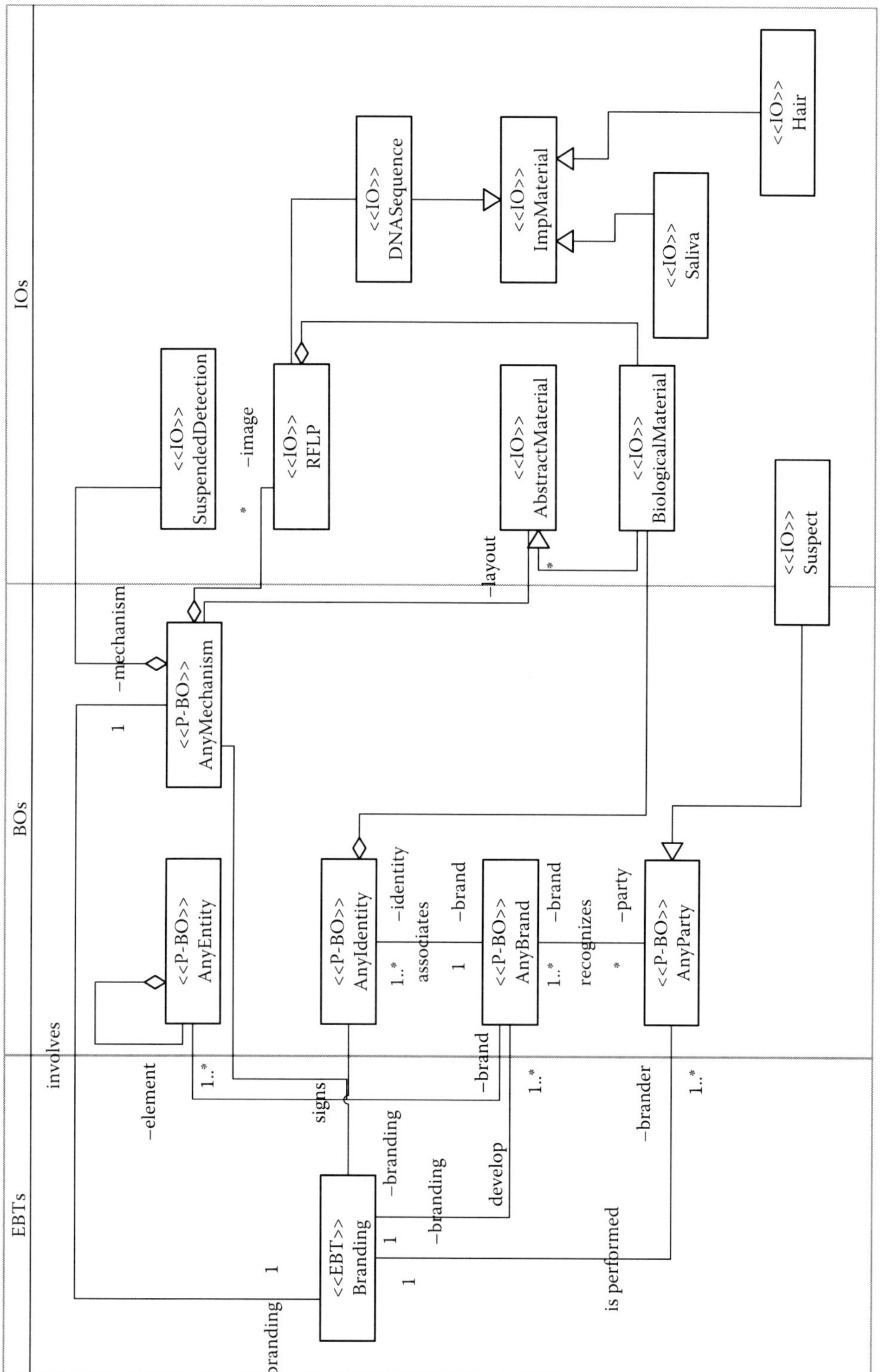

FIGURE 9.12 Using branding in biometric systems.

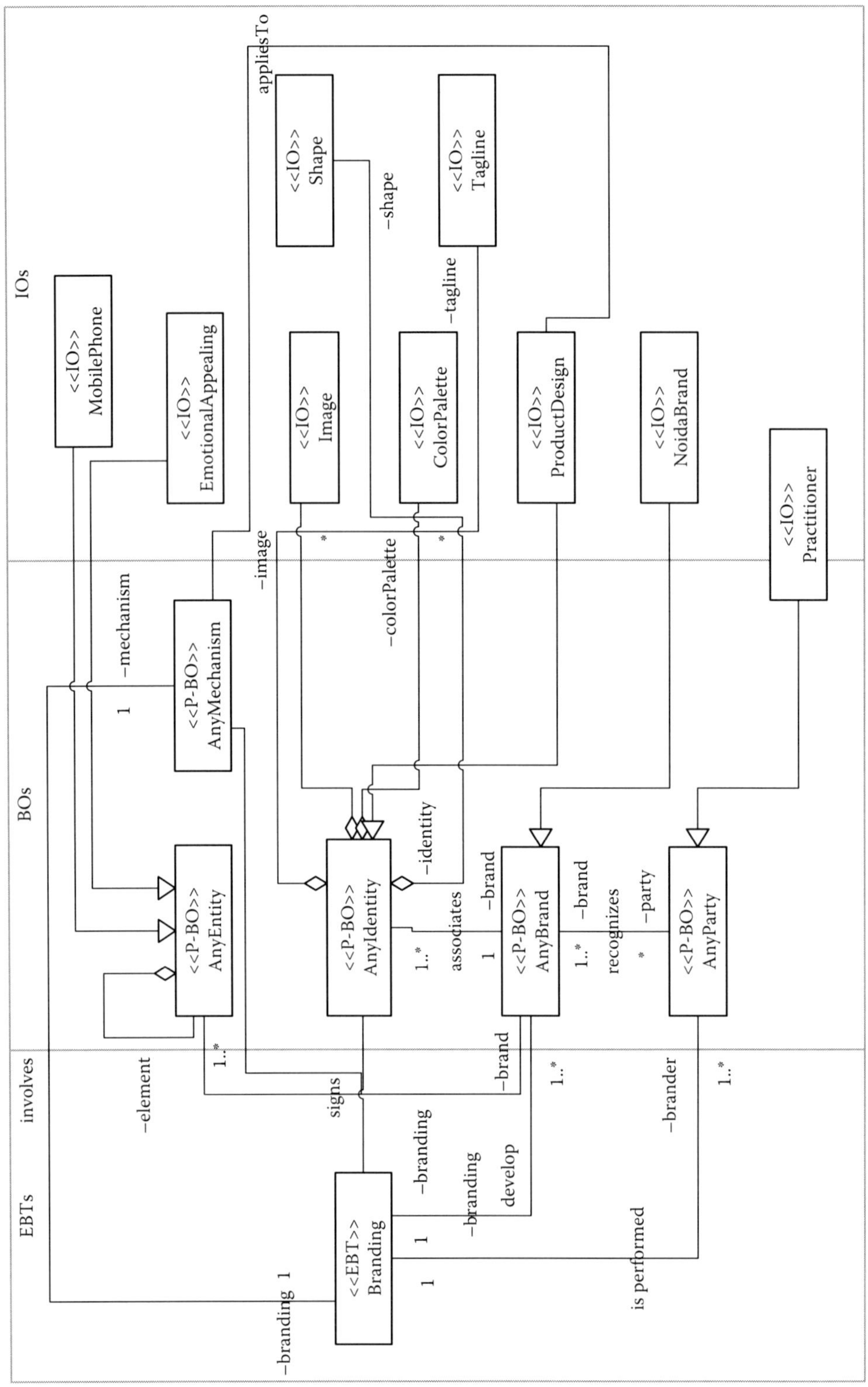

FIGURE 9.13 Using branding in marketing.

through 8, software practitioners will ensure that they have identified or collected the right requirements. The proper collection and identification of these requirements will guarantee a unified and accurate pattern topology, spanning all the building blocks of a particular domain or discipline.

9.3 KNOWLEDGE MAPS VERSUS TRADITIONAL PATTERN LANGUAGES

This section provides the results of a comparative study between existing pattern languages and the knowledge maps. The dimensions from where these two structures were compared are the quality factors of knowledge maps and the properties of both.

After an extensive perusal of traditional pattern languages, we obtained the following results:

1. Missing indicators and guidelines for assuring a focused nature or behavior for the provided patterns of the pattern languages with respect to a selected problem. There is no method for deciding which patterns to use and add in a particular pattern language. Therefore, practitioners are infected with a "keep adding what seems to satisfy my requirements" kind of inaccurate behavior. These actions at the end will provide more problems than visible benefits for software practitioners. For example, you may find out that there was indeed a better solution for your problem. Therefore, you will have to do a lot of refactoring or code maintenance.
2. No identifiable classification of patterns' rationale within a pattern language's structure. This also includes the lack of pattern languages concerns partitioning, which undermines the proper distribution of patterns' responsibilities in a pattern language. This will bring serious problems, like *macho class* problem (Fayad 2003), where all responsibilities are positioned in one class or high coupling and low cohesion problems, and so on.
3. Loss of traceability across all the different stages of software development, especially when tackling the deeper levels of patterns' implementation. There is a traceability problem in the instantiation processes of traditional pattern languages, which results in instances that cannot be traced back to their original design patterns (Hamza and Fayad 2003). The instances were deduced from the design pattern, but the internal structure of design patterns is invisible in traditional pattern languages (Manns et al. 2000).
4. No clear and systematic process to weave patterns together and create software architectures. This is done on an ad hoc basis; therefore most of the resulted solutions accuracy is questionable and open to serious contamination and ripple effects (Coplien 1996).
5. Loss of generality of pattern languages' patterns, because most of these patterns tend to be out of context, and usually, their implementation instances are lost within the pattern language's implementation (Wu, Hamza, and Fayad 2003). There is no way to pinpoint the pattern instances once the entire pattern language has been implemented (Cercone and McCalla 1987; Manns et al. 2000).
6. There is no distinction between associate and remote knowledge in pattern languages. This statement leads to the following questions: How can we associate two

or more pattern languages of a similar discipline? How can we intersect one pattern language with one or more pattern languages from a different (remote) discipline? The answer to these questions is that with traditional pattern languages, there is no systematic way to achieve that; usually, this is done on a ad hoc and temporary basis (Connelly et al. 2001, pp. 39–49).

7. Pattern languages are built for here and now, not for the future. Most of the patterns that they enclose have been built based on artifacts or components bound to specific business processes. We know that as business policies, mission, and so on change, the processes would change too; therefore, pattern languages would change as a consequence. As a result, there will be additional effort, cost, and time involved to maintain pattern languages in answer to new businesses' changes.
8. Pattern languages do not provide a retrospective of the subject matter's rationale. This disadvantage has two effects: First, there is a chance of building the wrong pattern language, due to its enclosed element design based on a wrong analysis of subject matter. Second, because the intent of the pattern language in a particular domain has been obscured by ignoring its original rationale, how the pattern language deals with the problem they addressed is neither straightforward nor easy.

Throughout the previous section, it was made clear that the knowledge maps overcome all the problems or shortages that traditional pattern languages always experience, especially when dealing with requirements, solution formulation, pattern composition, architecture generation, and system creation. The next section will illustrate some sample architectural patterns generated using the knowledge maps.

9.4 SAMPLES OF SArchPs

We will describe a couple of SArchPs that were generated by using the knowledge maps. We will also use a short template to document this architectural pattern. In this template, we will describe the essentials of the architectural patterns, while ignoring specific details of the solution's implementation. All the samples were extracted from previous work on the software stability concepts approach (Fayad, Islam, and Hamza 2003; Yavari and Fayad 2003).

9.4.1 Architecture 1

The first SArchP is called model-view-mapping (MVM) (Fayad, Islam, and Hamza 2003) and is shown in Figure 9.14.

- *Name.* Stable MVM architectural pattern
- *Problem.* How to build a high-level architecture model that can provide, for any application, flexible mapping between any abstract model (which could be either a passive model, such as text, frames, and diagrams, or an active model or alive model, which is a model returned by a specific application, such as animating data) to any abstract views and vice versa.
- *Solution and participants*
    - *Solution*

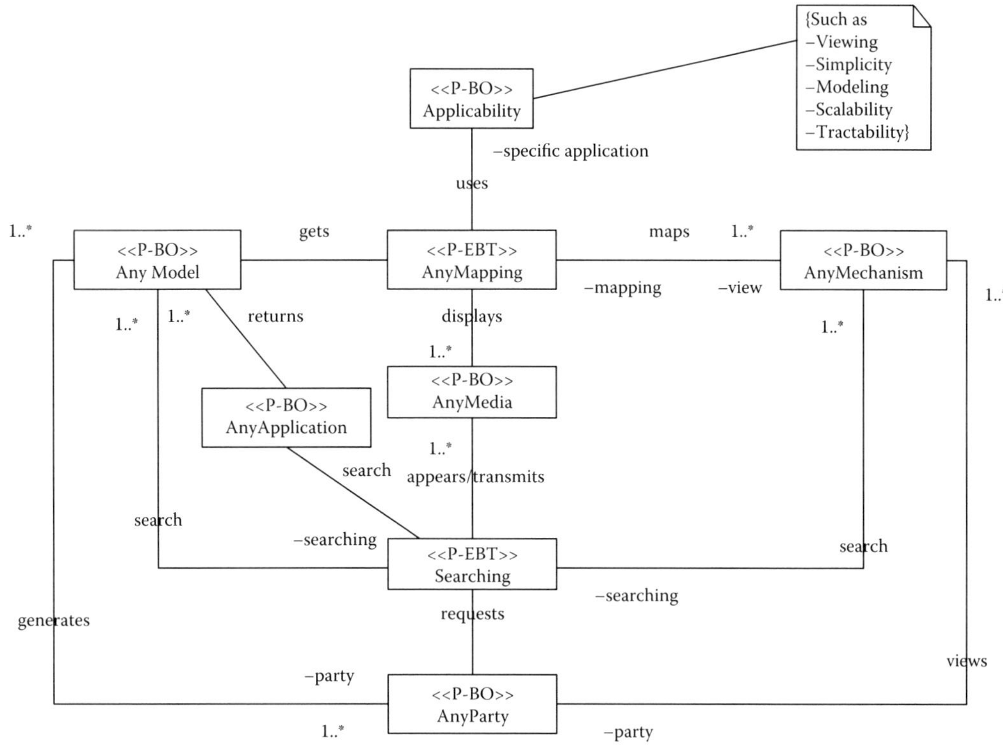

FIGURE 9.14 The stable MVM architectural pattern.

- *Participants*
  - *Classes*
    - *Applicability.* It describes the application and the purpose for which mapping is needed. For instance, in one application, a simple view that extracts part of the data from the original model is needed for the sake of simplicity.
  - *Patterns*
    - *AnyModel.* It describes the models within the application. The model is a representation of the data within the application.
    - *AnyView.* It represents the view of a collection of data (the model).
    - *AnyParty.* It represents both the modeler and the viewer. The modeler is responsible for building the data models in the appropriate abstract level. The viewer requests the model and the mapped view of that model.
    - *AnyMedia.* It identifies and defines the media upon which the models and views are mapped and transmitted. It also represents the media by which the views are to be displayed (e.g., devices and PCs).
    - *Mapping.* It defines the mapping rules between the models and their views. It also determines how this mapping will be performed.
    - *Searching.* It searches AnyMedia for the requested application, model, or view.
    - *AnyApplication.* It represents the application that is requested by AnyParty.

9.4.2 Architecture 2

The second SArchP is called magnetic resonance imaging (MRI) architectural pattern (Yavari and Fayad 2003) and shown in Figure 9.15.

- *Name*. Stable MRI architectural pattern.
- *Problem*. The MRI system provides two-dimensional images of soft cells of different organs. Each image shows a horizontal picture of a particular height. Any issue or problem in the organ would be presented in several consecutive pictures. So, the problem is how to build an accurate, fast, effective, and easy-to-use application for MRI visual analysis that performs all required processes and saves the results, so they are always available to be discussed and/or reviewed by the medical professionals.
- *Solution and participants*
    - *Solution*
    - *Participants*
        - *Classes*
        - *Patterns*
            - *Recording*. This Pattern-EBT indicates the purpose of using a log in a set of application areas.
            - *Interpreting*. It provides the means for explaining something, which is loose or fuzzy or is not obvious.
            - *Viewing*. It indicates the rules for using a set of distinct views of a specific subject (e.g., system model).
            - *AnyView*. It represents the view of a collection of data (a determined model).
            - *AnyParty*. It represents both the recorder and the interpreter. The modeler is responsible for building the data models in the appropriate abstract level. The viewer requests the model and the mapped view of that model.

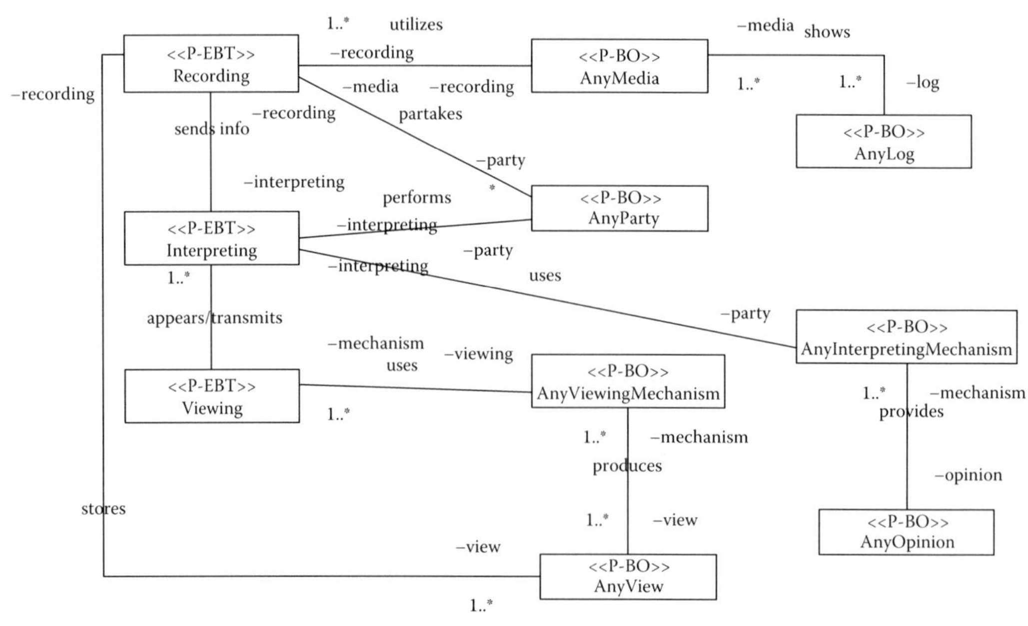

FIGURE 9.15 The stable MRI architectural pattern.

- *AnyMedia.* It identifies and defines the media, upon which the models and views are mapped and transmitted. It also represents the media by which the views are to be displayed (e.g., devices and PCs).
- *AnyLog.* It represents the canonical logging process itself.
- *AnyViewingMechanism.* It encapsulates the behavior and properties of mechanisms associated with viewing in one pattern.
- *AnyInterpretingMechanism.* It encapsulates the behavior and properties of mechanisms associated with interpreting in one pattern.
- *AnyOpinion.* It represents the beliefs about something that will be shared among a set of parties.

SUMMARY

Using knowledge maps is a new movement and it provides you a fresh vista and better opportunities; therefore, a few examples were defined under the principles in software stability concepts approach and used in our comparative study. The following represents how knowledge maps have overcome traditional pattern languages problems:

- *Systematic capture processes and full understanding of the domain where a targeted solution must exist.* This will provide a unified domain knowledge discovery and understanding approach, spanning all different categories for stable patterns.
- *Great reuse and on-demand context adaptation, via an identifiable stable core and its extension points.* IOs or transient aspects are attached to the core according to the context-specific needs via these extension points. These qualities will greatly reduce the cost of adaptation or upgrades.
- *Full traceability capability across the implementation of knowledge maps.* This is accomplished by describing a two-way mapping relationship between stable design patterns (SDPs) and their implementation instances. Therefore, traceability, maintainability, and stability become visible base for evaluation of your software architectures.
- *Unprecedented flexibility.* Knowledge maps provide an unprecedented flexibility by enabling a business to add or remove functionality from its system and on a real-time basis. Therefore, a rapid ROI will occur, because software architectures would become cheaper and they are faster to construct, as well as reuse.
- *Software architecture definition.* There is rapid software architecture definition and application through a systematic work flow pattern process and an identifiable pattern composition process.

OPEN RESEARCH ISSUES

As seen here, knowledge maps are the magical and golden recipes for building stable software applications, when we make use of reusable patterns. As this field is still emerging and in its nascent phase, there are some issues that have not been addressed yet.

1. *Knowledge map tool support.* There is a fool-proof methodology available to test and evaluate the knowledge map, which has been illustrated and highlighted in this chapter, but there are existing tools to test the created knowledge map. The second issue is that there are no standard tools that can be used to draw the structure of the

knowledge map. So, this might slow down the process. But we are working on two different tools:

a. *Knowledge map engine*
b. *Hook engine*

2. *Software economics.* It examines the application of economic factors at all levels of decision making in the management and development of software. The application of software stability impacts all economical aspects of software management and development. The knowledge map will provide several objectives that can be reached easily. They are as follows:
   a. A road map for software economics
   b. Empirical views of where money goes
   c. Understanding and controlling software costs and effort
   d. Identifying shortcomings in existing work
   e. Discussion of economic perspectives on software architecture promises
   f. Reviewing state of the art in software economics
   g. Better form of making decisions in managerial and technical aspects of all activities of software development.
   h. Financial evaluation and strategic analysis of software development based on knowledge maps and software stability

   The goal of future research is to study and enhance existing evaluation techniques with knowledge maps in mind and develop new ones to improve the effectiveness of decision-making processes that impact software economic parameters in software development.

   Knowledge maps also provide practical and intuitive views of many software economics issues, and it would be very useful to focus on software economics research by answering the following questions that are related to software quality:
   a. How to establish quality assurance of software projects and how much would it cost?
   b. How can quality data of the knowledge map and measurement of existing software projects help in evaluating future software projects?
   c. How do processes of the knowledge maps impact software economics?
3. *Software measurements.* Knowledge maps and SSM have many add-value and embedded qualities, such as stability over time, reusability, usability, scalability, understanding, customizability, extensibility, adaptability, configurability, integrability, traceability, testability, simplicity, and maintainability. Knowledge maps and the SSM can revolutionalize software measurements, impact existing metrics, and introduce new list of metrics: How fast can an application or a number of applications be produced? How easy to generate an application or a number of applications? How many architecture can be produced from a knowledge map? How many applications can be produced per architecture? Knowledge maps also include a build-in any additional quality and quantity aspects as part of the knowledge maps.
4. *Return on investment (ROI).* It is a ratio that is used to measure the performance of any investment or a number of investments in software developments by dividing the net profit (gain of investment/cost of investment) by the cost of investment. We also know that because of the stable and reusable nature of the knowledge maps, software

architectures will be easier and cheaper to construct, with a significant reduction in cost and innovation which means a faster and better ROI. We achieved over 76.3% ROI of early development and 84% ROI of later development. These are very promising findings and we will track and study the ROI in the future and new developments.
5. *Knowledge classification.* Knowledge maps and SSM provide a new and useful approach for knowledge classification. Knowledge maps and SSM classify knowledge into three categories: (1) EBT or goals or SAP; (2) BOs, capability, or SDPs; and (3) IOs or application objects. Points (1) and (2) are general knowledge that can be applied in any domain, and objects in (3) are domain- and application knowledge–specific, which is limited to the domain and application knowledge. It will be very useful to apply the knowledge maps and SSMs classification approach to as many software developments, and as many nonsoftware development projects as possible. This will also prove that the classification approach of the knowledge maps is more effective and efficient.
6. *Domain engineering, domain analysis, and domain modeling.* Domain engineering is an engineering approach of reusing domain knowledge during the development of new software systems. Domain analysis is the process of identifying domain boundary and scope and discovering commonalities and variability among the subdomains within the domain, and domain modeling is a process of creating a reusable model of the domain aspects or domain (see Sidebar 9.1). Both domain analysis and *domain modeling* are part of domain engineering. Knowledge maps and the SSM provide an established approach for domain engineering and produce more generic and domain-less patterns that can be reused in many different ways and in many different domains.

REVIEW QUESTIONS

1. Justify the following statement: Software patterns and patterns languages can be used to solve recurring problems.
2. Distinguish between software patterns and patterns language.
3. Provide an analogy for software patterns and patterns language and describe it.
4. Explain, why in spite of using patterns to devise application solutions, achieving short software development life cycles is not possible?
5. What are the shortcomings of the currently used methodology that makes use of patterns and patterns language for modeling a software problem?
6. What can be a solution for achieving shorter software life cycles for software development?
7. What is meant by knowledge map?
8. Why is it correct to say *Knowledge maps are a topology of software patterns*?
9. In what two ways can a system be built?
10. Briefly describe the structure of the knowledge maps.
11. Match the following column entries:

a. Concern	–	Represented by
b. Analysis	–	Quality factors

c. Design – IO
d. Implementation – Architectural patterns
e. Deployment – Capabilities
f. Knowledge – Goals

12. Explain what is meant by temporary and permanent collaboration of an element within the context of knowledge map.
13. The main targets for first perspective (structure notation) of a knowledge map are __________. However, ____________ can also use it to get a high-level idea of the domain.
14. How are main categories of knowledge map represented? Illustrate with a diagram.
15. What notation is followed for representing the elements of main categories of knowledge map? Use a diagram to explain it clearly.
16. How are the associations between categories and elements depicted? Draw a diagram.
17. Why do associations depart from the hexagon represented by SArchPs?
18. The UML perspective is target at __________________.
19. Can managers understand the UML perspective? How will it be useful to them?
20. What do the following concepts mean within the context of UML?
    a. Goals
    b. Capabilities
    c. Development scenarios
    d. Quality factors
    e. Associations

    Use diagrams wherever possible to depict the relationship.
21. Which two stereotypes are used to depict the classes that represent the goal of the application? What do they each of them represent?
22. When is the stereotype <<BO>> used? When is <<Pattern-BO>> or <<P-BO>> used?
23. What do stereotypes <<IO>> and <<Hook>> indicate?
24. What is the three-layer representation approach of knowledge maps?
25. How can knowledge maps provide within context solutions to recurring problems? Start with providing your understanding of the term *within context*.
26. Describe how each of the above listed quality factors can be achieved by using a knowledge map.
27. Enlist knowledge map properties.
28. Why do you think the following statement is correct?

    The proper implementation of quality factors in conjunction with a proper knowledge representation (structure of the knowledge maps) will make obvious the properties of the knowledge maps and the benefits that stem from them.
29. Justify. Knowledge map properties will help to streamline the process of building systems from patterns.
30. Contrast knowledge with traditional patterns languages and list the advantages of the knowledge map.
31. How does the knowledge map answer the shortcomings of patterns language?

EXERCISES

1. Read about patterns language and come up with some shortcomings which are not listed in this chapter.
2. Form a group of three people and debate on whether the structure notation used for representing knowledge maps is effective or not.
3. In a group of two, analyze the given MVM and MRI architectural patterns.

PROJECTS

1. Using knowledge maps structure's notation, create a knowledge map for the following:
   a. Tax calculation
   b. Data search engine
   c. Creating application log
2. Give a three-layer representation for the above applications.
3. Using the template given under Section 9.4, document the following architectural patterns:
   a. Recording
   b. Encapsulation
   c. Data hiding
   e. Abstraction
   f. Reusability
   g. Scalability
   h. Partitioning
4. Using knowledge maps structure's notation, create a knowledge map for the following film domains: Animation, Live-action, Filmmaking, Film criticism.
5. Using the knowledge map template, document the knowledge map for each of the domains in Project 4.
6. Using knowledge maps structure's notation, create a knowledge map for the following economics domains: Agricultural economics, Bio-economics, Consumer economics, Economic systems, Energy economics, Entrepreneurial economics, Experimental economics, Information economics, Islamic economics.
7. Using the knowledge map template, document the knowledge map for each of the domains in Project 6.
8. Using knowledge maps structure notation, create a knowledge map for the following visual arts domains: calligraphy, connoisseurship, creative arts, drawing, fine arts, painting, photography, sculpture.
9. Using the knowledge map template, document the knowledge map for each of the domains in Project 8.
10. Using knowledge maps structure notation, create a knowledge map for the following law domains: criminal law, Islamic law, Jewish law, civil law.
11. Using the knowledge map template, document the knowledge map for each of the domains in Project 10.
12. Using the knowledge maps structure notation, create a knowledge map for the following life sciences domains: biochemistry, bioinformatics, computational biology, cell biology, genetics, nutrition.

13. Using the knowledge map template, document the knowledge map of each of the domains in Project 12.
14. Using the knowledge maps structure notation, create a knowledge map for the following applied science domains: artificial intelligence, ceramic engineering, computing technology, electronics, energy, energy storage, engineering physics, environmental technology, materials science and engineering, microtechnology, nanotechnology, nuclear technology, optics, zoography.
15. Using the knowledge map template, document the knowledge map for each of the domains in Project 14.
16. Using the knowledge maps structure notation, create a knowledge map for the following information domains: communication, graphics, music technology, speech recognition, visual technology.
17. Using the knowledge map template, document the knowledge map for each of the domains in Project 16.
18. Using the knowledge maps structure notation, create a knowledge map for the following industrial domains: construction, financial engineering, manufacturing, machinery, mining, business informatics.
19. Using the knowledge map template, document the knowledge map for each of the domains in Project 18.
20. Using the knowledge maps structure notation, create a knowledge map for the following military domains: ammunition, bombs, guns, military technology and equipment, naval engineering.
21. Using the knowledge map template, document the knowledge map for each of the domains in Project 20.
22. Using the knowledge maps structure notation, create a knowledge map for following domestic domains: educational technology, domestic appliances, domestic technology, food technology.
23. Using the knowledge map template, document the knowledge map for each of the domains in Project 22.
24. Using the knowledge maps structure notation, create a knowledge map for the following engineering domains: aerospace, agricultural, architectural, biological, biochemical, biomedical, ceramic, chemical, civil, computer, construction, cryogenic, electrical, electronic, environmental, food, industrial, materials, mechanical, mechatronics, metallurgical, mining, naval, nuclear, optical, petroleum, software, structural, systems, textile, tissue, transport.
25. Using the knowledge map template, document the knowledge map for each of the domains in Project 24.
26. Using the knowledge maps structure notation, create a knowledge map for the following health and safety domains: biomedical engineering, bioinformatics, biotechnology, cheminformatics, fireprotection engineering, health technologies, pharmaceuticals, safety engineering, sanitary engineering.
27. Using the knowledge map template, document the knowledge map for each of the domain in Project 26.
28. Using the knowledge maps structure notation, create a knowledge map for the following transportation domains: aerospace, aerospace engineering, marine engineering, motor vehicles, space technology.

29. Using the knowledge map template, document the knowledge map for each of the domain in Project 28.
30. Form a knowledge map with stability in mind for the problem statement D1 which is titled "Ocean Resources Management System" (see Appendix D).
31. Form a knowledge map with stability in mind for the problem statement D2 which is titled "Dengue Fever Prevention and Outbreak Management System" (see Appendix D).
32. Form a knowledge map with stability in mind for the problem statement D3 which is titled "Organizing Cricket World Cup" (see Appendix D).
33. Form a knowledge map with stability in mind for the problem statement D4 which is titled "Pollution Management" (see Appendix D).
34. Form a knowledge map with stability in mind for the problem statement D5 which is titled "Natural Disaster Tracking System" (see Appendix D).
35. Form a knowledge map with stability in mind for the problem statement D6 which is titled "Global Warming Control System" (see Appendix D).
36. Form a knowledge map with stability in mind for the problem statement D7 which is titled "Circus" (see Appendix D).
37. Form a knowledge map with stability in mind for the problem statement D8 which is titled "Jurassic Park" (see Appendix D).

SIDEBAR 9.1 Domain Analysis

In the software engineering field, domain analysis is the intricate process of analyzing and evaluating related software systems in a given domain to seek their common and variable parts or modules. Experts suggest a number of methods for domain analysis. Each one of them produces domain models like feature tables, facet tables, facet templates, and generic architectures. A generic architecture system explains all available systems in a domain. It also provides an insight of proposed methodologies for domain analysis (Champeaux et al. 1993; Hjørland 1995; Arango and Prieto-Diaz 1989).

Domain analysis is "the process of identifying, collecting, organizing, and representing the relevant information in a domain, based upon the study of existing systems and their development histories, knowledge captured from domain experts, underlying theory, and emerging technology within a domain" (Kang et al. 1990).

The products, or *artifacts*, of a domain analysis are sometimes object-oriented models (represented with UML) and/or data models (represented with entity-relationship diagrams [ERD]). Software engineers employ these models as a foundation for the implementation of software architectures and applications.

REFERENCES

Arango, G. and R. Prieto-Diaz, eds. Domain analysis: Concepts and research directions. In *Domain Analysis: Acquisition of Reusable Information for Software Construction*. IEEE Computer Society Press, May 1989.

de Champeaux, D., D. Lea, and P. Faure. Domain Analysis, chapter 13, *Object-Oriented System Development*. Addison Wesley, 1993.

Hjørland, B. and H. Albrechtsen, "Toward a New Horizon in Information Science: Domain-Analysis," *Journal of the American Society for Information Science* vol. 46, no. 6 (1995): 400–425.

Kang, K., S. Cohen, J. Hess, W. Novak, and A. Peterson. "Feature-Oriented Domain Analysis (FODA) Feasibility Study." *Technical Report*. CMU/SEI-90-TR-021, Software Engineering Institute, November 1990.

SIDEBAR 9.2 What is the System of Patterns?

Pattern systems have been the focus of immense interest for the last few years and experts in the software engineering field have been using them recently, especially in the field of object-oriented and component-based software areas (see http://www.opengroup.org/architecture/togaf8-doc/arch/chap28.htm). Their usage is helping developers to extend the basic principles and concepts to a number of architectural domains. However, available literature on pattern systems indicates a conflict of terminology, as a number of experts like to use the term *architecture* to relate to software domains and a number of patterns as *architectural patterns*. In essence, patterns are useful tools of subjecting building blocks into beneficial context.

One example for this is describing a reusable solution to indicate a simple and easy to understand a problem. In fact, building blocks are what you use and employ, whereas patterns can inform you how you can employ and use them. They also indicate the *when*, *why*, and *what* parts of the problem at hand. With pattern systems, one can seek to identify various combinations of architectures and/or solution building blocks to arrive at effective and practical solutions.

A knowledge map is a system of patterns that includes SAPs + SDPs + SArchPs.

SIDEBAR 9.3 Knowledge Map of the CRC Cards

The knowledge map template consists of 6 sections as following for the knowledge map of the CRC Cards:

1. *Knowledge map name*. Class responsibility and collaborator (CRC) cards
2. *Knowledge map nickname*. None
3. *Knowledge map domain/subject/topic description*. The CRC cards are index cards that are utilized for mapping candidates to classes in predefined design scenarios, for example, use case scenarios. The objective of CRC cards is to facilitate the design process, while ensuring an active participation of involved designers. This chapter represents the first attempt toward a CRC card knowledge map or stable pattern language representation, via stable patterns, as a mean to discover, organize, and utilize CRC cards goals. Each stable pattern focuses on a distinctive activity and provides a way by which this activity can be conducted efficiently. The knowledge map or stable pattern language is a continuation of our early effort in improving the effectiveness of CRC cards and their role in the design process.
4. *EBTs/goals*. Name the EBTs of the *CRC cards* and provide a short description of each EBT and organize your answer in Table 9.1.
5. *BOs/properties*. Name the BOs of the *CRC cards* and provide a short description of each BO and organize your answer in Table 9.2.
6. *Knowledge map (core knowledge)*. Map each EBT to its BOs of the *CRC cards* and organize your answer in Table 9.3.

SIDEBAR 9.4 Knowledge Map of the Software Project Management

The knowledge map template consists of 6 sections as following for the knowledge map of software project management:

1. *Knowledge map name*. Software project management (SPM)
2. *Knowledge map nickname*. None
3. *Knowledge map domain/subject/topic description*. SPM is the art and science of planning and leading software projects (Stellman and Greene 2005). It is a subdiscipline of project

TABLE 9.1
EBTs of CRC Cards

EBTs/Goals	Description
Brainstorming	A group problem-solving techniques in which members sit around and come up with ideas and possible solutions to the problem. Current implementations of the brainstorming process are bound to a specific problem domain.
Engagement	We are always concerned about the quality of the involvement between participants involved in a particular activity, when interacting with each other. The act of sharing in the activities of a group.
Traceability	
Identification	
Modeling	The actual problems range from the overloaded generation of too many responsibilities per class to the lack of specific class roles, which defined the position of a class in a pre-animated scenario in accordance with certain responsibility.

TABLE 9.2
BOs of CRC Cards

BOs/Capabilities	Description
AnyClass	
AnyRole	
AnyResponsibility	
AnyCollaboration	
AnyService	
AnyAttribute	
AnyClient	
AnyForm	
AnyMedia	
AnyParty	
AnyContext	
AnyActivity	
AnySkill	

TABLE 9.3
Knowledge Map of CRC Cards

EBTs	BOs
Brainstorming	AnyParty, AnyForm, AnyContext, AnyMedia
Engagement	AnyParty, AnyCommitment, AnyDisposition, AnyActivity, AnySkill

management in which software projects are planned, implemented, monitored, and controlled (Stellman and Greene 2005; Pankaj 2002; Chemuturi and Cagley 2010).

4. *EBTs/goals*. Name the EBTs of the *SPM* and provide a short description of each EBT and organize your answer in Table 9.4.
5. *BOs/properties*. Name the BOs of the *SPM* and provide a short description of each BO and organize your answer in Table 9.5.
6. *Knowledge map (core knowledge)*. Map each EBT to its BOs of the *SPM* and organize your answer in Table 9.6.

TABLE 9.4
EBTs of SPM

EBTs/Goals	Description
Allocation	The process of parceling out a monetary budget, a technology budget, system requirements, software requirements, effort, or any other quantity that can be subdivided and assigned to the elements of a process or a system
Utilization	
Need	
Management	
Analysis	
Tracking	
Assessment	

TABLE 9.5
BOs of SPM

BOs/Capabilities	Description
Activity	An element of work performed during the course of a project. An activity normally has an expected duration, an expected cost, and expected resource requirement. Activities can be subdivided into tasks.
Assumption	A condition accepted as true but which cannot be verified at the current time or which would be too expensive to verify at the current time.
Authority	The power to make and implement decisions that must be made to fulfill one's responsibilities
Baseline	A work product that has satisfied its predetermined acceptance criteria and has been placed under version control. Baselines provide the basis for future work during software development and maintenance. Synonymous with baselined work product
Constraint	A limitation imposed by external agents on some or all of the operational domain, operational requirements, software requirements, project scope, monetary budget, technology budget, resources, completion date, and platform technology
Contract	A statement of understanding between two or more parties. A contract may be informal or legally binding (i.e., formal). See also Acquirer, Memo of understanding, and Statement of work
Event	Crisis—an event that halts or seriously impedes progress

(*Continued*)

TABLE 9.5
(Continued) BOs of SPM

BOs/Capabilities	Description
Critical path	The process of determining the set of (one or more) longest paths through a schedule network
Defect	A flaw in a work product that renders it incorrect, incomplete, and/or inconsistent. See also error and failure
Risk	
Scope	
Resource	
Project	
Milestone	
Task	
Party	

TABLE 9.6
Knowledge Map of SPM

EBTs	BOs
Tracking	Party, Actor, Criteria, Mechanism, Data, Entity, Media
Assessment	Party, Actor, Criteria, Mechanism, Outcome, Log, Reason, Type, Entity, Media

REFERENCES

Chemuturi, M., and T. M. Cagley Jr. *Software Project Management: Best Practices, Tools and Techniques*. Boca Raton, FL: J.Ross Publishing, 2010.

Pankaj, J. *Software Project Management in Practice*. Boston, MA: Addison-Wesley, 2002.

Stellman, A., and J. Greene. *Applied Software Project Management*. Sebastopol, CA: O'Reilly Media, 2005.

10 Development Scenarios
Setting the Stage

Always design a thing by considering it in its next larger context—a chair in a room, a room in a house, a house in an environment, an environment in a city plan.

Eliel Saarinen
Bederson and Shneiderman, 2003

10.1 INTRODUCTION

So far, we have learned and understood how we can identify underlying goals of any domain, the capabilities needed to fulfill or achieve them, and the synergy between them to generate necessary knowledge maps or stable analysis, design, and architectural patterns. Now, we will discuss the context or the situation wherein these knowledge maps will be deployed.

Context is an important entity that is useful in providing a precise message to help us arrive at a valid conclusion. Context is the notable environment, ambience, or scenario where one can expect something to occur or happen. Our brain employs the concept of context to administer a large amount of details culled from a number of different scenarios, like society, work space, family ambience, and community surroundings. By using the principles of context, one can instantly find out and decipher what type of information is useful and relevant for a given situation.

People can easily recognize the contexts or situations that they are in, and they would also know what information is applicable to each of those contexts and later derive or deduce more precise information from the context. Processing a given context is very much essential to humans, because a perceived paucity in analyzing the given context may likely to result in serious health problems-related cognition. Context is also the most important mental technique for humans.

Before initiating this discussion, we will need to set the stage and define the required concepts for this chapter, so that all of us will have a common understanding of the topic at hand. We will start with the actual definition of context. After an exhaustive search of context definitions, we have found one that nearly matches, to some extent, with what we are looking for when referring to contexts. According to Dey (2001):

> Context is any information that can be used to characterize the situation of an entity. An entity is a person, place, or object that is considered relevant to the interaction between the user and an application, including the user and the application themselves.

This definition allows software developers to associate a given application with a set of design scenarios. In fact, it also provides software developers with a mental construct that defines elements that are/are not part of the problem domain. Think metaphorically about context as an imaginary fence between relevant information to your problem and information that does not have anything to do with it.

From a conceptual point of view, this fence allows us to generate a pervasive core or base, focused and generic enough, which can host a set of countless fine-grained solutions. From a practical point of view, this fence allows us to utilize this core, via extension points or hooks, and specify their new target environment in a cost-effective manner. Likewise, this fence also allows us to keep on track, as we implement this core, along with its always changing target environments, thereby keeping us away from getting bogged down with irrelevant information—neither promising more for the solution nor requiring less.

Now, we have a clear and concise idea of the benefits provided by having a focused investigation of the core knowledge of a discipline, along with its *always changing* applications. We must now proceed with the illustration on how we can implement and adapt this core, as well as its target environments, in a way we can obtain the expected quality and consistency.

Just for the sake of record, you should know that most of the guidelines and coding standards that will be illustrated in this chapter are based on the software framework described in Chapter 8. Moreover, you should also realize that it is also quite important to know that the foundations for implementing this framework are critical too and they are provided in this chapter. Therefore, a recursive association exists between Chapters 6 and 8.

From a developmental point of view, the implementation of the selected core formed by the enduring business themes (EBTs) and business objects (BOs), along with the target environment formed by the industrial objects (IOs), requires certain clarifications, especially with the terms *classes* and *types*, *hooks specifications*, *user model specifications*, *coding*, *packaging*, and the solution's *encapsulation* that form a component. We will provide these clarifications in the following sections.

10.2 IMPLEMENTATION ISSUES OF THE KNOWLEDGE MAPS

Now, we will debate and discuss several issues related to the structure and behavior implementation of the knowledge maps (e.g., EBTs, BOs, and IOs implementation). The overall goal of this section is to create, among software developers, a more accurate idea of the coding guidelines they will use to implement partial or whole knowledge maps (some snippets will be provided). For example, we will start with *type versus class* issue within the type-oriented paradigm (TOP) (Fayad and Arun 2003). Then, we will proceed with the impact of using contracts in the partial or whole development of our knowledge map. Moreover, we will also provide the specification of hooks in the knowledge maps. Finally, we will describe the use of storytellers, packages, and components as a whole in the implementation of the knowledge maps.

10.2.1 Type versus Class within TOP

Within TOP, types are names that denote an interface of an individual (object), and classes are the implementers of the type. In other words, classes encapsulate the implementation of a type (Fayad and Arun 2003). Software developers often overlook this issue and thus tend to diminish the potential of types, such as the benefits of decoupling a class's specification from its implementation.

The relationship between types and classes is sometimes complex and leads software developers to confusion and antagonism, especially when a class has the same name as its type. To avoid this confusion and antagonism, we must step back and retract a bit and understand, by using examples, what types and classes really are in the context of pattern topologies and in TOP.

Take one of the capabilities of the knowledge maps: *AnyContent stable design pattern* (Fayad 2015). This stable pattern consists of five classes: *AnyContent* (*BO*), *Knowledge* (*EBT*), *AnyActor* (*BO*), *AnyForm* (*BO*), and *AnyMedia* (*BO*). Each one encapsulates the behavior specification of its respective types. At the same time, some of the classes' behavior specification can be of more than one type. In other words, one individual (object) may have one portion of its interface defined by one type and the portion implemented by another type (Fayad and Arun 2003). Let us now look at the Snippets 10.1 through 10.3 of the AnyMedia class and the BO type. In this class, we will implement two operations defined by the BO type.

SNIPPET 10.1 BO TYPE SPECIFICATION

```
//:~begin
public interface BO {
...............//Some code goes here
void    collaborates (EBT goal);
void    connects    (BO    collaborator);
Boolean disconnects    (BO    collaborator);
BO loads    (String extension);
void    disables    (String extension);
..................//Some code goes here
}//:~end
```

SNIPPET 10.2 ANYMEDIA'S BO TYPE IMPLEMENTATION

```
//:~begin
public class AnyMedia extends AbstractBO implements BO{
................//some code goes here
private Map boRegistry = Collections.synchronizedMap(new
    HashMap());
public void connects(BO collaborator){
if(capability! = null){
      ...../some code goes here
((AbstractBO)collaborator).determine (this, collaborator);
      .....//some code goes here
}
}
....................//some code goes here
private Map ebtRegistry = Collections.synchronizedMap(new
    HashMap());
public void collaborates(EBT goal){
      .....//some code goes here
      ebt   = goal;
      .....//some code goes here
}
................//some code goes here
}//:~end
```

Snippets 10.1 and 10.2 provide us with enough information to make our point straight and clear about types and classes, along with their relationship. First, within TOP, a class implements a type. A class can implement more than one type. Types can be associated with other types. Types are represented as interfaces in Java. More information can be found at Bloch (2001) and Oestereich (1999).

Within the realm of the knowledge maps, we will provide a template for describing this relationship, so software developers can have, at hand, the facility to distinguish between types and classes and make their software analysis and design more efficient and ordered.

10.2.2 Design Type Specification Template

To ensure accurate communication among software practitioners (e.g., analysts, designers, architects, and developers) throughout the software design and development process, we must facilitate the right means to convey design decisions. These right means are assisted by a list of rules that have been determined to represent designing outcomes, such as the following:

1. To ensure that we have at least seven operations per class or type within context (Fayad and Arun 2003)
2. To avoid the utilization of *get and set* methods (Fayad and Arun 2003)
3. To ensure that classes and types will follow a *camel casing* naming convention
4. To ensure that operations or services per class and type are not repeated in other classes and types

The following example will make the above rules clear. It uses three classes of the AnyContent stable design pattern (capability). These are AnyMedia (BO), AnyContent (BO), and Knowledge (EBT) within the design type specification template (see Tables 10.1 through 10.3).

TABLE 10.1
Type Specification 1

		EBT to EBT	
EBT Name	**Type Name**	**Interfaces/Services**	**Implementation**
Knowledge	EBT	release(), ignore(), enlist(), etc.	AbstractEBT
	Knowledge	synthesize(), recognize(), related(), etc.	Knowledge

TABLE 10.2
Type Specification 2

		BO to BO	
BO Name	**Type Name**	**Interfaces/Services**	**Implementation**
AnyMedia	BO	connects(BO), attaches(IO), collaborates(EBT), etc.	AbstractBO
	AnyMedia	broadcast(), captures(), switch(), etc.	AnyMedia
AnyContent	BO	connects(BO), attaches(IO), etc.	AbstractBO
	AnyContent	add(), remove(), denotes(), scope(), etc.	AnyContent

TABLE 10.3
Type Specification 3

		BO to IO	
BO Name	**Type Name**	**Interfaces/Services**	**Implementation**
AnyMedia	AnyMedia	broadcast(), captures(), switch(), etc.	Cellphone, Internet, Computer, etc.
AnyContent	AnyContent	add(), remove(), denotes(), scope(), etc.	Text, Diagram, Image, AudioFile, Movie, etc.

10.2.2.1 Design Type Specification Template 1: EBT

The design type specification template 1 is given in Table 10.1.

10.2.2.2 Design Type Specification Template 2: BO

The design type specification template 2 is given in Table 10.2 and the design type specification template 3 is given in Table 10.3.

In order to ensure that a type's behavior is doing what it is expected to do, it is possible to apply a set of restrictions to the operations' signatures defined by a type. These restrictions are commonly known as *contracts* (Bacvanski and Graff n.d.; Bruegge and Dutoit 2003; Findler and Felleisen 2000; Lackner, Krall, and Puntigam 2002), an overview of which is provided in the next section.

10.2.3 The Uses of Contracts in the Specification of Design Types

Contracts are the useful means to guide the proper use of objects' operations. They would ensure that an object's operations comply with the semantics defined by the objects' behavior. Likewise, they also establish a set of restrictions or constraints for each one of the operations' signatures defined by a type's interfaces. Constraints or restrictions of this sort include the following:

- *Preconditions.* They represent the conditions that need to be fulfilled prior to the operation's execution.
- *Postconditions.* These conditions must be satisfied after any type of operation has ended.
- *Invariants.* They describe the conditions of an object that must be always kept.
- *Exceptions.* They represent what an object does in exceptional circumstances or when the aforementioned conditions are violated.

In Java, especially when we are implementing stable patterns, these contracts are written within the /** **/symbols by using @Pre for preconditions, @Post for postconditions, and @Invariant for invariants; more details about defining contracts, especially in the object constraint language, can be accessed at Oestereich (1999) and Bruegge and Dutoit (2003). Then, they will be realized within a class's code using a simple *contract* application programming interface (API) facility. Snippet 10.3 shows API contract facility where AnyMedia class is used to show how contracts are written.

SNIPPET 10.3 INCLUDING CONTRACTS IN THE ANYMEDIA CLASS

```
//:~begin
Import core.vnv.contract.*;
/**
* @Invariant boRegistry.size() > = 0
* @Invariant ebtRegistry.size()> = 0
* @Invariant ultimateGoal ! = null
**/
public class AnyMedia extends AbstractBO implements BO{
................//some code goes here
/**
* @Pre capability ! = null
* @Post boRegistry.size() = self@pre.boRegistry.size()+1
**/
public void connectsTo(BO capability){
     ..............//some code goes here
     Assertion.requires (capability! = null);//pre
     ..............//some code goes here
     Assertion.ensures(boRegisty.size()>0)//post
     ..............//some code goes here
}
}//:~end
```

The next section deals with the description of the hook specification template. Take this template as the blueprint for defining hooks as well as their implementations.

10.2.4 The Hook Specification Template

The hook specification template consists of nine elements. As a whole, these elements provide the blueprint for adapting the core knowledge formed by EBTs and BOs to specific contexts. This template was extracted from the book *Building Application Frameworks* (Fayad, Schmidt, and Johnson 1998). Table 10.4 illustrates this template using an example from the stable pattern that has been used throughout this chapter—the AnyContent stable design pattern (in specific, we will use the class AnyMedia).

This template will be used to specify how your core knowledge will be seen in contexts that are more specific. The next section will provide a description of the core of the AnyContent stable pattern from a user's point of view. To do that, we will use user models or storytellers.

10.2.5 Specification of the User Model or Storyteller

The user model is just an object diagram that tells a story to a stakeholder. It also tells you what the overall artifacts involved in the software design are and the interactions that stem from them. These artifacts include model name, class names, attributes and operations (sometimes), roles, relationships, contracts, and the sequence of events that happens between the included classes (Fayad, Hamza, and Sanchez 2005). In short, this template is simply

TABLE 10.4
AnyMediaHook's Specification

Name. AnyMediaHook
Requirement. A new media is generated based on a selected media type
Type

Type	Option	Supported Pattern	Open ended
	Level of Support		
Enabling a Feature			
Disabling a Feature			
Replacing a Feature	✓		
Augmenting a Feature			
Adding a Feature			✓

Area. AnyMedia
Uses. N/A
Participants. AnyMedia, MobilePhone
Changes
New subclass MobilePhone of AnyMedia
MobilePhone.broadcast() extends AnyMedia.broadcast()
MobilePhone.captures(object) extends AnyMedia.captures(object)
Repeat as necessary
New property MobilePhone.signal where
Reads signal maps from the AnyMedia.object
Write signal maps into AnyMedia.object
Constraints. All media used in the AnyMediaHook must be derived from AnyMedia
Comments. All the derived media that are using this hook must be related to a specific type of media

a data dictionary of the elements included in your solution. We will use the AnyContent stable design pattern to illustrate this user model (see Figure 10.1).

The remaining sections deal with the facilities for packaging and composing the stable architectural patterns. In order to do that, we must provide a brief description of methods to create packages and processes to turn these packages properly into components.

10.2.5.1 Packages Overview

According to Oestereich (1999, p. 218), "packages are collections of model elements of arbitrary types, which are used to structure the entire model into smaller, clearly visible units." They represent some sort of a logical and physical boundary for the elements they contain. This boundary also defines a unique *namespace* or *home* for a set of model elements, such as classes, types, libraries, subsystems, and other packages.

From a modeling perspective, packages are represented as *folders*. This folder has a name written inside of it, and sometimes, even stereotypes to clarify its purpose. In case this folder has other model elements embedded in it, the package name will be written in its left top as shown in Figure 10.2. Table 10.2 shows as an example of a set of packages.

From a coding perspective, especially in Java, packages representation starts with the keyword *package*. Then, this keyword is followed by a list of packages (a hierarchy of packages). The above packages are programmatically represented in Snippet 10.4.

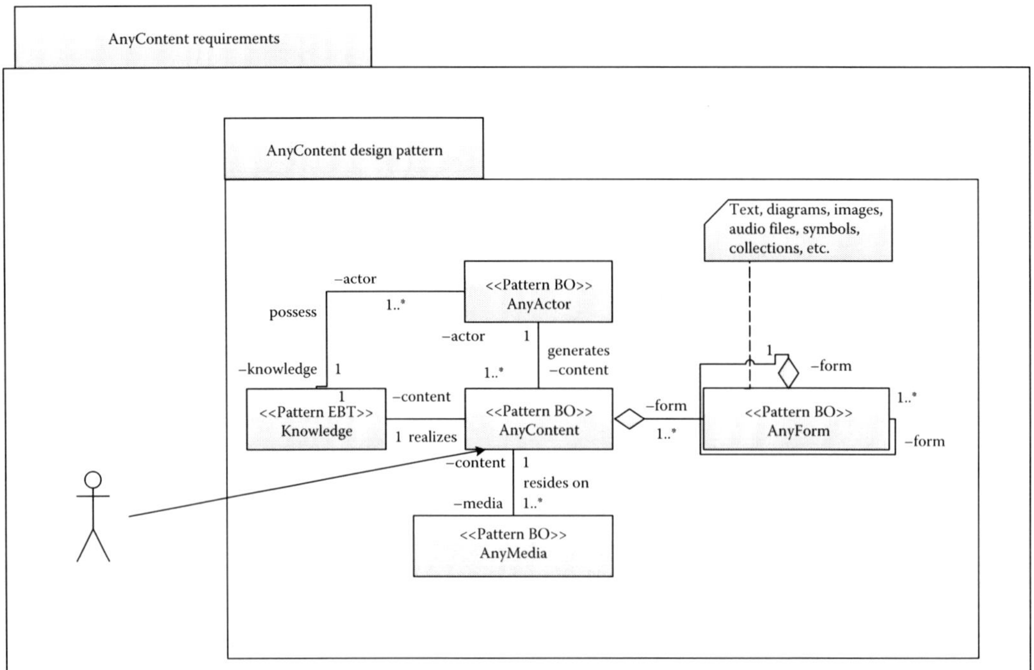

Term	Description
AnyActor	The user initiated the generation of determined content
Knowledge	The user has gained enough knowledge to be able to generate content
AnyContent	This content includes text, diagrams, movies, images, and audio files This class encapsulates the implementation of behavior and properties defined by the type AnyContent and BO
Knowledge	Knowledge is placed explicitly in the generated content
AnyContent	During the generation of a content, the form of content will be automatically selected
AnyForm	The generated content encloses a set of different form of content that will make the knowledge provided more visible and accessible
AnyMedia	This class is where content will reside on. During the generation of a content, this media will be automatically selected by the actor

FIGURE 10.1 The AnyContent stable design pattern's user model.

SNIPPET 10.4 PACKAGES REPRESENTATION

```
//:~begin
package net.hsanchez.km.core.*;//or
package net.hsanchez.km.core.hf.*;//or
package net.hsanchez.km.patterns;//or
package net.hsanchez.km.vnv.*;//and so on

........//import statements and comments go here
public class ...........{
................//some code goes here
}//:~end
```

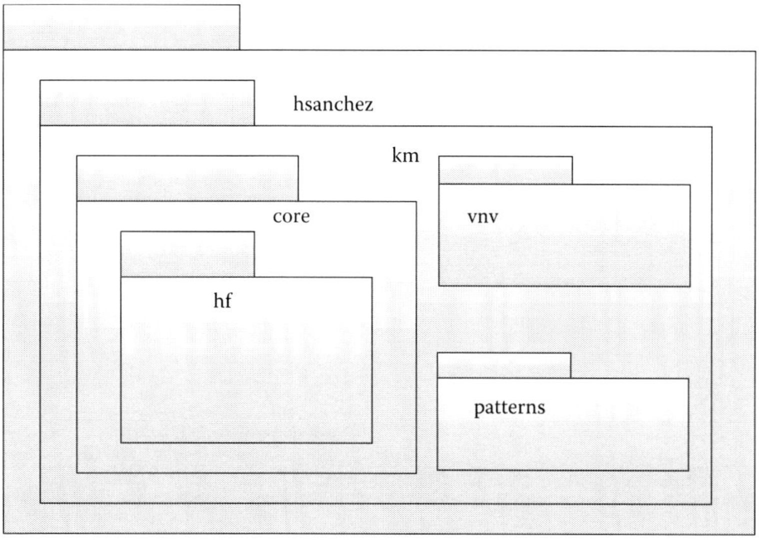

FIGURE 10.2 The packages representation and their relationships in knowledge maps.

One can perceive knowledge maps as the large system that contains a set of building blocks. In order to manage all these building blocks properly, you may need to group and organize each of them within its own boundaries. Likewise, some of them can be in the middle of overlapping boundaries. You may also access more information about packages at Oestereich (1999) and Bruegge and Dutoit (2003).

10.2.6 Components

Components are not that different from packages. For instance, they both realize some kind of boundary (Oestereich 1999) for a set of elements. They also define the means for structuring and grouping this set of elements as a unit (Oestereich 1999). The main difference between packages and components is that components are executable and interchangeable software units (Oestereich 1999) with well-defined external interfaces.

There are several platforms for components, such as Enterprise JavaBeans (EJB) and CORBA. The option for selecting a specific component environment will be left to the software developer's desire. From a more basic perspective, in the case discussed throughout this book, we will simply compose stable patterns, as well as stable software architectures, by using the EJB (Enterprise JavaBeans Technology) (Monson-Haefel 2000) and Java Archive (JAR) (Files, Packaging Programs in JAR). The JAR files will allow us to bundle our knowledge maps, individual stable patterns and their respective specific contexts, and stable software architectures, as a single executable file. More information about implementing EJB and JARs together are at Monson-Haefel (2000). For the purpose of simplicity, here we would provide only its unified modeling language (UML) representation. We will also provide a picture to clarify or provide the pertinent representation of components in UML (see Figure 10.3).

The combination of all the elements described in this chapter will allow us to implement robust, within-context, and reliable knowledge maps (partial or whole).

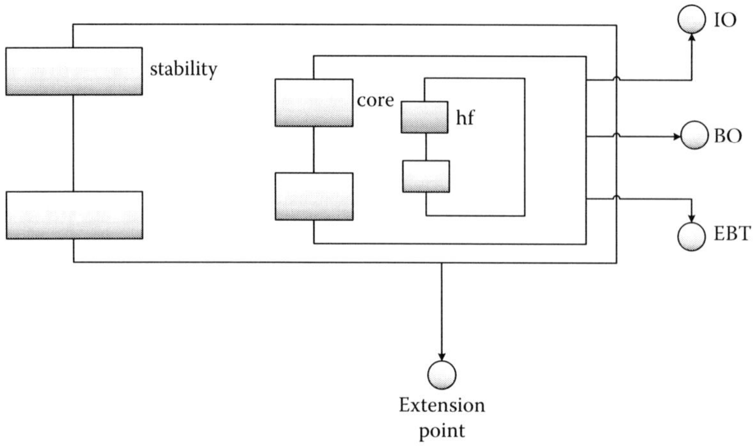

FIGURE 10.3 Representation of the components in UML.

10.3 ASPECT-ORIENTED MODELING

Software stability modeling = Aspect-oriented modeling

Software stability modeling and aspect-oriented programming (AOP) are both emerging paradigms in the software development field. Software stability concepts are based on identifying the core goals of a system and creating a layered model (of EBTs and BOs). The IOs come at the periphery of the model and are application specific. These are like leaves in the model and can be replaced for different application, but the core goal and the means to achieve the goals (represented by the EBTs and BOs) change very little and hence the model is a stable one. Aspect-oriented design aims at identifying functionality that is desired by many classes—referred to as a crosscutting concern and modularizing it as an aspect. Aspects help in increasing the maintainability and hence stability of the system. This means the purpose of the two approaches is similar, but they lead to the same goal differently. We have tried to understand the similarity and differences between the two paradigms in this chapter.

The main goal today, in the development of complex systems, is to achieve software *extensibility*, *reusability*, and *adaptability*. These goals must be explicitly engineered into the software without regard to the software technology applied (Fayad and Cline 1996). This will make the system more stable. The advantage of incorporating stability is that the models become easily extensible to add other features and adaptable to different scenarios without requiring to change the basic model of the system. This is not, however, a straightforward task due to certain aspects that are part of the problem analysis representing the core knowledge of the problem domain and are not explicitly stated in the problem statement.

New approaches are emerging looking for the ability to discover the core knowledge of the problem domain in software development projects. Their main benefits are a shorter time to market and the easy adaptation of systems to the continual evolution of new technological and client requirements.

In this chapter, we will focus on two of the most novel software technologies: *software stability model* (SSM) (Fayad 2002a, 2002b; Fayad and Altman 2001; Fayad

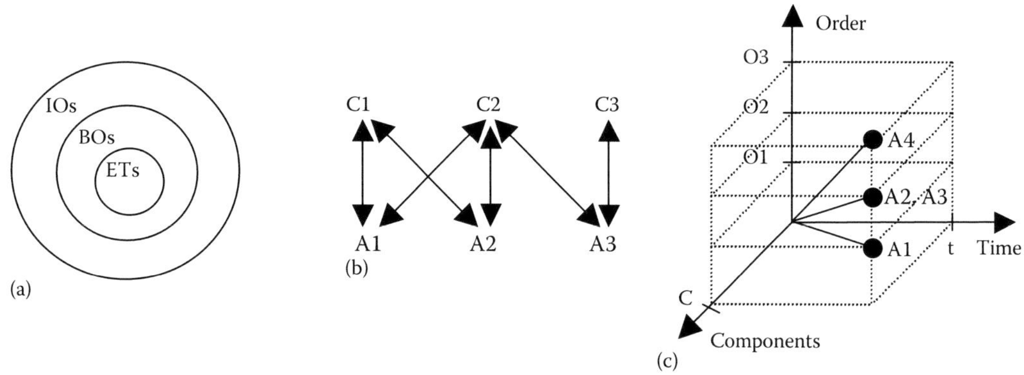

FIGURE 10.4 (a) SSM architecture, (b) AO architecture, and (c) aspect-oriented component: the execution time and order of the application or time and order of application.

and Wu 2002) and *AOP* (Constantinides, Elrad, and Fayad 2002), which analyze the relationship between both technologies and the differences and similarities between the *aspects* in AOP and *EBTs* and *BOs* in the SSM.

These technologies model the system in different structural ways, but share the same motivation, that is, making the components in the final system more *stable*. SSMs try to model the core of a system based on those *concepts*, named *EBTs* and *BOs*, which are stable over time. The AOP approaches separate different entities based on *concerns*, named *aspects*, that are spread along other components in the system and can change or evolve independently from them. This separation of concerns increases necessarily the stability of the system because it rises as first-order entities, properties that otherwise are implementation details.

In the SSM architecture (Figure 10.4a), the components are stratified in three layers according to the level of stability provided as shown in Figure 10.5. The EBTs layer represents all the concepts that remain stable both internally and externally over time. The BOs layer contains those concepts that are externally stable, but internally adjustable or adaptable. The IOs layer is formed by those entities that are not stable at all.

There are several possible aspect-oriented architectures (AOA) because the aspect–component relationship is bidirectional (Figure 10.4b). We can state for each component which aspects will be applied and also for each aspect which components are affected by that aspect. In addition, in the AOA, we must take into account two new dimensions, the *time* and the *order* of application. Considering these dimensions, an AOA might show for each component and execution time which aspects must be applied and the order of application (Figure 10.4c). Both the number of aspects and the order of application could change over time.

10.3.1 Aspects as Core Design Features in a System

Aspects are often goals or a matter of interest in a system. These goals could be at the topmost level representing the core concerns of the system as a whole. These are the functional goals of the system. The goals could be of system levels that are essential for the system to achieve its core goals. In an e-commerce example of an online store, the key goals are that of security and the presentation of information for convenience. The system level concerns would be that of the synchronization, failure handling, and so on. If one were to model the

EBTs	BOs	IOs

FIGURE 10.5 Software stability model template.

same example by using SSM concepts, one would recognize the core goals of the system, which would be modeled as EBTs and the system level concerns as BOs. In effect, it shows that the aspects of a system will be the EBTs and BOs put together from an SSM model. The properties modeled as aspects in AOP are part of the core design of a system in the same way than EBTs and BOs are in the SSM. Both paradigms model as first-order entity concepts that are implicitly present in the system due to either the domain application or the computational model.

An example of the stability model for e-commerce is shown in Figure 10.6. We could recognize the classes trading, convenience, and security as the EBTs, because they form the core goals in this example. EBTs are conceptual, and hence, we need the more concrete BOs to achieve these goals. For example, the EBT of security is achieved through an authorization class that encapsulates and has the authorization procedure methods. Then at the periphery, we have two IOs that are application specific. From the above model, it is easy to see that security (which is in the innermost layer of EBTs) has an effect on the classes (Authorization, CreditCard) in the layers above it. So, security has an effect on one or more classes and hence can be modeled as an aspect. The modeling using the software stability concepts helped us recognize that security was a concern and needed representation, which was not the case in the traditional model. Moreover, the authorization class cleanly encapsulates the needed procedure for security in the example.

10.3.2 Difference between the Aspect-Oriented Design Approach and SSM

In more traditional software modeling, the approach is not centered on recognizing the goals of the system that will not change over time and creating a model with these core concepts as the foundation. This is what the SSM enforces to increase stability in software. AOP, however, focuses on identifying the properties or functions that will affect one or more classes and how they will. These features, being of a crosscutting nature affecting various aspects, will decrease maintainability and therefore stability. An AOP approach

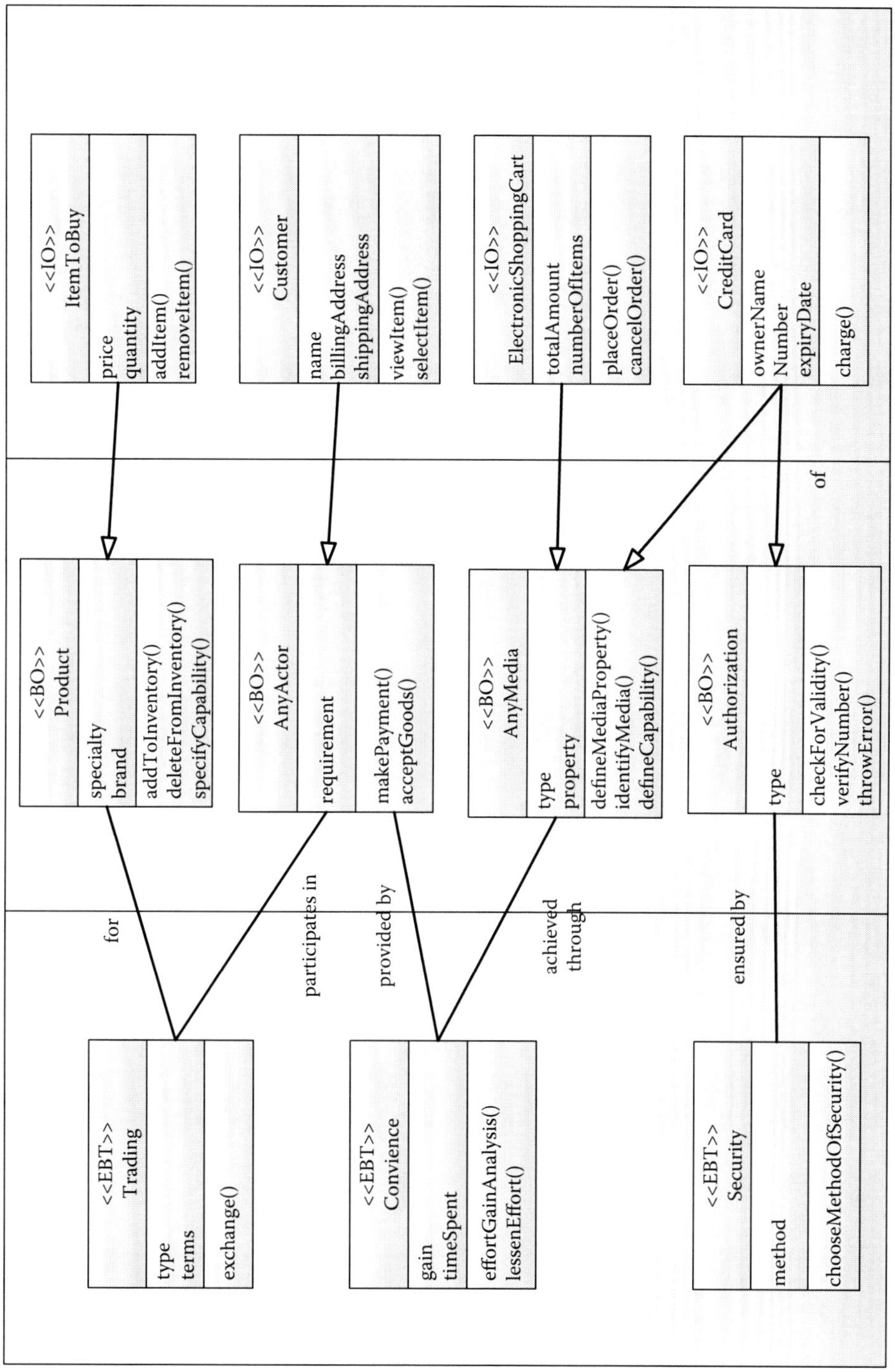

FIGURE 10.6 Stability model for e-commerce example.

involves modularizing these crosscutting concerns into an abstraction called an aspect. The aspects can be independently maintained, so that changes can be incorporated easily and in one place without involving the re-implementation of the other classes it affects.

Though the approaches of AOP and SSM are different, we want to highlight here that the SSM comes up with the same aspects as an AOP. All the aspects in an AOP can be modeled as BOs or EBTs in the SSM.

10.3.3 Identification of Whether Aspects Will Be All the EBTs and All the BOs in a System

Analyzing the other direction in the relationship among aspects and EBTs or BOs, we were convinced that all EBTs and BOs in the SSM are aspects. However, it is not easy to perceive all core concepts in an SSM as aspects. Using the example of a collaborative environment (CE), we will try to show the difficulty in understanding this relationship. A CE is a common shared environment that supports collaborative work and social interaction.

First, EBTs and BOs are concepts that are part or complement the functionality of a system, but have entity by their own and this is exactly the definition of aspects. Figure 10.7a shows partially a CE interface to (1) initiate a session to collaborate with other users, (2) show the awareness list to know who is connected and their state, (3) add new friends to the awareness list, and (4) initiate a communication with selected users. This interface hides some concepts present across the components in CE systems as *collaboration*, *integration*, *synchronization*, *distribution*, and *awareness*. By applying the SSM, we can clearly consider these concepts EBTs and BOs applied to an IO *room* (Figure 10.7b). The room component is where communication and collaboration among users is performed. Looking into them from an AOP perspective, these concepts can be seen as aspects composed with the room component to complete its functionality. This led us to affirm that all the EBTs and BOs can be modeled as aspects.

Second, tangible elements that should be modeled as IOs in the SSM are sometimes modeled as BOs. This decision might be taken because of a misunderstanding of the problem statement, but not always. System analyzers might include as part of the core design relevant elements of the problem domain they think are essential to model the system correctly, though they are tangible concepts and are easier to understand as IOs. For instance, in Figure 10.7c, a new analyzer considers the necessity to include *place* as a BO, whereas

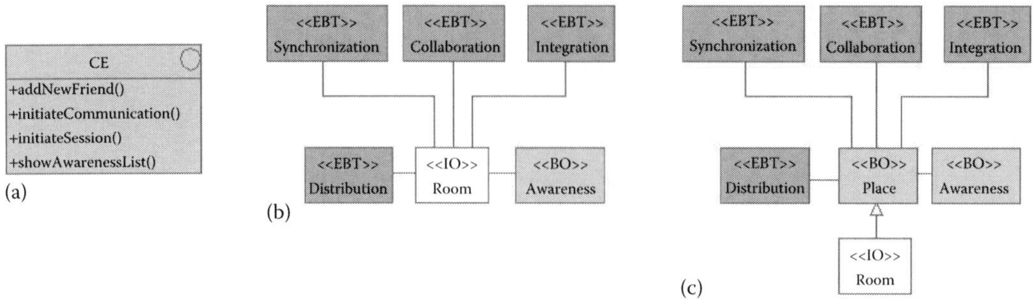

FIGURE 10.7 CE models (illustration). (a) Collaborative environment (CE) partial interface, (b) EBTs and BOs are apsects, and (c) EBTs layer → BOs layer → IOs (right order).

the one in Figure 10.7b shows that the concept of *integration* was enough to represent the users' accessibility to all the resources in the environment. Though both models are correct, it is difficult to see *place* as an aspect, because unconsciously we imagine a real-world place. However, the concept of *place* is broadly used in the researching of CEs to express *integration*. In this context both *integration* and *place* are stable concepts and should be aspects in AOP.

10.3.4 Understanding Whether Aspects in AOP Will Represent either EBTs or BOs of SSM in a System

All the aspects in an AOP approach are EBTs or BOs in the SSM. We also think that most of them will be BOs and only a few will be represented as EBTs. Further, we believe so, because though aspects represent concepts stable over time that you cannot take away from the system, internally they will be able to evolve or manifest in different ways. One may think that EBTs do not fit well with the definition of aspect given in this column, where we said "aspects can change or evolve independently of the components they affect." But they do, because EBTs are really representing *aspect frameworks* (AFs), each one covering a set of related aspects (or BOs) that are then applied to components (or IOs). In Figure 10.8, the *communication language*, *communication media*, and *transport protocol* BOs represent the mechanisms through which distribution is implemented in the system. So, these will be the aspects applied to the components or IOs in the system, whereas the *distribution* EBT is an AF that encloses all these concepts.

10.3.5 Comparison between Extraction of Aspects in AOP, EBTs, and BOs in SSM

We think that the same reasoning is not followed, when we apply AOP or SSM. When a system is engineered by using SSM, the developer will have in mind the goal of *stability* and will look for those concerns (EBTs and BOs) that are immutable over time determining

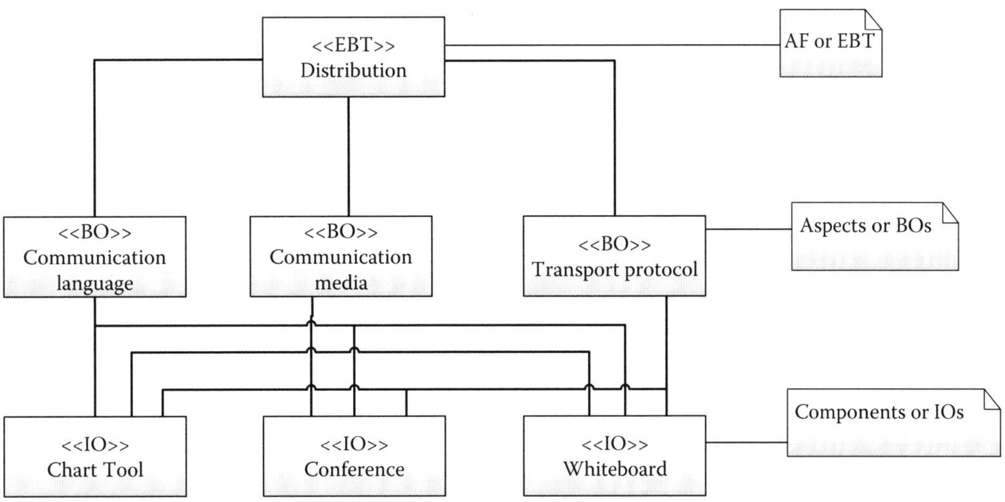

FIGURE 10.8 AOP and SSM entities relationship (illustration).

the core system. In AOP, the main goal is *separation of concerns*, and the developer is worried about the identification of those *concerns* that are present across multiple components in the system, decoupling them in different entities. This means that achieving *stability* is not the primary goal in the extraction of aspects. What we want to emphasize here is that despite of the difference in reasoning, both paradigms share the requirements of a deep knowledge in the application domain and a change of thinking process—in the system analysis process. These requirements will constitute the main difference with many traditional paradigms.

Following the above discussion, AOP and SSM should come up with the same core concepts. However, both models could not be completely equivalent, because the reasoning process is different and so might be the final design of the system. In order to take advantage of both *stability* and *separation of concerns*, we will consider the application of both techniques in different stage of the software life cycle. First, the SSM should be applied in the analysis and design phases, by analyzing the problem statement and defining the EBTs and BOs that constitute the core design of the system and those IOs that are in the periphery. Then, AOP should be applied during the implementation phase. All EBTs and BOs in the design will be implemented as aspects that will be composed with the rest of the components in the system. The main advantage of this combination is that by following this process, we are explicitly engineering stability in the whole system.

10.3.6 Modeling EBTs and BOs as Aspects

SSM concepts are centered on the goals of the system. Once the goals of the system to be modeled are identified, they are put at the center of the model as EBTs; the means to achieve these goals are represented by the BOs that form the next layer. Then IOs would arrive, which are application specific and are at the periphery of the model. One can clearly see that EBTs functionality cuts across the BOs and IOs associated with it in the next two layers. Therefore, the EBTs and BOs can be modeled as aspects. Adopting this kind of approach can have many advantages. Any changes that have to be made to the EBTs functionality due to an earlier flaw in modeling can have a rippling effect on the classes in the layers above it. Instead, if the EBT is modeled as an aspect, it will not be scattered in the classes associated with it. Enhancements can be made to it in a single place without affecting other classes and all the changes can be made in a single place. All this will make the EBTs and BOs more maintainable.

SUMMARY

The objectives of this chapter are twofold. First, we provided the coding recipes for implementing partial or whole knowledge maps. Second, we provided their visual representation according to UML standards. We also implemented these objectives by providing a set of specification templates, some snippets, and proper examples. Chapter 8 provides additional information. Third, the SSM concepts and aspect-oriented approach work toward the same goal of stability in a system. The advantages of both these approaches can be exploited by employing the SSM concepts in designing systems and by using the AOP concepts in the implementation phase. The SSM concepts can also be very useful with aspect-oriented design, where creating a stable model will make aspects in the system evident at an early stage. In all, both the paradigms can work to the advantage to one another.

OPEN RESEARCH ISSUES

Because implementation of stable design patterns and hence knowledge map is carried out using Java, which is a robust object-oriented language, knowledge maps inherit this robustness, if designed and implemented correctly. Only some guidelines of package creation and creation and implementation of interfaces must be taken care of.

REVIEW QUESTIONS

1. What do you understand by the terms *context* and *entity*?
2. Why is it necessary to define a context when investigating the core knowledge of a discipline?
3. Explain in your own words the relation between type and class in type-oriented paradigm.
4. Is the following statement true or false? A class must implement one unique type.
5. Enlist rules that one needs to follow during designing, so that right design decisions are conveyed.
6. Explain what are contracts and why are they used?
7. What are some commonly used contracts in the realm of designing types to ensure proper use of object's operations?
8. How do you specify contracts while implementing patterns?
9. One can use ________________ template to specify how core knowledge can be applied in specific contexts.
10. Explain user model.
11. Is the following statement true or false? Storytellers' specification is a data dictionary of the elements.
12. What are packages? Why do you need them for knowledge map?
13. How are packages represented in knowledge maps.
14. What is meant by component? How are they different from packages?
15. How are knowledge map bundled together. Explain.
16. Describe and depict representation of a component using UML.

EXERCISES

1. Research type-oriented paradigm and find its usage in knowledge map.
2. Explore use of contracts in object constraint language.
3. Using hook template, create blueprint for the following:
   a. AnyContent
   b. AnyHook
   c. AnyPartition
   d. AnyEntity
   e. AnyCollection
4. Create user model for the above scenarios.

PROJECTS

1. Create a knowledge map for collaborative virtual environment (CVE) and come up with the following:
   a. The traditional model of CVE
   b. The stability model that includes the (EBTs + BOs) of CVE
   c. Describe briefly three different scenarios of the stable model of CVE.
   d. Discuss briefly the functional requirements of the stable model of CVE.
   e. Discuss briefly the nonfunctional requirements of the stable model of CVE.
   f. Discuss briefly 12 of the challenges of implementing the stable model of CVE.
   g. List 25 constraints of the stable model of CVE.
   h. Create CRC cards for all the EBTs and BOs of the stable model of CVE. (Make sure to come up with unique responsibility per class, list the collaborators, five services provided by the CRC class, and seven of its attributes.)
   i. Create two case studies, one of them is the CVE of the stable model, and include a significant use case, class diagram that include the IOs, and a sequence diagram per case study.

11 Deployment, Verification and Validation, and Quality Factors

Quality is not an act, it is a habit.

Aristotle

Deployment, verification, and validation are the intricate processes of guaranteeing that software being developed or modified will satisfy almost all functional and other sundry needs and requirements (which is called validation), and ascertain that each and every step in the ensuing process of designing the software product provides the right and correct type of products in the end (which is called verification). The proponents of the software design and development process use the term *V&V* (verification and validation) just to refer to all those critical activities that focus at making sure that the software products designed will act and behave in the way required.

V&V aims at a systematic and calculated technical evaluation of software products involved in the development and maintenance processes. At the end of these exhaustive processes, designers will review and test all the developmental processes just to make sure that all the requirements and needs are complete, exhaustive, and completely testable. Testing is a process of operating a software product under real or simulated conditions and inputs to prove or demonstrate that the manufactured product completely satisfies all the requirements and needs (Eushiuan 2008; Tsai et al. 1999).

11.1 INTRODUCTION

Deployment and V&V focus specially on the definition of robust pattern topologies (knowledge maps), by means of using a set of quality factors that are pertinent to the discipline of interest and concern. Creating a definition of these robust patterns topologies is a big challenge, especially when dealing with disciplines with such unique characteristics and behavior, and with a lack of a complete and systematic process to support their creation.

Chapters 3 through 6 of this book described various methods and means for identifying, defining, and evaluating the functional requirements of the subject that has been determined, such as knowledge maps. Chapter 7 discussed a different set of systems requirements, which is the commonly called *-ilities* (Larman) of the system—the nonfunctional requirements. These nonfunctional requirements are the systems operational constraints that are not related to a system's functional specification (Bruegge and Dutoit 2003). Within the realm or domain of knowledge maps, these nonfunctional requirements will indicate the deployment expectations of knowledge maps. Take the deployment or deployment stage as the effective retrospective of the sharing and execution abilities of the knowledge maps. This retrospective deals with the formulation of a set of goals or quality factors that will drive a robust deployment of a created patterns' topology, as well as the building blocks within it.

Similar to a discipline, quality factors are embodied and represented as goals, but with a central focus on the intended use of the discipline in question. By using knowledge maps as this discipline, we may encounter a number of quality factors, such as reliability, scalability, traceability, extensibility, usability, and others, none ending with *-ility*, such as performance and accuracy. The selection of these quality factors will be determined by examining the knowledge maps as a well-defined and operational unit.

The process for identifying, filtering, and evaluating these quality factors is the same as the one that was described in Chapter 3. The only thing that software practitioners need to pay close attention to is the fact that the subject of interest will be a knowledge map, and not its internal structure, such as rationale. This process can be assisted by means of using heuristics, evaluation/filtering questions, and V&V processes that were used to assess and evaluate the underlying goals of a discipline. We will go further with the V&V processes throughout this chapter.

11.2 DEPLOYMENT OF KNOWLEDGE MAPS

The deployment stage of the knowledge map represents the set of interrelated activities (e.g., release, installation, and acceptance) that will make knowledge maps available and ready to use. These interrelated activities represent, as a whole, a systematic process that can be executed and customized according to a determined set of requirements (e.g., expected response time and throughput) that are defined by stakeholders. To understand this stage, we must tackle and approach it from both a conceptual and practical point of view. The conceptual point of view will give us a clear notion of what terms and semantics are used to understand and realize what deployment of the knowledge maps is or consists of. The practical point of view will give us the proof of the concept and benefits of deployment of the knowledge maps. Together, they will allow us to have the same frame of reference in terms of how to deploy a knowledge map.

11.2.1 The Conceptual Point of View of Deployment

Let us begin by defining the requirements that drive the usage of the knowledge maps. These requirements are known as nonfunctional requirements (Bruegge and Dutoit 2003; Jacobsen, Kristensen, and Nowack 1999; Malan and Bredemeyer 2001). They consist of a set of quality factors and constraints related to the operation of the knowledge maps. Quality factors, for example, are the essential properties that stakeholders will care about while using the knowledge maps. In other words, they are the factors that will drive stakeholders satisfaction and contentment when using the knowledge maps. Constraints are the nonnegotiable restrictions imposed on the use of the knowledge maps. Some examples of qualities and restrictions include an intuitive graphical user interface's navigational style as a quality and a prohibited transformation of the knowledge maps to text files as a restriction. This quality and constraint will determine the overall scope of the usage of the knowledge maps. Therefore, the proper awareness of qualities and restrictions will guide us to properly use the knowledge maps.

For the knowledge maps, especially because they can be associated with other adjacent maps (within a determined discipline) and with remote maps of a different discipline, their deployment stage involves a proper coordination of a set of determined quality factors. These qualities will change according to each discipline of interest; however, some may

TABLE 11.1
Goals of Deployment of the Knowledge Maps

Quality Factors	Description
Traceability	This quality deals with the ability of knowledge maps to be traced back to their original goals after their implementation
Stability	This quality deals with the ability of knowledge maps to handle future requirements and changes
Generality	This quality refers to the ability of knowledge maps to be reused elsewhere and create partial or complete solutions via patterns
Adequacy	This quality is the ability of knowledge maps to satisfy the intended and established requirements, along with their consequences[a]
Scalability	This concentrates on the ability of knowledge maps, along with their enclose elements, to adapt to evolving needs and insights without unnecessary effort[b]
Verification and validation	This concentrates on the ability of knowledge maps to be self-verified and validated according to their context of application

Sources: [a] Fayad, M. E. et al. "Enterprise Frameworks Characteristics, Criteria, and Challenges." *Communications of the ACM* 43, no. 10 (2000): 39–46. [b] Fayad, M. E. et al. "Towards Scalable and Adaptable Software Architectures." Paper presented at the IEEE International Conference on Information Reuse and Integration, Las Vegas, NV, 2005.

commonly appear in all of them. These qualities were already described in Chapter 5. Table 11.1 summarizes them.

11.2.2 The Practical Point of View of Deployment

It is well known that nonfunctional requirements have the tendency to be conceptualized by using loose or fuzzy terms that lead to different types of interpretations (Malan and Bredemeyer). Accordingly, they provide little tidbits or snippets of advice or guidance to software engineers, especially when making tough trade-off decisions or when meeting functional goals (Malan and Bredemeyer). The deployment stage of the knowledge maps seeks to alleviate, from a practical point of view, this shortage by providing the necessary mechanisms to guide the identification and implementation of nonfunctional requirements. This practical view represents a combination of best practices, expectations, actual qualities, and promises of what knowledge maps can possibly do, such as the qualities that were described in Table 11.1, and the properties of the knowledge maps as described in Chapter 5. These elements are then encapsulated in the form of stable software patterns, which are a retrospective of best practices and lessons learned in software development.

The encapsulation and use of these qualities as stable software patterns is considered as the proof of the conceptual view of the deployment of the knowledge maps. Clearly, in order to make deployment activities accessible, complete, and accurate, we must explain or note how the context of the knowledge maps is set, and what the purpose of its use is. One way to represent this notion or perception is to represent these activities as quality factors (see Table 11.1). Then, we can proceed with the indication of their proper capabilities, so that there will be a clear idea of what exactly is required to use the knowledge maps. We may assist this process by attaching a set of IOs that will streamline the functionality of their capabilities.

TABLE 11.2
Common Stakeholders Quality Factors

Quality Factor	Description	Potential Capabilities	Pattern Provided?
Reliability	This deals with quality or consistency of knowledge maps services	AnyService, AnyMetric, AnyMechanism, etc.	No
Assembling	This quality deals with the ability of knowledge maps to allow systems to be assembled from different parts/components (each component may enclose one or many software patterns)	AnyEntity, AnyMechanism, AnySequence, etc.	No
Portability	This quality refers to the ability of knowledge maps to be used in different environments to that, for which it was originally created (portability)	AnyEnvironment, AnyMedia, AnyUnit, AnyMechanism, etc.	No
ROI	This quality reduces the software development lifecycle and meet time to market with less cost, time, and effort involved	AnyLifecycle, AnyMechanism, AnyMetric, etc.	No
Usability	This denotes how easy it will be for any software practitioner to use knowledge maps to generate software solutions (nonfunctional requirements)	AnyMechanism, AnyContext, AnyActor, etc.	No

ROI, return on investment.

As along with goals and capabilities, quality factors and their respective capabilities can form insightful synergies. These synergies provide the basis for the generation of an unimaginable number of stable software architectures (see Chapter 5), which in our case, are translated as the realization of specific qualities that stakeholders are after when using a knowledge map, such as faster return on investment, assembling, and usability. Table 11.2 complements the aforementioned deployment goals, as illustrated in Table 11.1, and provides more practical (stakeholder-oriented) qualities. Again, the existence of these qualities is bound to each discipline of interest. In other words, these qualities may change from discipline to discipline.

The next section will provide a description of additional quality factors that will streamline the deployment of the knowledge maps.

11.3 THE QUALITY FACTORS OF DEPLOYMENT

Chapter 5 and Table 11.1 provide a set of canonical quality factors by driving the internal structure and behavior organization/management of the knowledge maps. In Table 11.2, via stable software patterns, we also described an additional set of quality factors dealing with the operational process of the knowledge maps as a well-defined and self-contained software unit. (Some software patterns may repeat from Chapters 4, 5, 7, and 8.) We also specified that these quality factors may change or can be modified based on the domain knowledge of the subject being explored, as is the case of the knowledge maps as the

TABLE 11.3
Quality Factors of a Common Target Environment

Deployment Activity	Description	Pattern Provided
Configuration	This process prepares knowledge maps for their new target environments. It can be considered the preparation work that developers must perform before installation or release	No
Installation	This process is defined the same way as if we were talking about any type of software. It is the initial physical deployment into a determined customer's site. It requires prior configuration	No
Releasing	According to nonfunctional requirements and V&V, releasing activity is the interface between development and deployment processes	No
Activation	According to nonfunctional requirements and V&V, activation is the activity starting the knowledge maps or a knowledge maps' tool as an executable unit	No
Acceptance	This is the quality that deals with the formal process of accepting a new developed knowledge map or a deliverable one	No

subject of interest. For example, in the domain of class responsibility and collaboration (CRC) cards, these quality factors are Blueprint, Groupware, Adaptability, and Bookkeeping. (Fayad and Cline 1996; Fayad, Sanchez, and Hamza 2004), or in the domain of performance evaluation, such as compatibility, flexibility, and pattern relatedness (Fayad and Pradeep 2004). Now, we need to be aware of certain activities that are more oriented to the target environment where the knowledge maps will be deployed.

The proper implementation of the quality factors illustrated in Tables 11.1 and 11.2 is not the entire piece of the puzzle with respect to the deployment of knowledge maps. We may also need to be aware of certain activities that deal more with the target environment where knowledge maps will be deployed. These activities are part of the software development life cycle, and they usually include software, knowledge maps in our case, configuration, installation, releasing, and so on (Bruegge and Dutoit 2003). Table 11.3 will provide these environment-oriented quality factors.

11.4 DEPLOYMENT—V&V PROCESS

In order to make a knowledge map's solution valid, complete, accessible, operational, and, most importantly, focus on the discipline or subject's specification, we must use a proper V&V process during the entire realization of the knowledge maps—from goals identification to actual deployment.

In V&V, a definition of verification and validation concepts is provided. In this definition, the target subject is software/system. To assure that we all are on the same page, let us use or imagine that the term *knowledge maps* replace the term software/system. The following are the definitions of V&V:

> Verification is the process of determining if a system meets the conditions set forth at the beginning, or during previous activities of the software development lifecycle, correctly. These conditions are set forth in software requirements, which are usually formally documented.
>
> Validation is the process of evaluating the system to determine whether it satisfies the specified requirements and meets customer needs.

V&V processes are of prime importance and critical significance during the formulation of knowledge maps, especially because they determine that knowledge maps are actually portraying an accurate and focused solution. In terms of the quality of knowledge maps, V&V processes are not the entire recipe to achieve that objective. In fact, to assure the quality of the knowledge maps, we use V&V processes in combination with other tools found in software engineering, such as configuration management, planning (V&V), good analysis, and good design.

From a practical point of view, V&V processes of knowledge maps can be perceived as follows.

Verification processes in knowledge maps usually include the following methods:

1. Documentation reviews of the knowledge map (after documenting each one of the generated elements in the knowledge maps)
2. Inspection of the knowledge maps (evaluation of the past knowledge maps' conformance)
3. Auditing of the knowledge maps, as well as stable patterns (evaluation of the current and future knowledge maps' conformance)
4. Validation of records (in the design scenarios)
5. Simulations of the knowledge map (especially in prototyping and final release of the knowledge maps)
6. Demonstration measurement of the knowledge maps

Validation processes in knowledge maps usually include the following methods:

1. Test case generation (created after the formulation of each of the elements in a knowledge map) of the knowledge maps (including each individual pattern)
2. Test suites generation of the knowledge maps (the collection of all test cases generated in the knowledge maps)
3. Scenario testing of the knowledge maps

To show a typical testing cycle that can be used in knowledge maps (to support the above methods), we would provide the following list:

1. *Goals identification phase.* Testing should start in this phase with the purpose of guaranteeing an accurate solution.
2. *Capabilities identification phase.* The capabilities found are evaluated, so that designers and developers have some sort of accurate interface or same frame of reference between design and implementation.
3. *Test planning.* The formulation or selection of the testing strategy and approach to evaluate the development scenarios of the knowledge maps.

4. *Test development*. The proper generation of test cases, test suites, test scripts, and scenario testing.
5. *Test execution*. Developers run all the testing elements that are already defined.
6. *Test reporting*. Based on the previous execution of tests, developers record each one of the generated results, so that they can create metrics and final reporting the total effort spent during testing, and if they consider that knowledge maps are ready for deployment.
7. Continue steps 5 and 6 until necessary.

More work has to be done to make the V&V process automatic throughout the conceptualization, development, and deployment of the knowledge maps. This is part of what we call dynamic analysis. This activity will be part of our future work.

SUMMARY

In general, the rationale of providing a deployment concern in the knowledge maps is to guarantee and ascertain a proper forming of core sets, along with the enclosed patterns that are pertinent to the discipline of interest. In short, this stage implies guaranteeing quality and authority of the knowledge maps with respect to the mastering of a particular discipline.

In this chapter, we also highlighted the respective notions of what the knowledge maps deployment's activities are. The actual documentation process, as a pattern of these activities will be a part of our future work.

OPEN RESEARCH ISSUES

The deployment and V&V methods described for knowledge map are still in their evolutionary phases and so there are many gaps. It is still not used in mainstream development, and hence, no automated V&V tools are available. However, by using the concept of stability, such tools can be developed and will have a great potential.

REVIEW QUESTIONS

1. What do you mean by deployment in the context of software application?
2. What is the difference between verification and validation? Are they related?
3. What does deployment and V&V focus on? Why is it challenging to define for knowledge map.
4. Describe deployment in the context of knowledge map.
5. What are the goals for deployment of the knowledge map? Describe each of them.
6. Explain the conceptual point of view of deploying knowledge map.
7. What do you mean by nonfunctional requirements? What do these requirements encompass?
8. Describe the practical point of view of deployment of knowledge map.
9. List the common stakeholder's quality factors. Describe each of them briefly.
10. What are the quality factors of a general target environment? Describe each of them briefly.

11. What are the methods typically employed in the verification process of knowledge maps?
12. Describe briefly the methods employed in the validation process of knowledge maps.
13. Describe the typical testing cycle used in knowledge maps.
14. What is meant by dynamic analysis?

EXERCISES

1. Discuss the differences and similarities between traditional deployment and knowledge map deployment process.
2. Discuss how the V&V process differs in the traditional software development cycle and knowledge map.
3. Discuss the role of stakeholders in the domain of quality factors.

PROJECTS

1. Create a knowledge map for verification and validation (V&V) and come up with
   a. The traditional model of V&V.
   b. The stability model that includes the (EBTs + BOs) of V&V.
2. Describe briefly three different scenarios of the stable model of V&V.
3. Discuss briefly the functional requirements of the stable model of V&V.
4. Discuss briefly the nonfunctional requirements of the stable model of V&V.
5. Discuss briefly 12 of the challenges of implementing the stable model of V&V.
6. List 25 constraints of the stable model of V&V.
7. Create the CRC cards for all the EBTs and BOs of the stable model of V&V. (Make sure to come up with unique responsibility per class, list the collaborators, five services provided by the CRC class, and seven of its attributes.)
8. Create two case studies, one of them in the V&V of the stable model, and include a significant use case, a class diagram that include the IOs, and a sequence diagram per case study.

Section V

Case Studies of the Knowledge Maps

The significant goal of the knowledge maps is driven by the special motto *divide and conquer*, and this principle is applied here throughout the structure of knowledge map. The knowledge maps are basically a system of dividing a domain into a number of different levels of fineness to enable easy management of domains. A knowledge map engine is the architecture of different patterns driven forward by the software stability concepts approach as espoused by Fayad (2002a, 2002b) and Fayad and Altman (2001).

Section V discusses knowledge map engine: initial work that consists of a number of interfaces, implementations, heuristics, and a hooking facility for context adaptation. The interfaces are the types that the knowledge map engine supports. These are enduring business theme; business object; industrial object; ExtensionPoints and existing software patterns, such as Gang of Four patterns (Gamma et al. 1995); Siemens group (Buschmann 1996); and others, and case study about knowledge map or stable pattern language for CRC cards and future work and conclusions. Section V contains three chapters.

Chapter 12, titled "Knowledge Map Engine: Initial Work," defines engines-supported interfaces and interface implementations, types, structures, engine construction heuristics, and the knowledge map engine's hooking facility. This chapter concludes with a brief summary and open research issues. This chapter also provides review questions, exercises, and projects.

Chapter 13 is titled "CRC Cards Knowledge Map," and it discusses the case study of a knowledge map for CRC cards that includes what makes an effective CRC cards; CRC cards knowledge classification, which includes its goals and capabilities; detailed approach toward a knowledge map for CRC cards; and the actual knowledge map for CRC card that includes a family of stable analysis and design patterns. This chapter concludes with a summary and some open research issues. This chapter also provides review questions, exercises, and projects.

Chapter 14 is titled "Future Work and Conclusions," and it discusses in brief any future work that can be carried out. This chapter concludes with a summary and also provides review questions and exercises.

12 Knowledge Map Engine
Initial Work

Real genius of moral insight is a motor which will start any engine.

Edmund Wilson

Knowledge map is a tangible and typical representation of the underlying concepts and relationships of a given set of knowledge. The catalog of knowledge is an efficient navigational or steering aid that assists a user to detect and seek the most desired concept or idea and then retrieve or cull out correct and relevant knowledge sources.

The significant goal of knowledge maps is driven by the special motto *divide and conquer*, which is applied here throughout the structure of knowledge maps. Knowledge maps are basically a system of dividing a domain into a number of different levels of fineness to enable easy management of domains. A knowledge map engine is the architecture of different patterns that are driven forward by the software stability concepts approach as espoused by Fayad (2002a, 2002b) and Fayad and Altman (2001) (http://www.kmglobe.com).

12.1 INTRODUCTION

The knowledge map engine specifically consists of a number of interfaces, implementations, heuristics, and a hooking facility for contextual adaptation. The interfaces are the types that the knowledge map engine supports. These are enduring business theme (EBT), business object (BO), industrial object (IO), ExtensionPoints, and existing software patterns, such as gang of four (GoF) patterns (Gamma et al. 1995), Siemens group (Buschmann 1996), and others. The implementations here are the classes that provide a concrete version of these types. Heuristics are the Java coding best practices that will be used to write effective Java codes, such as "prefer interfaces to declare types, avoid prefer static class members over private class members, provide a skeleton implementations of types using abstract classes, etc." (Bloch 2001). The hooking facility is a set of classes that allow us to adapt specific BOs to any application contexts by means of associating any type of IO with the BO.

12.2 INTERFACES SUPPORTED BY THE KNOWLEDGE MAP ENGINE

The main interfaces of this engine are EBT, BO, IO, and ExtensionPoint including existing software patterns, such as GoF design patterns. There are other interfaces, such as Analysis, Design, and Verifiable, which are used more with the engine's implementations and deployments. For this sole reason, we will not describe and explain them in this section. Figure 12.1 provides a model that represents these main interfaces.

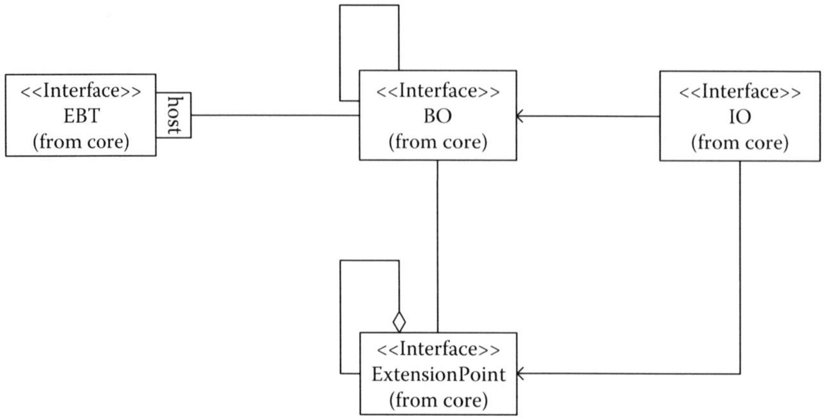

FIGURE 12.1 The engines' interfaces or types.

We must get familiar and acquainted with these types, because they are the core of the implementation of goals, capabilities, development scenarios (as well as their adaptation to new contexts), and deployment goals of the knowledge maps.

Each one of the types forming the core of this engine, in isolation, is very important and critical to the user; however, it is the synergy or harmony between them that makes the knowledge maps distinct from other approaches (see Chapters 2 through 6). The type EBT is the *what* of the system. It realizes the analysis concerns of a perused problem. This type is associated with a list of classes implementing the type BO, which is the type that represents the *how* of the system. Each BO can be associated with several EBTs; however, there is only one EBT that represents the ultimate goal of the BO. The type IO is a more tangible version of the type BO*; it represents the concrete application of the BOs. IOs are associated with the BOs by using associations, aggregations or compositions, and inheritance. These forms of association generation are handled by the ExtensionPoint type. The ExtensionPoint type is the one responsible for externally adapting a BO to a set of always-changing requirements (IOs) or application contexts at an on-demand basis. Examples of extension points are hooks and hot spots.

12.3 IMPLEMENTATIONS OF THE INTERFACES

In the previous section, we provided the interfaces that form the core of the knowledge map engine. Now, we will explain the basic implementations of these interfaces—abstract implementations. This technique is the *skeleton implementation* and it is described in Bloch (2001). In summary, this technique allows us to implement a default behavior for the supported types and provide the choice, for future specializations, to override the provided behavior. Figure 12.2 displays implementations of the interfaces.

The utilization or use of these default implementations, also known as skeleton implementations (Bloch 2001), allows us to change and transform the implemented behavior from

* This is not the generalization–specialization problem; see Chapters 4 and 6 for more information.

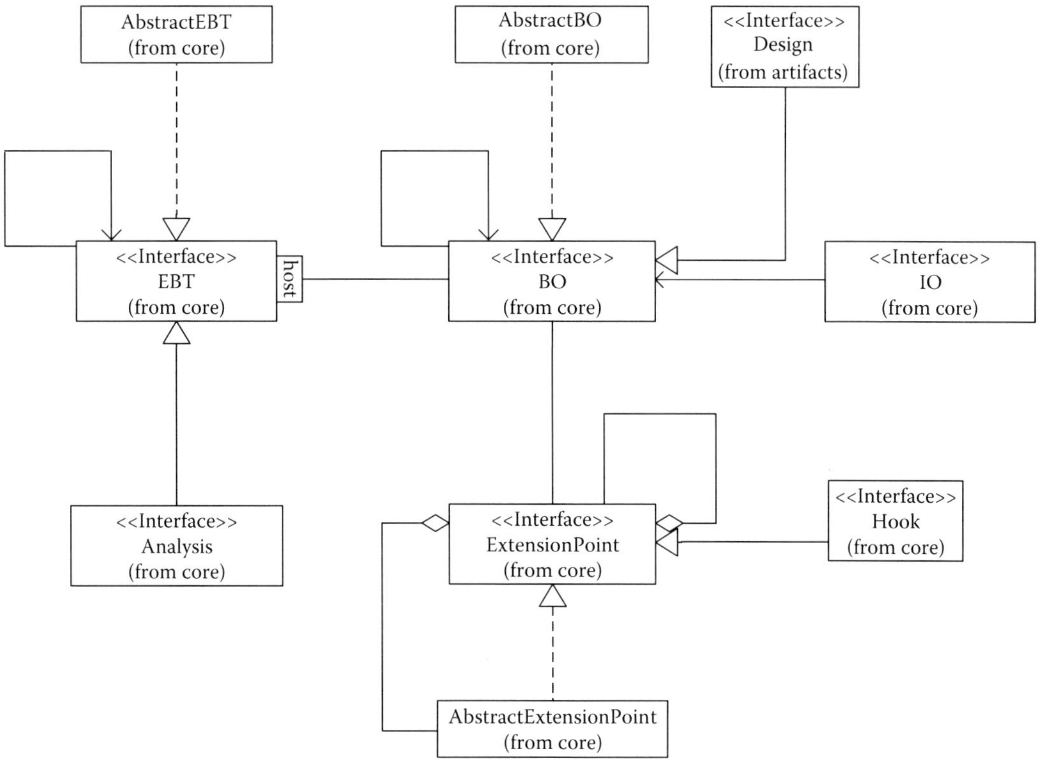

FIGURE 12.2 The engines' interfaces implementations.

the interfaces without affecting the rest of classes that are implementing a specific type and extending the skeleton implementation. Figure 12.3 shows a good example, where we show three elements of knowledge maps: goal, capability, and hook. We will also show how they implement, inherit, or aggregate the behavior of the EBT, BO, and ExtensionPoint types with a purpose of granting some sort of default functionality that will be used throughout the engine.

Table 12.1 shows a summary of the skeleton implementations of this engine's interfaces, as well as their concrete implementation.

Besides the implementation of this engine's main interfaces, a set of supporting classes complement the responsibilities of the main interfaces. These supporting classes, in partnership with the main interfaces and implementation, provide immense benefits to the utilization of the engine like, the following:

- The on-demand adaptation of BOs via hooks
- The loading of classes (IOs) not necessarily located in the same classpath as the application being developed
- Pre- and post-conditions and invariants assessment by using a *design by contract* facility

These classes are illustrated in the complete engine's class diagram (see Figure 12.4).

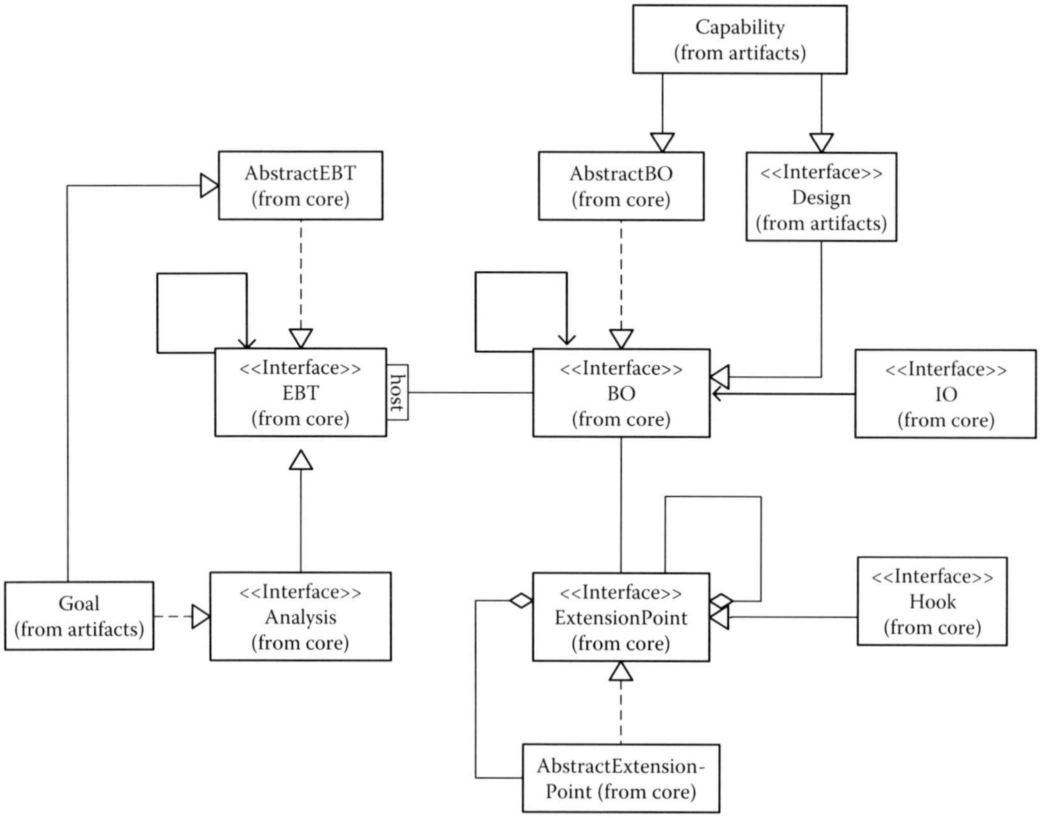

FIGURE 12.3 The engines' additional interfaces implementation.

TABLE 12.1
Summary of Interfaces Implementations

Types or Interfaces	Subtype	Skeleton Implementation	Concrete Implementation
EBT	Analysis	AbstractEBT	Goal
BO	Design	AbstractBO	Capability
IO	n/a	n/a	As many as necessary
ExtensionPoint	Hook	AbstractExtensionPoint	As many as required

12.4 STRUCTURES OF TYPES

Each one of the types mentioned here, as well as their skeleton implementations, defines a set of methods that realize their behavior or actions. Some of the methods that implement such behavior are given in Tables 12.2 through 12.6. For the sake of simplicity, we will provide only a summary of the methods of the main types. The rest of the methods, as well as the implementations' methods, are provided in Appendix B.

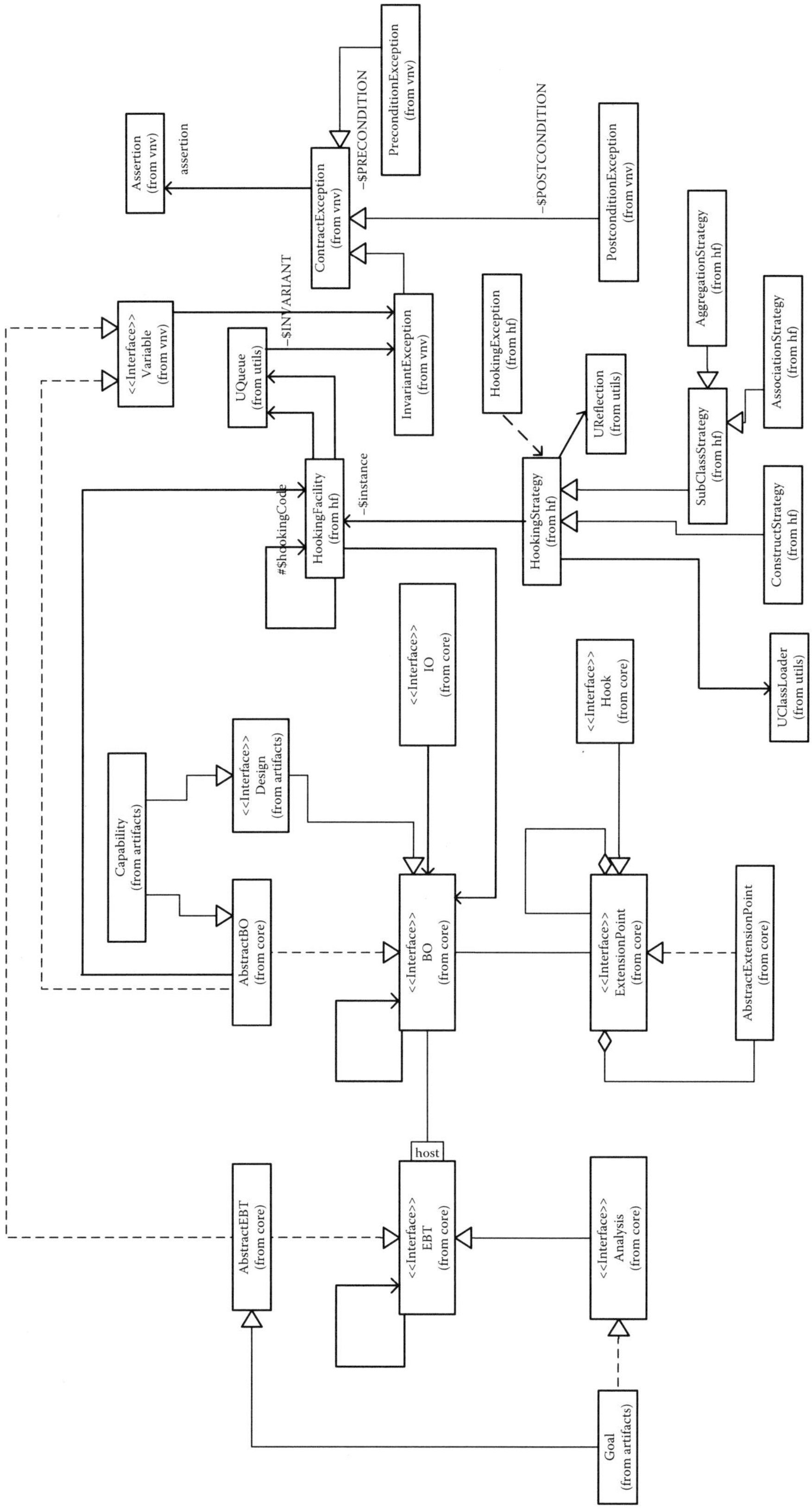

FIGURE 12.4 The engines complete class diagram.

TABLE 12.2
Structure of EBTs

EBT		
Method	**Version**	**Description**
introduces(): String	1.0	Gives a name to an EBT
releases(BO): boolean	1.0	Removes a BO from an EBT's structure
enlist(BO)	1.0	Add BOs to an EBT
divides(EBT)	1.0	Divides an EBT into a set of EBTs or subgoals
ignores(EBT): boolean	1.0	Ignores a subgoal from the set of the EBT's subgoal registry
....	....	.....

TABLE 12.3
Structure of BOs

BO		
Method	**Version**	**Description**
collaborates(EBT)	1.0	The BO is associated with an EBT
connects(BO)	1.0	A BO is connected to another BO
loads(String): BO	1.0	Loads an ExtensionPoint
disables(String)	1.0	Disables an ExtensionPoint
....	....	.....

TABLE 12.4
ExtensionPoint's Structure

ExtensionPoint		
Method	**Version**	**Description**
describes(): String	1.0	Gives the name of the ExtensionPoint
binds(BO)	1.0	Binds the ExtensionPoint to a BO
enables()	1.0	Enables an ExtensionPoint
disables()	1.0	Disables an ExtensionPoint
add(IO)	1.0	Adds new IOs to the ExtensionPoint
....	....	.....

TABLE 12.5
Structure of IOs

IO
IO represents a tag that the engine will use to recognize transient aspects or application-specific objects

TABLE 12.6
Classes of the Hooking Facility

Hooking Facility's Elements		
Class	**Version**	**Description**
HookingFacility	1.0	Controls all calls to the HookingStrategy class
HookingStrategy	1.0	Represents the parent of all the hooking strategies
ConstructStrategy	1.0	Constructs IO classes
SubClassStrategy	1.0	Creates SubClasses of a selected BO
AssociationStrategy	1.0	Creates associations between IOs and a selected BO
AggregationStrategy	1.0	Creates aggregations between IOs and a selected BO

12.5 CONSTRUCTION HEURISTICS OF ENGINES

The implementation of this engine was supported by a set of heuristics derived from best practices in Java development (Bloch 2001; Forman and Forman 2004; Simmons 2004; Stelting and Maassen 2001), as well as our prior Java programming experience. For simplicity, we have included some of the heuristics that we used in the implementation of this engine. This list is as follows:

1. The implementation of the types interfaces is unsynchronized to streamline the engine's performance. However, some actions in the types skeleton implementation are synchronized, such as the methods that deal with the singleton *coding* pattern.
2. Some types interfaces are driven by the concept of optional methods, so that if a type's implementation does not support a particular action, it will have to throw an UnsupportedOperationException.
3. Use interfaces to define types and classes to implement these types
4. Try to use static member classes over the nonstatic ones. In other words, if you are creating an inner class, always try to create it as a static class. This will allow members of the same sort (classes containing the inner class) to share the functionality provided by this static member class.
5. Favor protected constructors over private, when creating singletons. This will allow you to extend a class and prevent its instantiation from classes of a different sort. This is a good idea for creating extendable singletons.
6. When implementing methods, do not try to control the checking of parameters validity, within your method implementation. Instead, try use exceptions to check parameters validity, such as IllegalArgumentException and NullPointerException.
7. Disable assertions in production code; otherwise, it will degrade the performance of your developed application.
8. Always ensure that you are using Java's Reflection Application Program Interface (API) with extreme care. Even though there are so many benefits of using Java's Reflection API, such as the ability to ignore the source code of components and simply extract their interfaces, uncontrolled use of Java's Reflection API might actually degrade the performance of your developed solution.

12.6 THE HOOKING FACILITY OF THE KNOWLEDGE MAP ENGINE

Once we have obtained a fair idea of the types that are supported by this engine, along with their skeleton implementations and internal structure, we need to look now at how our engine allows both the external adaptation of BOs to ever-changing requirements and the loading of these adaptations into our solution, so that we are able to use them.

To understand how the BOs are externally adapted to new contexts, we may also need to take a closer look at the classes that realize the distinct hooking strategies in our engine. These classes are Hook (ExtensionPoint's subtype), HookStrategy, SubClassStrategy, ConstructStrategy, AssociationStrategy, and AggregationStrategy. Table 12.4 provides a description of the tasks performed by these classes.

An overview of the entire process to adapt externally, the BO of interest, by using the hooking facility as provided in Figure 12.5.

As one can note from Figure 12.5, the hook is attached to the BO of interest. If the user of the BO decides to adapt this BO to new contexts, he or she has three hook options to choose from—extend the BO to help specialize the BO's functionality, associate any IO to the BO of interest, and aggregate any IO to the BO of interest. The ConstructStrategy class is used only to create *dummy* classes (IOs without implementation) to be later implemented by a developer.

To adapt the BO to a new context, via the first option, the user needs to pass as input the BO to be adapted and the output name of the new class. Then, the hook selects that right type of strategy to create the BO's external adaptation. After that, the hooking facility will take this BO and extract implement the BO's constructors, fields, and the function

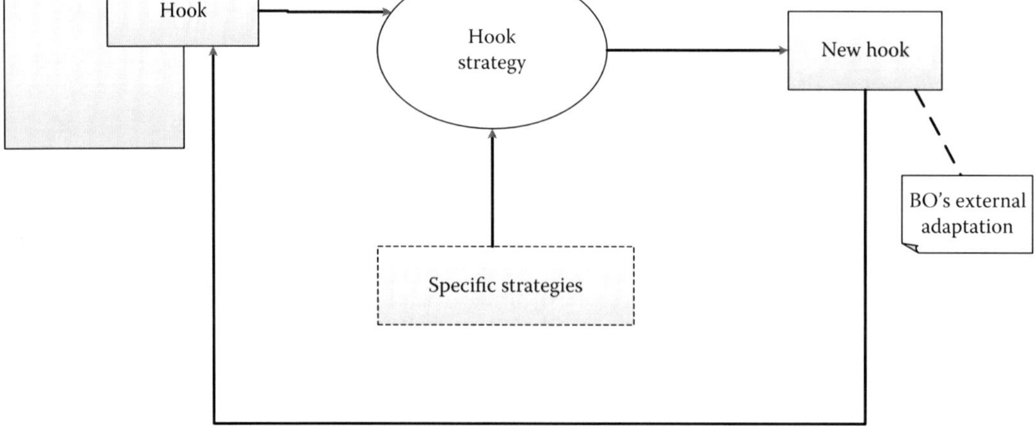

FIGURE 12.5 The hooking facility's process overview.

members using reflection. This is possible by using the Reflection API. In other words, instead of focusing on implementation details (source code) of the BO, we will focus on its available fields and methods signatures. Snippets 12.1 through 12.6 show the hooking method responsible for creating the BO's identity subclass.

The use of the UClassLoader.search(….) method allow us to dynamically load constantly modified classes into memory, that is, it loads and compiles the BO's hooks modified previously by the hooking facility, so that one can return them to the BO of interest. Snippet 12.2 shows a snippet of the URLClassLoader.

In the second option, the user provides as inputs the BO of interest, the IO to be associated with the BO, the association's multiplicity, and the name of the hook that needs generation. The process is quite similar to the previous option. The only difference now is that the hooking facility will also extract the fields and methods of the IO to be selected. Then, it will implement the respective methods that create an association between the IO and the resulting hook. Snippet 12.3 displays partial implementation of the method that uses the AssociationStrategy.

SNIPPET 12.1 CREATING A BO's SUBCLASS

```
public void adapt(BO cls, String output) {
Class newBO = UClassLoader.search ("net.hsanchez.km.repository."
                                    + output) ;
          setFirstInputObject (cls.getClass ()) ;
          setOutputClassName (output) ;
          setPackageName ("net.hsanchez.km.repository");
          Class nBO = new SubClassStrategy().createClass
               (this);
          if(nBO = = null)
          throw new HookingException("Unable to adapt the
               given BO..");
     }
```

SNIPPET 12.2 DYNAMICALLY LOADING CLASSES AND/OR MODIFIED CLASSES

```
.....
     //why? because we can modify the class and reload it again
     //without restarting the application
     ClassLoader cl = new URLClassLoader(urls);
     cls = cl.loadClass(name);
......
```

SNIPPET 12.3 ASSOCIATING IOs AND BOs

```
....
    this.setPackageName ("net.hsanchez.km.repository");
    this.setMultiplicity (multiplicity);
    Class nBO = new AssociationStrategy().createClass(this);
          if (nBO ! = null) {
                hook  = (BO)instantiates (nBO);
          }
....
```

SNIPPET 12.4 AGGREGATING AN IO TO A BO

```
....
    this.setPackageName ("repository");
    this.setMultiplicity (multiplicity);
    Class nBO = new AggregationStrategy().createClass(this);
          if (nBO ! = null) {
                hook = (BO)instantiates (nBO);
          }
....
```

The third option is similar to the previous option. The only difference is that it uses the AggregationStrategy. Snippet 12.4 shows partial implementation of the hook's method calling the AggregationStrategy.

The following section provides an example using the AnyLog BO and the Receipt IO. The goal here is to create an association between these two classes. Snippets 12.5 and 12.6 provide the input code and the resulted Java file.

12.6.1 An Example: Creating Associations between BOs and IOs

The following snippets illustrate how BOs and IOs will be associated by using the hooking facility. For the sake of simplicity, we are providing a rudimentary example.

SNIPPET 12.5 THE INPUT CODE

```
.....
          Hook hey = new Hook();
    AnyLog log = new AnyLog();
    Receipt receipt = new Receipt();
    hey.associate(log, receipt, "AnyLog");//the hooking
          facility will
    //add the postfix Hook to the output file's name
.....
```

SNIPPET 12.6 THE RESULTING HOOK

```
......
public class AnyLogHook
extends AnyLog
implements Hook{
//= = = = = = = = = = = = = = = F I E L D S = = = = = = = = = = =
private Receipt _receipt;
//= = = = = = = = = = = = C O N S T R U C T O R S ==== = = =
public AnyLogHook()
{
super();
_receipt = new Receipt();
}
//= = = = = = = = = = = = = = = M E T H O D S = = = = = = = = = =
public Receipt getReceipt(){
      return _chair;
}
public void setReceipt(final Receipt ipCls) {
if(ipCls = = null)
      throw new IllegalArgumentException("Argument must not be
            null.");
_receipt = ipCls;
}
//= = = = = = = = = = = N E S T E D C L A S S E S = = = = = =
}
......
```

One can use the generated file (a hook) within the AnyLog (BO) of interest to streamline its functionality, by using any functionality from the Receipt (IO).

SUMMARY

The main objectives of this chapter were to provide a concise view of the knowledge maps' engine and to show with examples the hooking facility structure. This first version of this engine attempts to facilitate a platform for developing the concerns (stable patterns) of the knowledge maps in a systematic manner. The main benefits of having a built-in hooking facility for the BOs to use are numerous. For instance, developers will reduce their efforts in adapting their software to new contexts; they will simply use and adapt the hooking facility and employ their BOs the way they wanted by means of extensions, associations, and aggregations of hooking strategies. Another benefit is that one of faster return on investment; developers will be able to accelerate their coding deliverables in lesser time.

OPEN RESEARCH ISSUES

The first prototype for the hook engine by providing easy hooking of IOs to BOs has already been developed based solely on stability model. The second version of this hook engine is currently being developed. This engine would be useful for novice users, as well as advanced developers to hook their IOs to the core knowledge map. The second version also includes more than 1000 existing software patterns, such as the GoF (Gamma et al. 1995) and Seimens Group patterns (Buschmann 1996). The second version of the hook would have many capabilities, such as extend, adapt, customize, configure, modify, and change any existing systems (see Sidebar 1.2).

REVIEW QUESTIONS

1. Describe the model that represents the knowledge map engine's interface.
2. Describe the technique *skeleton implementation*.
3. Describe the knowledge map engine's interfaces implementation.
4. Give the subtype, skeleton implementation, and concrete implementation for the following interfaces:
   a. EBT
   b. BO
   c. IO
   d. ExtensionPoint
5. What is the role of supporting classes in knowledge map engine?
6. Draw a class diagram for the knowledge map engine and describe it.
7. Provide a list of methods that implement the behavior of each of the types.
8. What do you mean by heuristics? State the heuristics for knowledge map engine.
9. BOs are externally adaptable to new contexts. Do you agree with this statement? Explain.
10. Name the classes that are used to realize distinct hooking strategies in the knowledge map engine.
11. Describe the responsibilities of each of the classes that provide hooking strategies for the knowledge map engine.
12. What options are available to the user to extend BOs?
13. Explain with the help of diagram the process to adapt BO externally.
14. Explain how to extend BOs by creating subclass of BO.
15. Explain how to extend BOs by associating IOs with BO.
16. Explain how to extend BOs by aggregating IOs with BO.

EXERCISES

1. Come up with an exhaustive list of EBT, BO, and ExtensionPoint's structure.
2. Create hooks for each of the following using any of the three methods described in this chapter for extending BOs:
   a. AnyActor and musician
   b. AnyCollection and stamp collection
   c. AnyEntity and laptop
   d. AnySkill and communication
   e. AnyPresentation and charts

PROJECTS

1. Design an engine that can generate hooks automatically for an application.

13 CRC Cards Knowledge Map

Quality is everyone's responsibility.

W. Edwards Deming

The class responsibility collaborator (CRC) cards are index cards that are utilized for mapping candidates classes in predefined design scenarios, for example, use case scenarios. The objective of CRC cards is to facilitate the design process, while insuring an active participation of involved designers. This chapter represents the first attempt toward a CRC card knowledge map or stable pattern language representation, via stable patterns, as a mean to discover, organize, and utilize CRC cards endured knowledge. Each stable pattern focuses on a distinctive activity and provides a way by which this activity can be conducted efficiently. The knowledge map or stable pattern language is a continuation of our early effort in improving the effectiveness of CRC cards and their role in the design process.

13.1 INTRODUCTION

The notion of CRC cards was first introduced in 1989 at the annual Object-Oriented Programming, Systems, Languages, and Applications conference. The acronym CRC stands for class, responsibilities, and collaboration, and, while they are not formally used in Unified Modeling Language, they can offer valuable insights during the early stages of development (Pressman 2001). They are primarily used as a brainstorming technique to rapidly and thoroughly explore design alternatives by identifying the classes and their associations within a system.

CRC cards are index cards utilized for mapping candidates classes in predefined design scenarios, for example, use case scenarios. They provide a simple alternative for a collaborative design environment, where analysts, designers, and developers try to simulate the system behavior, that is, role-play-driven approach. This process ends up with a set of collaborative classes represented by index cards, along with their roles, which are played by the members of the development team in a pre-animated design scenario (Biddle et al. 2009). Figure 13.1 shows the original CRC cards and illustrates in light gray color, current changes incurred over its original format.

Aside from its original purpose that was to teach programmers the object-oriented paradigm (OOP), CRC cards have been redefined to become valuable beyond the educational purpose. For their simplicity and flexible form, this tool can be applied to different domains' purposes, such as teaching OOP, a mean for not only documenting and identifying relevant classes of a system, serving as a methodology or as a front-end for other design methods, but also solving modeling problems, and engaging the entire development team through effective brainstorming sessions.

Some of those applications of CRC cards, as such, specify the underlying goals and expectations of CRC cards' use. The processes of accomplishing these goals, especially in software development, throughout iterative sessions are done at ad hoc, not precisely knowing when, how, and where to apply them to successfully accomplish the expected goals. This limitation

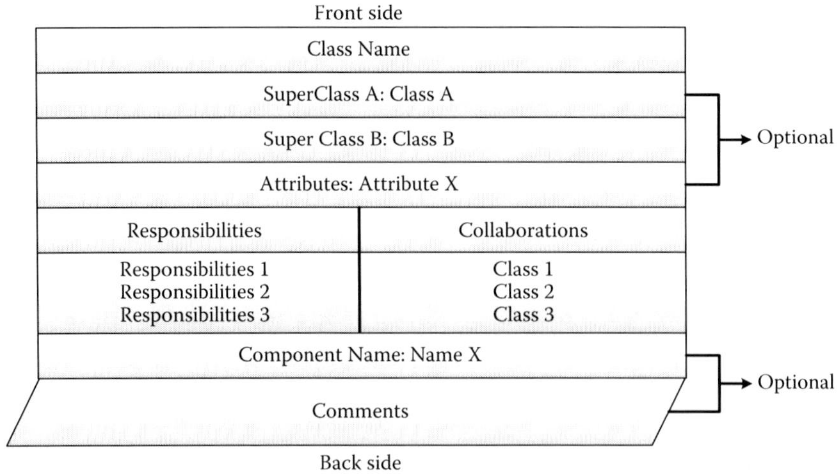

FIGURE 13.1 Original CRC card and current changes in its structure.

calls software practitioners for a set of suitable guidelines and stable knowledge to answer those former inquiries. This chapter illustrates the first stable pattern language for CRC cards and communicate this stable knowledge and suitable guidelines called knowledge map for CRC cards. This stable pattern language will enhance the way we commonly see and use CRC cards, turning it from simple index cards to a valuable knowledge repository. Our idea of this chapter is not to provide a specific approach on how to define and deploy a CRC card, but to provide a straightforward conceptualization and understanding of the CRC card's domain knowledge and fundamental processes to deploy one. To do this, we relied on the formed synergy of two methodologies: the software stability concepts paradigm (Fayad and Hamza 2003) and its subelements, stable analysis and design patterns (Fayad and Hamza 2003) and pattern language methodology (Fayad, Sanchez, and Goverdhan 2005). The reader, in this case, would truly visualize the distinct elements that composed the CRC card's domain knowledge and how they relate with each other to cope with a determined area of application.

The rest of the chapter is organized as follows. Section 13.2 provides an overview of the essential characteristics of effective CRC cards. Section 13.3 focuses on a bird's-eye description of the CRC card's implicit goals and capabilities. Section 13.4 discusses the rationale of the pattern language for CRC cards and the set of stable patterns involved in the CRC card usage. Section 13.5 provides the detailed description of CRC cards pattern language. Section 13.6 concludes this chapter.

13.2 WHAT MAKES EFFECTIVE CRC CARDS

CRC cards, regardless of which format is used (Halbleib 1999; Beck and Cunningham 1989), embody a particular set of characteristics that transcend across any context of applicability. Each of these characteristics possesses essential semantics that must be taken into consideration when applying CRC cards across domains. The focusing on those semantics, along with the utilization of CRC card quality factors, would be key factors in improving CRC card utilization within software development life cycle.

Regardless of structure's simplification, current CRC cards follow the original structure and share same basic elements (i.e., class name, responsibility, and collaborators).

Nevertheless, it may include more elements, if necessary (Halbleib 1999; Beck and Cunningham 1989). In the long run, these current structures might not always attain the essential aspects needed in future development stages (i.e., building system class diagrams). Therefore, for CRC cards to aid system development, main essential quality factors need to be satisfied. The fulfillment of these quality factors will have a high impact on the CRC card's characteristics realization. These quality factors are provided herein:

1. *High level of understandability.* Illustrate its sections in an orderly and specific manner, showing an efficient distribution of its elements.
2. *Accurate identification of class elements.* Assure proper *identification* of artifact/class and its elements. This will prevent any confusion during class assignation.
3. *Well-defined role.* Include a well-defined role, within context, for the artifact/class being developed. This role has to strongly reference the artifact's assigned responsibility. Each class may have multiple roles according to a specific simulated design scenario (Humans, for example).
4. *One cohesive responsibility per class.* Assign a unique, cohesive responsibility, within context, to each class. This responsibility must match the class's defined role. Avoiding overlapping/redundant responsibilities will prevent complex class interactions, when applying them in design scenarios.
5. *Self-descriptive services.* Provide a descriptive, straightforward definition of the services per class. These services will sustain a strong correlation with the class's responsibility. Otherwise, it would be difficult to know which services to invoke to fulfill an artifact's specific job.
6. *Explicit notion of the external collaborators.* Identify the class's collaborators.

A class needs to know which artifacts/classes are its collaborators for the achievement of responsibilities.

Based on the effective CRC card format, described in Fayad et al. (2003), we provide Table 13.1 listing a summary of CRC cards relevant characteristics.

13.3 CRC CARDS KNOWLEDGE CLASSIFICATION

Our primary focus is on the higher level patterns that conform to the underlying goals found in CRC card utilization: Patterns that help us to develop more understanding of CRC cards and their effective utilization.

It is worthwhile to mention that with the product of the association of higher level patterns (goals) and semitangible patterns (capabilities), we will be able to foresee a generic skeleton for myriad of development scenarios of CRC cards. Figure 13.2 illustrates a general view of the CRC card knowledge stratification in relation to its goals and capabilities. However, the domain knowledge from which CRC cards are made up may be enormous; therefore, for simplicity purpose we include few of them.

In the following section, we would provide a more specific view of the distinct elements involved in the CRC card domain knowledge, by the means of using and describing a map representation. This map representation focuses on the realization of the dissimilar artifacts, quality factors, and how they are associated with each other within the CRC cards domain boundary. It is worthwhile to mention that each element or artifact represented in this map would come to represent a stable pattern.

TABLE 13.1
Explicit Characteristics of CRC Cards and Solutions

Characteristic	Solution
Portable	No computers are required, they can be used anywhere, from the tranquility of your home to a very important meeting.
Reviewable	You can go back and review these index cards anytime after a long period of time without being concern of information deterioration.
Simple	It possesses a simple structure and is easy to read, learn, and understand by any person without a previous experience on CRC cards.
Multipurpose	Due to its simplicity and its portability, CRC cards may be utilized in different application domains, for example, education, software analysis, and design, as a teaching technique.
Accessable	The set of CRC index cards are highly available during the sharing decision process done by analysts, designers, and developers in the early stages of the software development phase.
Implementable	From the CRC card blueprint to its implementation, there is a short path to follow. Due to well-defined classes within the self-described structure of CRC cards, developers are able to implement these cards (classes) with ease.
Traceable	These cards can be traced throughout the entire specification of animated scenarios by the explicit exhibition of the classes and their different roles and behavior to which it represents.
Mapping ability	These index cards represent an exact match and definition of a class, and its elements, in a particular design scenario.
Reusable	After a project is completed, this does not mean that we cannot use our already defined index cards in another project. The classes were defined with a stable and reusable core in mind; therefore, they may be utilized within several applications that share the same domain knowledge. Think about this as a piece of knowledge that may be shared in other contexts of applicability that possesses similar rationale and capabilities, and hence, the effort in coming up with a stable system design would be reduced, and quality enhanced.

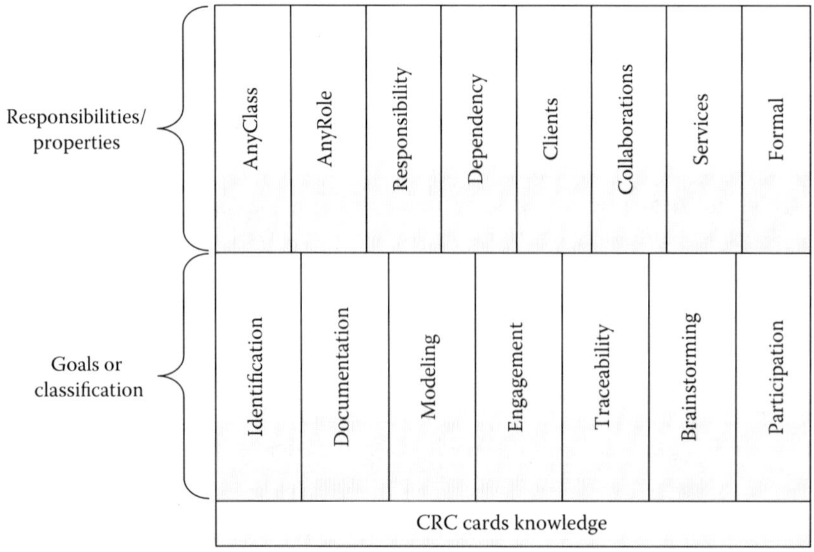

FIGURE 13.2 CRC card knowledge classification.

13.4 TOWARD A KNOWLEDGE MAP FOR CRC CARDS

CRC cards' goals and capabilities embody a set of related stable patterns that build CRC cards rationale and usage over a myriad of application contexts. When related stable patterns are interlaced together, they form a family of patterns that will cover multiple domains; this family of patterns is called pattern language.

The objective of the overall pattern language is to cover the essential aspects related to the process of conceiving, understanding, writing, and utilizing CRC cards. Concretely speaking, this pattern language will come to surface the behind-the-scenes endured knowledge of CRC cards. The process of defining Fayad's CRC cards endured knowledge involves four main steps, which can be perceived as the basis for our pattern categorization: (1) *goals or classification*, (2) *capabilities/properties of CRC cards*, (3) *development or scenario development of CRC cards*, and finally (4) *deployment of CRC cards across multiple domains.* The outcome of these main steps will be a set of interrelated patterns that interact together to serve a particular purpose, within CRC cards usage and rationale. As a whole, this pattern language will embody the core insights as a set of stable patterns and their interactions among them.

If we delve into the pattern language definition that is presented in Coplien and Schmidt (1995), our stable pattern language for CRC cards is far from being just a decision tree of patterns. On the contrary, it is a network of patterns (not hierarchy of patterns), where each generated interaction among its set of patterns come to represent a distinctive route or path serving a particular purpose or goal. Therefore, the number of distinctive routes that can be orderly navigated to satisfy distinctive purposes can be a very large number.

Before getting started with the family of patterns, let us go through the number of patterns that would be presented in this chapter, along with the ones that will be included in future versions. Also, let us mention the undertaken methodology and how patterns will be allocated accordingly to their target purpose via a knowledge map. Currently, our pattern language for CRC cards proposes 22 patterns covering CRC cards' domain knowledge; however, only four patterns are documented in this chapter. More patterns will be added or documented in the future version of this chapter. Additionally, as stated above, these patterns would be discovered and organized by means of applying the aforementioned four main steps, each one of them addressing a particular objective within the definition of our CRC cards language.

13.4.1 Goals or Classification

This step is concerned with surfacing the implicit goals hidden within the CRC cards core knowledge. This process requires the capture and full understanding of the context where our solution would be laid down. That includes describing the goals not from its tangible side, but focusing more on its conceptual side. In Fayad, Sanchez, and Goverdhan 2005, they are named enduring business themes (EBTs). Examples of the resulted patterns represented within this main step are documentation, identification, and brainstorming goals.

13.4.2 Capabilities or Properties

The second step concentrates on the discovery of those recipes that are required to fulfill the stated goals and purposes of the CRC cards. Without those concepts or stable patterns, there will do a vague understand (almost none) on how these goals will be achieved.

These stable patterns are known in Fayad, Sanchez, and Goverdhan 2005 as business objects (BOs). For instance, within our language for CRC cards, we have AnyClass, AnyRole, Responsibility, Services, and so on.

13.4.3 Development Scenarios

The third step relates to the following: (1) the myriad of development scenarios, where the CRC cards can be involved. These development scenarios are realized through the distinct routes or paths taken due to the interactions among the involved patterns (EBTs associated with BOs). The product of this association is known in Fayad, Sanchez, and Goverdhan 2005 as architectural patterns. Each one of the complete routes taken will represent a distinct application domain or context where the CRC cards would be used. For instance, teaching scenario and groupware and (2) how the CRC cards language would be implemented across dissimilar domains, based upon the utilization of tangible artifacts that would confirm the domain-specific patterns. These patterns are known in Fayad, Sanchez, and Goverdhan 2005 as industrial objects (IOs). An example of these patterns would be bookkeeping.

13.4.4 Deployment

The last step, as its name states, deals not only with how the CRC cards knowledge would be deployed across different application domains, but also with the representation of the artifacts or patterns that will aid the actual deployment process. For instance, we have the blueprint pattern.

The rationale of discovering and stratifying our pattern language, by means of using a systematic approach, which involves four main steps and ends up with four categories, is to facilitate the findings, execution order, and description of the stable patterns embodying the concepts or building blocks of the CRC cards domain.

Figure 13.3 depicts the overall pattern language structure. In the given figure, the four important steps are presented in light gray boxes with circle inside. The light gray boxes represent the generic aspects/recipes or stable patterns that are related to those steps. For instance, the first step relates to the analysis of the domain that is looking for hidden goals weaving the domain under discussion. Each aspect is then interconnected with other set of patterns through the CRC card pattern language.

In summary, the routes taken, when defining a new architecture would provide us the established road map to fulfill particular goals expected from the CRC cards. The outcome of these interconnections represents self-supported aspects of the generated framework that will define the order of employing CRC cards in dissimilar domains.

13.4.5 Family of Patterns—Bird's-Eye View

The proposed family of patterns, as stated above, contains nearly 22 patterns. These patterns as a whole describe the basis for defining, understanding, deploying, and embodying a stable knowledge of the CRC card across domains. Table 13.2 references in a concise way the artifacts included in this patterns language. For simplicity purpose, only the patterns provided in this chapter would be described. Future versions of this chapter will include more related patterns to the CRC cards domain knowledge and these patterns' description.

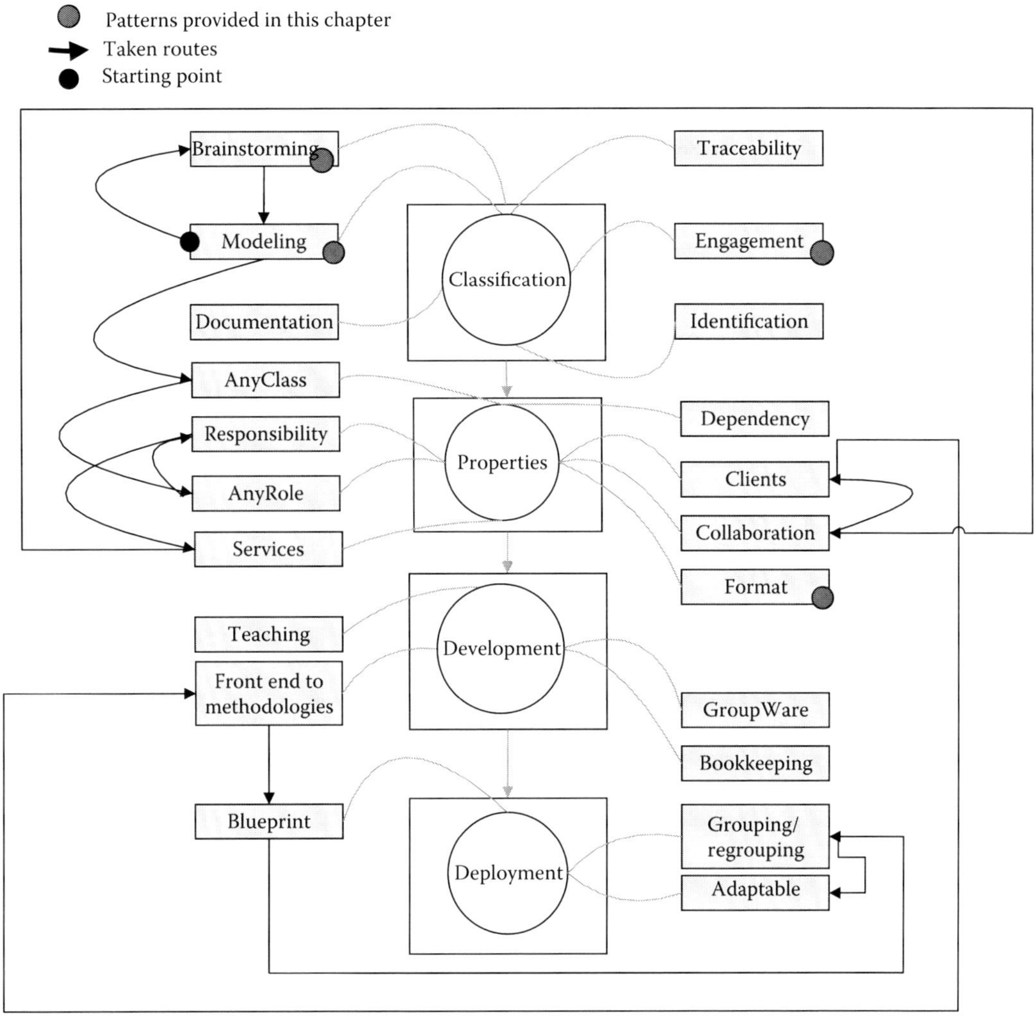

FIGURE 13.3 A pattern language for CRC cards—achieving the modeling goal scenario.

13.5 KNOWLEDGE MAP FOR CRC CARDS

A pattern language is not a formal language, rather a family of interrelated patterns organized in such a way that it facilitates a vocabulary that guides their application when solving standard problems (Schmidt, Fayad, and Johnson 1996). In our case, it would be a pattern language for communicating the underlying knowledge of CRC cards' conception, understanding, and application.

In this section, we will also describe only those patterns that are marked as filled or a solid circle in the knowledge map shown in Figure 13.3. The rest of the patterns would be described in the future versions of this chapter.

13.5.1 The Main Step in the Classification

This step relates to surfacing the implicit goals hidden within the CRC cards' core knowledge. These goals form the basis for stable analysis patterns representation.

TABLE 13.2
Summary of the Pattern Language of the CRC Cards

Category	Pattern	Problem	Solution
Goals or classification	Documentation		
	Brainstorming	Current implementations of the brainstorming process are bound to a specific problem domain.	Brainstorming stable analysis pattern
	Engagement	We are always concerned about the quality of the involvement between participants involved in a particular activity, when interacting with each other.	Engagement stable analysis pattern
	Traceability		
	Identification		
	Modeling	The actual problems range from the overloaded generation of too many responsibilities per class, to the lack of specific class roles, which defined the position of a class in a preanimated scenario in accordance with certain responsibility.	Modeling stable analysis pattern
	Participation		
Capabilities properties	Any Class		
	Any Role		
	Responsibility		
	Dependency		
	Clients		
	Collaboration		
	Service		
	Format	Current CRC cards lack some essential qualities that might affect the effectiveness of the developed system that uses them.	Effective format pattern
Development	Teaching		
	Front-end to methodologies		
	Groupware		
	Bookkeeping		
Deployment	Blueprint		
	Adaptable		
	Grouping/regrouping		

In this chapter, we present three stable analysis patterns: brainstorming, modeling, and engagement stable analysis patterns.

13.5.2 Pattern 1—Brainstorming Stable Analysis Pattern

Brainstorming is an informal way of generating ideas or solutions to write about, or points to generate a particular solution based on some engaging activities.

13.5.2.1 Context

Brainstorming is an informal way of generating ideas or solutions to write about, or points to generate a particular solution. It can be done at any time during the writing process or during certain interactions in particular meetings. In the case of CRC cards, brainstorming process is intended to generate ideas with respect to the simulation of system behavior. Such process encloses several benefits, such as the treatment of several argument issues, the achievement of agreement, engaging a particular group of people in the process of identifying candidate classes, their responsibility, and collaborations according a distinctive context or predefined scenarios through the utilization of heterogeneous media. This technique is open to a *freestyle* generation of ideas and is not a technique for idea evaluation. For instance, in a CRC card session, the modeling team, aided by the use of index cards, will try to identify and understand the requirements for the application they are building. However, because there is no idea-evaluation technique, the application of this technique brings to surface some trade-offs, such as a high uncertainty whether the generated idea is correct or not. Nevertheless, it sure helps practitioners to explore a vast set of possible solutions that may reflect the correct system behavior.

13.5.2.2 Problem

Brainstorming process can be done at any time, individually or collectively; it can target one particular subject or multiple ones; it can be limited to one particular context or contexts; it can be done synchronously and asynchronously. Currently, brainstorming solutions are limited to cope with one context at a time, not allowing certain grade of flexibility to allow practitioners expand their target context scope(s) of determined problem of discourse. Additionally, current brainstorming solutions strive, when practitioners deal with several problem domains at the same time, especially when switching from different contexts back and forth, in a random order, and trying to keep certain topic's discussion state, where it was previously left by practitioners. Therefore, current brainstorming solutions are unsuitable for and across problem domain application.

In the case of CRC cards use and understanding, brainstorming process seems to be bound to the ability of a moderator to engage practitioners in collaborative environment and the willingness of practitioners to truly be part of the session. For such a case, the process of engaging practitioners is detached from the brainstorming process itself. Bringing as a consequence, a poor running session, where the session-generated results would have higher degree of failure due to the fragile link of attention experienced between the practitioners running the brainstorming process with respect to a topic of discourse.

13.5.2.3 Solution

The following model will represent the proposed solution of the brainstorming analysis pattern, using the software stability concepts approach (see Figure 13.4).

Participants.

The participants of the brainstorming analysis pattern are as follows:

Classes

Brainstorming. Represents the brainstorming process itself. This class contains the characteristics and behavior that initialize the brainstorming process.

Engagement. See engagement analysis pattern.

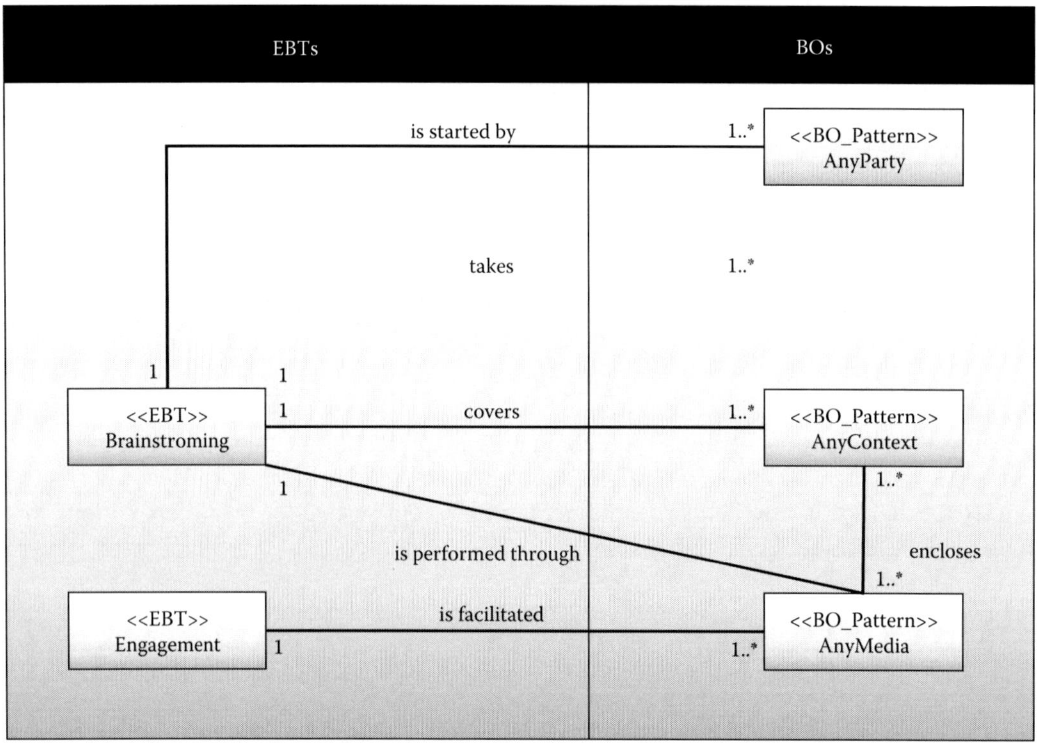

FIGURE 13.4 Brainstorming analysis pattern—stable object model.

Patterns

AnyMedia. Represents the media through which the brainstorming process will take place. For instance, one can brainstorm the candidate classes of the system being developed through the utilization of CRC cards and use case scenarios, by utilizing the role-play brainstorming process. Others might use simply the CRC cards, in a model, to represent all the candidate classes and then brainstorm how these classes interact with each other in a high-level representation of the system.

AnyParty. Represents the brainstorming inducers or practitioners. It models all the parties that are involved in the brainstorming process, including the facilitator, who is the person that rules the entire brainstorming process. Party can be a person, organization, or a group with specific orientation and organization.

AnyForm. Represents the forms and how the brainstorming process can be performed. It models all the forms that may be employed when carrying out a brainstorming process.

Form can be a CRC card modeling, a role-playing and use case form, or writing brainstorming form (Durfee 1998).

AnyContext. Represents the brainstorming topics, problem, or subjects to be brainstormed with. It models all the contexts that may be covered to generate possible solutions by using a brainstorming process. It includes the scope or boundaries of the context and the specific points to be discussed. AnyContext can be class identification step, responsibility identification, or just writing an action novel for a local magazine.

13.5.2.4 Example

In order to illustrate the use of the brainstorming pattern in different application areas, one example is presented: CRC cards are one of the methods, where brainstorming is present, by interacting in a role-playing and use case scenarios to discover, with users, the real-world objects that make up a system.

They are meant to assist in mapping the collaborations among classes. Because the purpose of this example is to demonstrate the usage of the proposed pattern, and for simplicity, this example does not present the complete model for the problem. Instead, they focus on the part that involves the brainstorming process.

13.5.2.4.1 Example 1: Role-Playing and Use Case Scenario Brainstorming

Role-playing and use case scenario brainstorming process requires a list of predefined scenarios, a group of CRC cards, and the practitioners of the brainstorming session (analyst, designers, developers), including the leader or facilitator of the session. However, it is almost impossible to make all the practitioners participate openly in those types of sessions, because they are afraid to give a bad idea or opinion. Here is where the facilitator has to come to shine, he needs to bring all the practitioners into a complete engagement by distributing the CRC cards to the team members, so that team members *play* one of the classes (preferably not so that possible collaborating classes are assigned to the same individual). This example models a simple solution to interact in an interesting brainstorming process called role-playing and use case scenario and generates certain points or ideas that refer to the possible candidate classes, their responsibility, and collaborations at a specific scenario. Figure 13.5 shows the stability model of the brainstorming used in *role-playing* and *use case scenario*. Classes that are not in the original *brainstorming* pattern are colored in gray.

13.5.3 Pattern 2—Engagement Stable Analysis Pattern

Engagement represents the process for candidates (participant) to meaningfully involve in a particular activity(s), through interaction with others and worthwhile activities. Such engagement could be accomplished without the use of technology; however, technology may facilitate engagement in ways, which are difficult to achieve otherwise (Kearsley and Shneiderman 1998).

13.5.3.1 Context

By engagement process, we meant the sense of concern with and curiosity about a particular activity, in which all participants are immersed in an iterative environment, based on their level of commitment and disposition regarding a particular set of tasks.

Engagement concern is based upon the idea of creating a collaborative environment for its participants, where the act of sharing activities and knowledge is increased.

The context may be summarized based on the following elements:

1. Collaborative teams or participants
2. Participants' proficiency in certain skills
3. Strong commitment toward the activity to be performed
4. Be truly involved in particular activity
5. Activities within scope

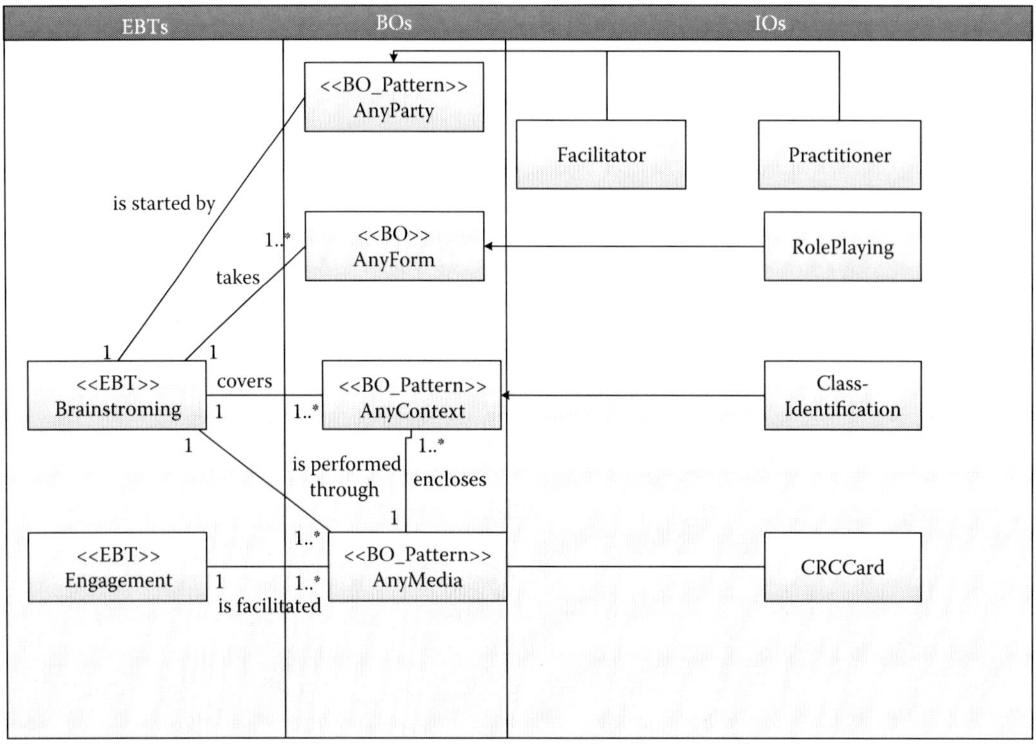

FIGURE 13.5 Brainstorming analysis pattern—stable object model applicability.

13.5.3.2 Problem

We are always concerned about the quality of the involvement between participants who are involved in a particular activity, when interacting with each other. Those concerns become more noticeable from time to time. This is especially so, when we are trying to engage participants with different level of proficiency toward a particular task. A number of barriers interfere with such an *effort*. Many involve a lack of commitment or interest from participants. These, in turn, may delay or halt the completion of the assigned set of tasks. Other obstacles relate to the process of defining the context boundaries of these activities. Therefore, accomplishing engagements may be hindered, when proficiency on particular skills and activity's context are not in consensus (not related).

13.5.3.3 Solution

Figure 13.6 represents an abstract representation of the solution that deals with the engagement concept.

13.5.3.4 Example: Conceptual Map Creation

In order to illustrate the use of the engagement pattern in different application areas, one example is presented: the creation of a conceptual map (Armitage and Cameron 2006). Because the purpose of this example is to demonstrate the use of a proposed pattern, and for simplicity, this example does not present the complete model for the problem. Instead, they focus on the part that involves the engagement process as shown in Figure 13.7.

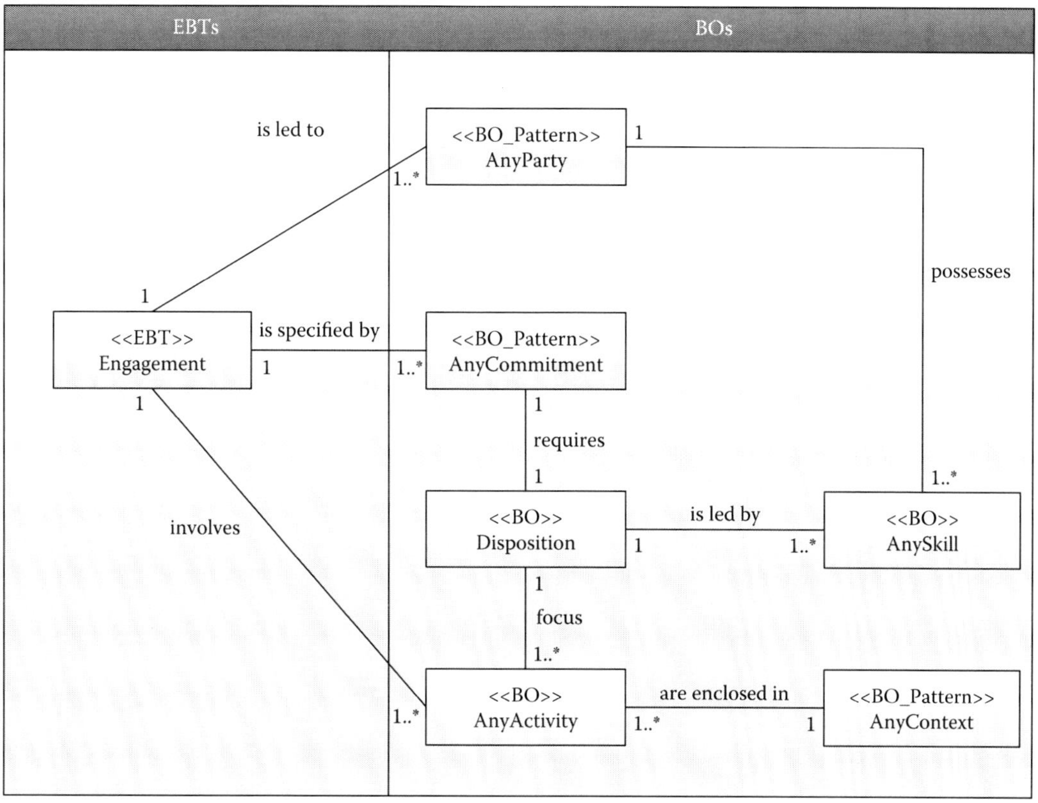

FIGURE 13.6 Engagement analysis pattern—stable object model.

13.5.3.4.1 Example 1: Creating a Conceptual Map to Predict a Constructive Engagement

The creation of a conceptual map (Armitage) should involve reshaping, adding personal links, keywords, and conceptual areas. Note that these activities are considered, when undertaken by a single user. Therefore, a user must be familiar with the semantics of creating conceptual maps. Only activities associated with the use of the navigation aid are considered, rather than activities associated with other aspects of the hypermedia content.

13.5.4 Pattern 3—CRC Card Modeling Stable Analysis Pattern

The main intent is to represent a collection of index cards and their interconnection with each other, to provide a big picture of the system functionality.

13.5.4.1 Context

A CRC card model is a low-tech method that implies a collection of index cards arranged in a client/server order to visualize how the system, being developed, will function. We also mean by *client/server order* as bringing interlocked classes or classes collaborating with each other closer in the model (having one or more server classes with their clients). This session is an interactive process, which brings together analysts, designers, developers, and/or any other professional that accedes to participate in this pre-animated exploration of the

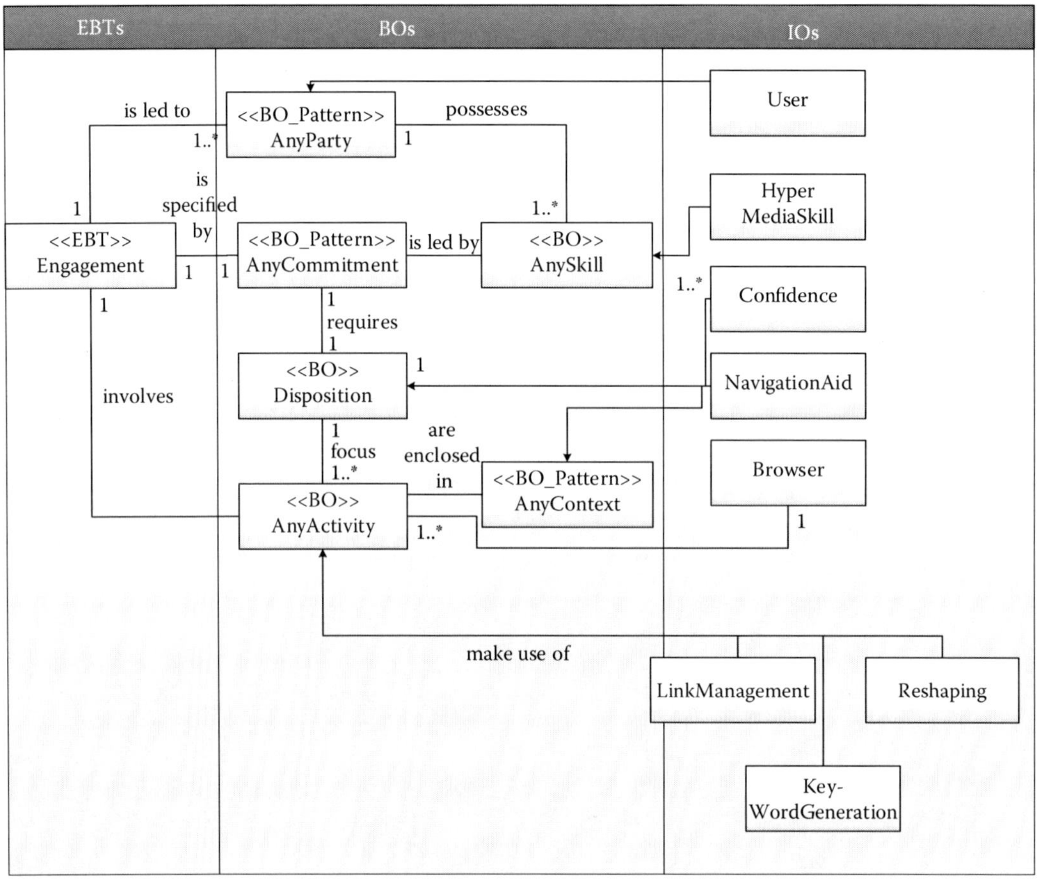

FIGURE 13.7 Stable conceptual map creation.

understanding and identification of the business requirements. As a low-tech, simple, yet powerful method for object-oriented analysis, this technique provides us great opportunities for a better understanding of the system throughout its entire development cycle through a responsibility-driven analysis.

Usually, this CRC card model is created by a group of domain experts, who are led by the CRC card session moderator or also known as *facilitator* (Ambler 1998). This moderator is in charge of organizing and performing the CRC card session, providing clear information and background to the participants about the CRC card techniques and how it should be performed. This moderator is usually assisted by one or two scribes who are in charge or recording the logical perspective on how to fulfill the business requirements product of a constant analysis of the pre-animated scenarios executed in the CRC card session.

13.5.4.2 Problem

During the execution of the CRC modeling session, domain experts initiate a looping process in which, based on brainstorming and several other methodical processes (Fayad et al.), all the candidate classes, their responsibilities, and collaborations are identified, and then, start filling their respective index cards, creating use case scenarios, and arranging these index card on certain table. However, current CRC card formats are bound to enclose limited information about the classes being defined. These actual problems range from the overloaded generation of too many responsibilities per class to the lack of specific class roles, which defined the position of a class in a pre-animated scenario in accordance with certain responsibility (Fayad et al.). These inconsistencies found on current CRC cards formats will restrain the fluency with which the execution of the CRC modeling session will be held, because of the increment in the number of the times a process will be carried out to assure an accurate definition of a class or an accurate match to the business requirements. Our problem consists of many major issues: the first one is how to express the process of running the CRC modeling technique by using the quality factors (see Section 13.2) already built in the proposed CRC card format to better understand the business requirements, and the second one is how to accurately identify the artifacts that would be modeled that include accurately defining the artifacts' enclosed elements. So in the end, practitioners would feel confident about the experience of understanding the system behavior that is being simulated during a pre-animated session.

13.5.4.3 Forces

- Before starting a CRC card session, the practitioners must have a clear understanding on how to perform this session. Usually, this is done by the facilitator, prior to the beginning of the session. Therefore, some sort of workflow may be useful for practitioners to enhance their understanding of the process.
- How to verify the identified artifacts or candidate classes, along with their enclosed elements, are accurate and later express the expected behavior from the system being simulated when being modeled as whole.
- How to include the CRC card quality factors into this workflow? Common workflow for CRC card session might lack this element. Therefore, based on the proposed CRC card format (Fayad et al.), those quality factors must be added to the defined process when running a CRC card session.

13.5.4.4 Solution

For CRC card modeling, there are six steps that need to be followed (Ambler 1998), which are as follows:

1. Put together the CRC modeling team
2. Organize the modeling room
3. Do some brainstorming
4. Explain the CRC modeling technique
5. Iteratively perform the steps of CRC card modeling
6. Perform use case scenario testing

Throughout the process, these steps should be constantly supervised by the CRC card moderator to make sure that the session is efficiently performed. In some cases, this moderator would allow the entrance of certain observers to the session. Typically, this clearance is for training purposes. These observers, however, cannot actively participate in the CRC card session.

At the time of identifying candidate classes, their responsibilities, and their collaborators, and then turn them into a CRC card, we will make use of a new look at the CRC card format that is described in Fayad et al. and the methodical process to fill it. This new format has a lot more to offer than the currently used formats (Ambler 2001; Maciaszek 2001; Pressman

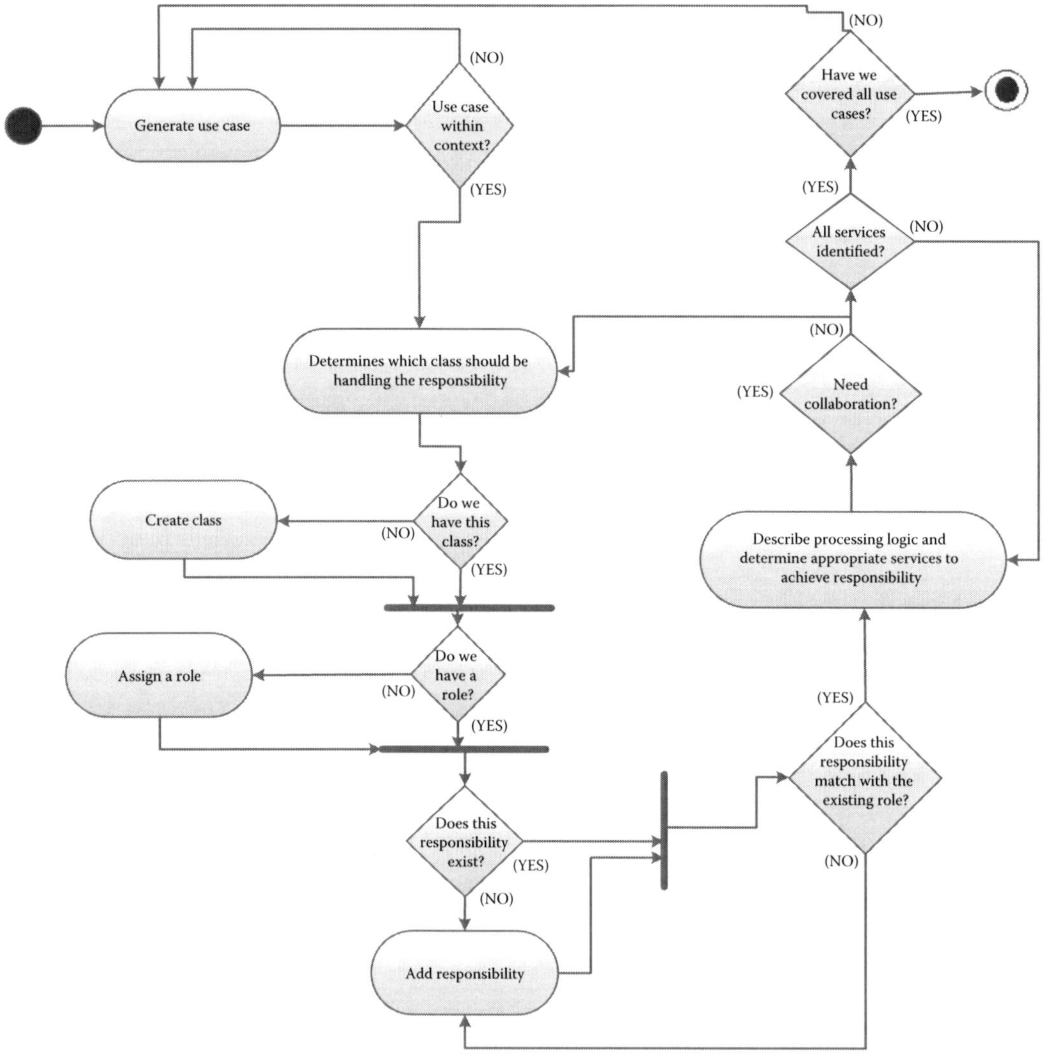

FIGURE 13.8 CRC modeling process.

2001), due to the avoidance of all the problems described in Section 13.5.2, and provides a clear blueprint of their respective candidate classes.

The following process flow, shown in Figure 13.8, aims to represent the appropriate steps to model candidate classes, previously identified using some brainstorming and methodical process (Fayad et al.), embodied as CRC cards, to see if they match the business requirements of the entire system. This new process flow is an adaptation of the process flow described in Ambler, but solely focusing on the new CRC card format proposed in Fayad et al., avoiding, for example, the inclusion of many responsibilities per class and the appearance of distinctive role per class. For more information about this new CRC card format, please refer to Fayad et al.

It is important to note that the success of this modeling process relies on how good and accurate your model mirrors the problem domain from where your application is being developed.

13.5.5 The Main Step in the Capability

This step emphasizes the discovery on those recipes that are required to fulfill the stated goals and purposes of the CRC cards. These recipes are embodied as stable design patterns. In this chapter, we will present only one stable design pattern: effective CRC card format pattern.

13.5.6 Pattern 4—Effective CRC Card Format Pattern

13.5.6.1 Context

The CRC cards technique can be used in either practical software development or in object-oriented (OO) education. In short, it embodies a higher grade of reusability, proportional to the number of problems it modes. For instance, in development, CRC cards can offer valuable insights during the early stages of development. Another benefit of CRC cards is in teaching object concepts in programming language courses. Several reported case studies have demonstrated the effectiveness of CRC card as a tool for introducing OO programming concepts and for improving the understanding of objects/classes (Börstler 2002).

13.5.6.2 Problem

Current CRC cards lack some essential qualities that might affect the effectiveness of the developed system that uses them. The main problems in current CRC cards can be summarized in the following points:

1. *Possibility of low cohesion and high coupling.* Because there are no limitations on the number of responsibilities allowed for a given class, it is possible to overload a class with too many responsibilities eventually resulting in low cohesion within the model. A cohesive class should ideally have just one responsibility (Fayad 2000–2003; Fayad et al.). Excessive responsibilities could also lead to a large number of collaborators required to support them.

   Too much collaboration between classes will also produce high coupling and they needlessly increase the complexity of the system. In software design, we strive for just the opposite—high cohesion and the lowest possible coupling.
2. *Macho classes.* Multiple responsibilities can also result in the creation of macho classes. Macho classes will instantiate an object that performs most of the work, leaving all minor operations to a set of essentially useless classes (Fayad 2000–2003). Ideally, the system intelligence should be distributed as evenly as possible across the application and the work shared uniformly. When all of the intelligence is concentrated in one or two classes, it also increases the difficulty of making changes.
3. *Exclusion of services.* Another problem with these CRC cards is the exclusion of the services provided by the class (Fayad, Hamza, and Sanchez 2003; Fayad et al.). By including the services on the CRC card, the classes can be checked for duplicate functionality (Fayad 2000–2003). By identifying duplicate functionality, it may be possible to combine or consolidate classes that perform similar functions. In addition, because the responsibility of a class is merely a summary of its operations, explicitly providing the services performed by a class may help verify that the responsibility is properly defined.

4. *No clear role is defined.* The absence of a class role may lead to assigning wrong, useless, or even missing responsibilities. Although the role seems fairly insignificant, it serves a very important purpose. If a class performs more than one role, it is possible that a generalization exists, where each role is actually a subclass of some superclass (Fayad 2000–2003). Humans provide a good example of performing multiple roles; a woman could be a mother, a wife, a daughter, and so on. By defining the role, generalizations and specializations can be explored early in the process.
5. *Difficulty in defining responsibilities.* Coming up with class responsibilities can be a difficult task, especially with the absence of a clearly defined role (Fayad et al.). It is easy to get off track and assign responsibilities that are either ambiguous or irrelevant. It is not until the developer begins to actually map the CRC cards to various use case scenarios or the class diagram that these extraneous responsibilities are realized.
6. *Difficulty in mapping.* Multiple responsibilities can make it difficult to map the classes identified by the CRC cards to the actual class diagram and use case scenarios. When numerous responsibilities are assigned to one class, the interactions can become complex (Fayad 2000–2003; Fayad et al.). This complexity carries over when the CRC cards are used to map use cases. The use cases are provided to determine if the class model provides the necessary functionally to support all possible scenarios. Because the responsibilities of a class are actually a summary of its functionality, mapping can be greatly complicated, when the class's operations are tied to multiple responsibilities.

13.5.6.3 Forces

For CRC cards to enhance the development of systems, main essential quality factors should be satisfied (Fayad et al.). Yet, satisfying these qualities is not too easy. The following summarizes the main points that show writing an effective CRC card is not straightforward:

- A major advantage of the CRC cards tool resides in its simplicity to understand. Current CRC cards consist of three elements: name of the class, its responsibilities, and its collaborations. However, such simplicity may not convey all the required information needed in the following steps in the development. For example, collaboration section in current CRC cards does not provide any information about the kind and type of collaboration. In other words, the card does not specify the services that its class offers to the other classes that collaborate with it. Such information is important in developing the class diagram and in verifying its accuracy. However, having too much information in a CRC card might preclude them from being widely applied. They might become difficult to understand or to use. This may scarify the simplicity of the technique. Therefore, compromising between completeness and simplicity should be considered while writing CRC cards.
- It is important to understand the role of each class in the system in order to identify its responsibilities (Fayad 2000–2003). A class within the system might have several roles; however, current CRC cards do not provide a way to differentiate between the different roles of the same class. How can we handle multiple roles for the same class in a simple way?

- Defining the class responsibility is crucial for developing an accurate class diagram and latter an effective system. A class might have several responsibilities within the system; however, identifying all these responsibilities in one CRC card might create great confusion for the developers (Fayad et al.). This is because different responsibilities for a class can result in different collaborations between this class and other classes in the system. By listing all the responsibilities and collaborations of a class in one CRC card, it becomes confusing to match a responsibility of a class to another collaborating class in the system. This complicates the deriving of the system class diagram from the written CRC cards.

13.5.6.4 Solution

In this section, we present an enhanced representation for CRC cards as a solution to some of the problems in the current CRC cards. The new version will include a clear role for each class, which will aid in the discovery of superclasses and their respective subclasses. This class role will also be useful, when defining the class responsibility. Each class will be allowed to have only one unique responsibility. If more than one responsibility is identified, additional classes should be formed. Limiting responsibilities will help prevent low cohesion and high coupling as well as reduce the possibility of macho classes. Finally, the revised CRC card will also include the services offered by each class. This will help verify the validity of the class responsibility, as well as ensure that overlapping functionality is avoided. The proposed CRC card format is shown in Table 13.3.

Creating CRC cards with the proposed format requires three main steps (see Figure 13.9) given as follows:

1. Class identification
2. Assigning roles and responsibilities
3. Discovering collaborators

Each step is concerned with filling one element of the proposed CRC card at a time. In summary, relevant classes of the system will be identified. Then, proper roles would be assigned to them based on a number of responsibilities that are found in particular design scenarios. Collaborators or clients of each class would be discovered to account for the certain inabilities of some classes to fulfill its assigned responsibility (Pressman 2001). This process will be aided by the explicit declaration of the external services offered by each relevant class.

13.5.6.5 Example

In this section, we will show an example of applying the effective CRC card on a simple problem.

TABLE 13.3
Proposed CRC Card Format

Class (Role)		
Responsibility	**Collaboration**	
	Client	**Server**

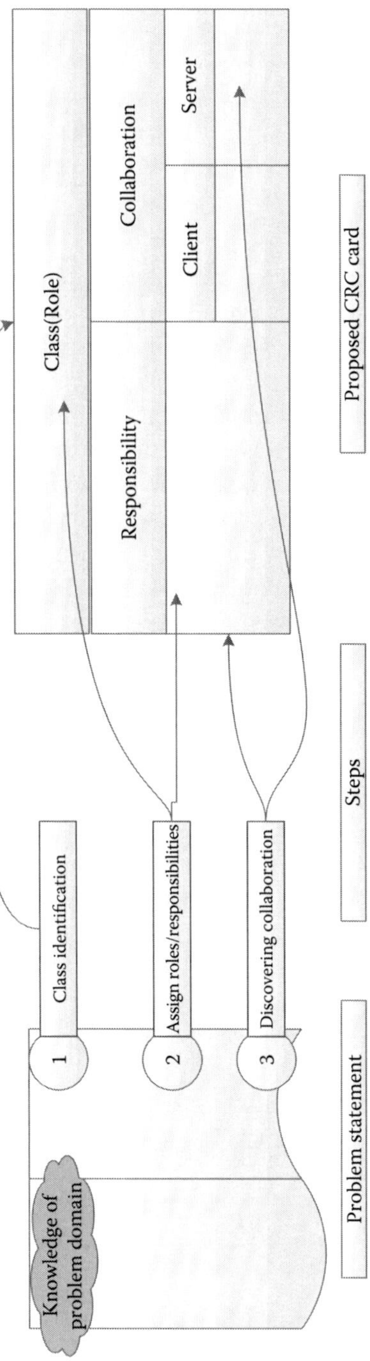

FIGURE 13.9 Applying the proposed CRC card.

13.5.6.5.1 Requirements

The services that are linked with the word *Genealogy* have been growing tremendously across the Internet landscape. This idea has been touched by several online businesses, such as Ancestry.com and MSN, but it is still in its infancy. We would like to propose a simple family tree design. This system will offer a central storage device, where all the information of current members will be stored. Each member will have full control of his/her information portrayed in the system.

The system will provide consistent updating, searching, and tracking mechanisms to facilitate an easy interaction between the system and the members throughout the entire family tree. An efficient user-friendly navigation mechanism will be presented as well. This will guarantee full access to all the features of the system. Members of the system will be able to share this experience by inviting new users to either start up their own family tree or enroll in a member's family tree. The latter would happen in the case that these potential members are relatives of an already enrolled member. Regarding guests of the system, there will be a section exclusively for them, allowing them to visit current family trees with certain limitations. They can also create their own family tree if they desire.

Step 1: Identifying classes
The class identification process does not vary in both approaches. Both approaches use similar techniques (Maciaszek 2001; Pressman 2001) along with the overall knowledge on the subject from analysts and designers. Similar naming conventions are applied, except they vary, when dealing with compound nouns. Compound nouns are treated as one word, and no spaces are allowed between these two words. This is sometimes called *camel casing*. For method declarations, the accepted notation is just as it is for classes *except* that the first letter is in lowercase. The following is an example of a possible class in the presented problem (Table 13.4).

Step 2: Assigning roles and responsibilities
The proposed CRC card format offers much more to the process of discovering responsibilities than the current one. The presence of a well-defined role makes things easier for the analyst, because each role is tightly bound to a unique responsibility (Fayad 2000–2014; Fayad et al. 2003). Therefore, the analyst can map the distinct responsibilities per class based on particular scenarios. Other techniques may be used to assist this process (Pressman 2001).

A class that contains multiple responsibilities will be partitioned into several classes (Fayad et al. 2003). No naming rules are required in this step. However, for understanding purpose, analysts need to define clear and cohesive responsibilities. Table 13.5 is an example of assigning one unique responsibility and a role.

TABLE 13.4
A Class Representation Using CRC Card

FamilyTree (Role)		
Responsibility	**Collaboration**	
	Client	**Server**

TABLE 13.5
Assigning a Role and Unique Responsibility

FamilyTree (Role)		
Responsibility	**Collaboration**	
Illustrate the bond of a group of People	**Client**	**Server**

Step 3: Discovering collaborators

The goal of this step is the definition of a set of methods that will help achieve a responsibility. The most common techniques used for services identification is the application of a grammatical parse over the problem statement, looking for verbs and/or verb phrases (Fayad 2000–2003; Pressman 2001). These verbs or verb phrases usually correspond to the methods used to fulfill certain functionality or unique responsibility of a particular class. They will be explicitly nested on the right compartment of the collaboration section of the proposed CRC card. This compartment is named *server*. They obey certain naming rules to assure a proper definition. The identifying classes section refers to these naming rules.

The inclusion of these services in the proposed CRC card will enhance the ability for responsibility identification, making it a very straightforward process. Along with this inclusion of services, several classes that communicate with one particular class will be placed on the left compartment of the collaboration section. This section is called *clients* because they collaborate with a particular class by requesting services from it. These requests are performed based on a message-wise action. These classes (clients) are usually identified at the moment of establishing an interaction between classes and its surroundings in a particular design scenario (e.g., use case scenario) (Fayad 2000–2003). This step is not just a common step. Having exhibited two more sections of this CRC card, we have granted the accomplishment of several quality factors at the same time (e.g., self-descriptive services and explicit notion of collaborators) (Fayad et al.). This result will also increase the level of understandability for analysts and designers toward the proposed CRC card structure.

Coming up with classes (clients) is a repeatable process done by analysts/developers, and it will be completed only when the analysts/designers feel that they have covered all the distinct design scenarios. The following Table 13.6 is an example of filling out the collaboration section of the proposed CRC card.

TABLE 13.6
Collaborators and Services Identification

FamilyTree (Role)		
Responsibility	**Collaboration**	
To show relationships between people	**Client**	**Server**
	Member	initiateTree() connectFamily()
	Family	joinToTree() search()

SUMMARY

Our objectives for defining this knowledge map or stable pattern language for CRC cards were concentrated in the offering of a suitable language for writing and applying CRC cards across domains. This language definition was aided by the addressing of CRC card's domain knowledge and their fundamental deployment processes from a conceptual and purpose-driven perspective. This conceptualization and understanding was laid out through *four main steps of stratification* and a set of stable patterns. The rationale of this stratification was to facilitate the findings, execution order, and description of the stable patterns embodying the concepts or building blocks of the CRC cards domain. The resulted work was possible due to the synergy of two methodologies: the software stability concepts paradigm (Fayad and Hamza 2003) and its subelements: stable analysis and design patterns (Fayad and Hamza 2003) and pattern language methodology (Fayad, Sanchez, and Goverdhan 2005). Patterns left without a description would be addressed in the future versions of the chapter.

OPEN RESEARCH ISSUES

Develop a unified Software Engine for CRC Cards to automate and fulfill the purposes and quality factors of the CRC Cards.

REVIEW QUESTIONS

1. T or F __________ CRC card stands for class responsibility and contracts.
2. T or F __________ CRC card is a dynamic model.
3. T or F __________ Fayad's CRC card stands for class responsibility and collaborations.
4. T or F __________ Fayad's CRC cards are utilized for mapping candidates classes in predefined design scenarios, for example, use case scenarios.
5. T or F __________ The objective of CRC cards is to facilitate the design process while insuring an active participation of involved designers.
6. T or F __________ Fayad's knowledge map is called stable patterns language.
7. T or F __________ Fayad's CRC cards representation, via stable patterns as a mean to discover, organize, and utilize CRC cards endured knowledge.
8. T or F __________ CRC cards are primarily used as a brainstorming technique to rapidly and thoroughly explore design alternatives by identifying the classes and their associations within a system.
9. T or F __________ CRC cards provide a simple alternative for a collaborative design environment, where analysts, designers, and developers try to simulate the system behavior, that is, role-play-driven approach.
10. T or F __________ Fayad's CRC cards knowledge map turns the CRC cards (index card) to a valuable knowledge repository.
11. What are the differences between Fayad's CRC cards and existing CRC cards (Biddle, Noble, and Tempero 2009)?
12. T or F __________ Fayad's CRC cards include a well-defined role.
13. T or F __________ Fayad's CRC cards provide a unique, cohesive responsibility, within context, to each class.
14. List the Fayad's CRC cards' quality factors and justify each one of these quality factors.

15. List and examine Fayad's CRC cards explicit characteristics and solutions.
16. Name the Fayad's CRC cards goals and describe each goal briefly.
17. Name the Fayad's CRC cards capabilities and describe each capability briefly.
18. Discuss the context of Fayad's CRC cards.
19. Name and describe briefly three different scenarios for utilizing Fayad's CRC cards.
20. Discuss briefly the process of defining Fayad's CRC cards knowledge map.
21. T or F__________ Brainstorming is a BO.
22. Explain what do you mean by the term *brainstorming*?
23. What is the usage of the brainstorming pattern?
24. Can the term *brainstorming* be used in any other context than what you thought of?
25. Can brainstorming pattern be used interchangeably with AnyDebate pattern? Explain your answer.
26. Can brainstorming pattern be used interchangeably with AnyDiscussion pattern? Explain your answer.
27. What problem does the brainstorming pattern solve?
28. In what context is the brainstorming pattern being applied?
29. Name a few scenarios for the application of brainstorming pattern.
30. What are the challenges faced in implementing the brainstorming pattern?
31. What are the constraints faced in implementing the brainstorming pattern?
32. Discuss briefly the functional requirements of brainstorming pattern.
33. Discuss briefly the nonfunctional requirements of brainstorming pattern.
34. Explain brainstorming pattern model with the help of class diagram and CRC cards.
35. What are the design and implementation issues for the given brainstorming pattern?
36. Provide some patterns related to the brainstorming pattern.
37. Explain usage of brainstorming pattern with two examples other than the ones provided in this chapter.
38. How does traditional model differ from the stability model? Explain using the brainstorming pattern model.
39. Enlist some of the business issues encountered for the brainstorming pattern.
40. Explain procedure for testing the brainstorming pattern.
41. Discuss some of the real-time usages of brainstorming pattern.
42. What are the lessons learned by you from the brainstorming pattern.
43. List some of the domains in which brainstorming pattern can be applied.
44. What is the trade-off of using the brainstorming pattern?
45. List some advantages of using brainstorming pattern in real applications.
46. Can you think of any scenarios where brainstorming pattern will fail? Explain each scenario briefly.
47. Describe how the developed brainstorming pattern would be stable over time.
48. Briefly explain how brainstorming pattern supports its objective.
49. Try to list few more business rules for the brainstorming pattern.
50. T or F __________ Engagement is a BO.
51. Repeat questions 23–49 for engagement pattern.
52. T or F __________ Modeling is a BO.
53. Repeat questions 23–49 for modeling pattern.

EXERCISES

1. Find a glossary or two of each one of the following domains/subjects/topics.
2. Use the knowledge map template and the glossary of each of the domains below to
   a. Identify and document all the EBTs/goals of your domain/subject/topic (Section II)
   b. Identify and document all the BOs/capabilities of your domain/subject/topic (Section III)
   c. Map each EBT (from Section I) to its BOs (from Section III) in your domain/subject/topic (Section IV)
3. Two or three of the following subdomains of visual arts:
   a. Art history
   b. Calligraphy
   c. Connoisseurship
   d. Creativearts
   e. Drawing (outline)
   f. Finearts
   g. Painting (outline)
   h. Filmmaking
   i. Photography (outline)
   j. Mixed media
   k. Printmaking
   l. Studio art
   m. Sculpture (outline)
   n. Art conservation
4. Two or three of the following subdomains of economics:
   a. Agricultural economics
   b. Behavioral economics
   c. Bioeconomics
   d. Complexity economics
   e. Computational economics
   f. Consumer economics
   g. Development economics
   h. Ecological economics
   i. Econometrics
   j. Economic geography
   k. Economic history
   l. Economic sociology
   m. Economic systems
   n. Energy economics
   o. Entrepreneurial economics
   p. Environmental economics
   q. Evolutionary economics
   r. Experimental economics
   s. Feminist economics
   t. Financial economics
   u. Financial econometrics
   v. Game theory

w. Green economics
x. Growth economics
y. Human development theory
z. Industrial organization
aa. Information economics
ab. Institutional economics
ac. International economics
ad. Islamic economics
ae. Labor economics
af. Law and economics
ag. Macroeconomics
ah. Managerial economics
ai. Mathematical economics
aj. Microeconomics
ak. Monetary economics
al. Neuroeconomics
am. Political economy
an. Public finance
ao. Public economics
ap. Real estate economics
aq. Resource economics
ar. Social choice theory
as. Socialist economics
at. Socioeconomics
au. Transport economics
av. Welfare economics

5. Two or three of the following subdomains of political sciences:
    a. American politics
    b. Canadian politics
    c. Civics
    d. Comparative politics
    e. Geopolitics (Political geography)
    f. International relations
    g. International organizations
    h. Nationalism studies
    i. Peace and conflict studies
    j. Policy studies
    k. Political behavior
    l. Political culture
    m. Political economy
    n. Political history
    o. Political philosophy
    p. Psephology
    q. Public administration
        i. Nonprofit administration
        ii. Nongovernmental organization (NGO) administration

    r. Public policy
    s. Social choice theory
6. Two or three of the following subdomains of engineering:
    a. Aerospace engineering
    b. Agricultural engineering
        i. Food engineering
    c. Architectural engineering
    d. Bioengineering
        i. Biomechanical engineering
        ii. Biomedical engineering
    e. Chemical engineering
    f. Civil engineering
        i. Geotechnical engineering
        ii. Engineering Geology
        iii. Earthquake engineering
        iv. Highway engineering
        v. Transportation engineering
    g. Computer engineering (outline)
    h. Control systems engineering
    i. Ecological engineering
    j. Electrical engineering (outline)
    k. Electronic engineering
    l. Instrumentation engineering
    m. Engineering physics
    n. Environmental engineering
    o. Industrial engineering
    p. Materials engineering
        i. Ceramic engineering
        ii. Metallurgical engineering
        iii. Polymer engineering
    q. Mechanical engineering
        i. Manufacturing engineering
    r. Mining engineering
    s. Nanoengineering
    t. Nuclear engineering
    u. Ocean engineering
        i. Marine engineering
        ii. Naval architecture
    v. Optical engineering
    w. Quality assurance engineering
    x. Petroleum engineering
    y. Safety engineering
    z. Software engineering (outline)
    aa. Structural engineering
    ab. Systems engineering
    ac. Telecommunications engineering

ad. Vehicle engineering
    i. Automotive engineering
7. Two or three of the following subdomains of law:
    a. Canon law
    b. Comparative law
    c. Constitutional law
    d. Competition law
    e. Criminal law
        i. Criminal procedure
        ii. Criminal justice (outline)
            A. Police science
            B. Forensic science (outline)
    f. Islamic law
    g. Jewish law
    h. Jurisprudence (Philosophy of Law)
    i. Civil law
        i. Admiralty law
        ii. Animal law/Animal rights
        iii. Corporations
        iv. Civil procedure
        v. Contract law
        vi. Environmental law
        vii. International law
        viii. Labor law
        ix. Paralegal studies
        x. Property law
    j. Tax law
    k. Tort law
8. Two or three of the following subdomains of space sciences:
    a. Astrobiology
    b. Astronomy (outline)
        i. Observational astronomy
            A. Radio astronomy
            B. Microwave astronomy
            C. Infrared astronomy
            D. Optical astronomy
            E. UV astronomy
            F. X-ray astronomy
            G. Gamma ray astronomy
    c. Astrophysics
        i. Gravitational astronomy
            A. Black holes
        ii. Interstellar medium
        iii. Numerical simulations in
            A. Astrophysical plasma
            B. Galaxy formation and evolution

            C. High-energy astrophysics
            D. Hydrodynamics
            E. Magnetohydrodynamics
            F. Star formation
        iv. Physical cosmology
        v. Stellar astrophysics
            A. Helioseismology
            B. Stellar evolution
            C. Stellar nucleosynthesis
    d. Planetary science (alternatively, a part of earth science)
9. Two or three of the following subdomains of life sciences:
    a. Biochemistry (outline)
    b. Bioinformatics
    c. Biotechnology (outline)
    d. Biology (outline)
        i. Aerobiology
        ii. Anatomy
            A. Comparative anatomy
            B. Human anatomy (outline)
        iii. Botany (outline)
            A. Ethnobotany
            B. Phycology
        iv. Cell biology (outline)
        v. Chronobiology
        vi. Computational biology
        vii. Cryobiology
        viii. Developmental biology
            A. Embryology
            B. Teratology
        ix. Ecology (outline)
            A. Agroecology
            B. Ethnoecology
            C. Human ecology
            D. Landscape ecology
        x. Genetics (outline)
            A. Behavioral genetics
            B. Molecular genetics
            C. Population genetics
        xi. Endocrinology
        xii. Evolutionary biology
        xiii. Human biology
        xiv. Immunology
        xv. Marine biology
        xvi. Mathematical biology
        xvii. Microbiology
        xviii. Molecular biology

xix. Nutrition (outline)
xx. Neuroscience (outline)
A. Behavioral neuroscience
xxi. Paleobiology
A. Paleontology
xxii. Systems biology
xxiii. Virology
A. Molecular virology
xxiv. Xenobiology
xxv. Zoology (outline)
A. Animal communications
B. Arachnology
C. Carcinology
D. Entomology
E. Ethnozoology
F. Ethology
G. Herpetology
H. Ichthyology
I. Oology
J. Ornithology
K. Primatology
L. Zootomy
e. Biophysics (outline)
f. Limnology
g. Linnaean taxonomy
h. Mycology
i. Parasitology
j. Pathology
k. Physiology
i. Human physiology
A. Exercise physiology
B. Systematics (Taxonomy)
10. Two or three of the following subdomains of business:
a. Accounting scholarship
b. Business administration
c. Business analysis
d. Business ethics
e. Business Law
f. e-Business
g. Entrepreneurship
h. Finance (outline)
i. Industrial and labor relations
i. Collective bargaining
ii. Human resources
iii. Organizational studies
iv. Labor economics
v. Labor history

   j. Information systems (Business informatics)
      i. Management information systems
      ii. Health informatics
   k. Information technology (outline)
   l. International trade
   m. Marketing (outline)
   n. Purchasing
   o. Risk management and insurance
   p. Systems science
11. Two or three of the following subdomains of psychology:
   a. Abnormal psychology
   b. Applied psychology
   c. Biological psychology
   d. Clinical psychology
   e. Cognitive psychology
   f. Community psychology
   g. Comparative psychology
   h. Conservation psychology
   i. Consumer psychology
   j. Counseling psychology
   k. Cultural psychology
   l. Differential psychology
   m. Developmental psychology
   n. Educational psychology
   o. Environmental psychology
   p. Evolutionary psychology
   q. Experimental psychology
   r. Forensic psychology
   s. Health psychology
   t. Legal psychology
   u. Media psychology
   v. Medical psychology
   w. Military psychology
   x. Neuropsychology
   y. Occupational health psychology
   z. Organizational psychology
   aa. Parapsychology (outline)
   ab. Personality psychology
   ac. Political psychology
   ad. Positive psychology
   ae. Psychometrics
   af. Psychology of religion
   ag. Psychophysics
   ah. Quantitative psychology
   ai. School psychology
   aj. Social psychology
   ak. Sport psychology

12. Two or three of the following subdomains of chemistry:
    a. Agrochemistry
    b. Analytical chemistry
    c. Astrochemistry
    d. Atmospheric chemistry
    e. Biochemistry (outline)
    f. Chemical engineering
    g. Chemical biology
    h. Cheminformatics
    i. Computational chemistry
    j. Cosmochemistry
    k. Electrochemistry
    l. Environmental chemistry
    m. Femtochemistry
    n. Flavor
    o. Flow chemistry
    p. Geochemistry
    q. Green chemistry
    r. Histochemistry
    s. Hydrogenation
    t. Immunochemistry
    u. Inorganic chemistry
    v. Marine chemistry
    w. Materials science
    x. Mathematical chemistry
    y. Mechanochemistry
    z. Medicinal chemistry
    aa. Molecular biology
    ab. Molecular mechanics
    ac. Nanotechnology
    ad. Natural product chemistry
    ae. Neurochemistry
    af. Oenology
    ag. Organic chemistry (outline)
    ah. Organometallic chemistry
    ai. Petrochemistry
    aj. Pharmacology
    ak. Photochemistry
    al. Physical chemistry
    am. Physical organic chemistry
    an. Phytochemistry
    ao. Polymer chemistry
    ap. Quantum chemistry
    aq. Radiochemistry
    ar. Solid-state chemistry
    as. Sonochemistry
    at. Supramolecular chemistry

au. Surface chemistry
av. Synthetic chemistry
aw. Theoretical chemistry
ax. Thermochemistry

13. Two or three of the following subdomains of earth sciences:
    a. Edaphology
    b. Environmental science
    c. Environmental chemistry
    d. Gemology
    e. Geodesy
    f. Geography (outline)
    g. Geology (outline)
    h. Geochemistry
    i. Geomorphology
    j. Geophysics (outline)
    k. Glaciology
    l. Hydrogeology
    m. Hydrology (outline)
    n. Meteorology (outline)
    o. Mineralogy
    p. Oceanography (outline)
    q. Pedology
    r. Paleontology
        i. Paleobiology
    s. Planetary science (alternatively, a part of space science)
    t. Sedimentology
    u. Soil science
    v. Speleology
    w. Tectonics

PROJECTS

1. Create a knowledge map for each of the sample requirements of Appendix D. [Hint: 3 ultimate EBT each and approx. 12–20 BOs.]
2. Draw a class diagram for each of the generated Knowledge Map.
3. Generate a CRC card and as discussed in Section 13.5.6 (Pattern 4—Effective CRC Card Format Pattern) and as shown in Tables 13.3 through 13.6 for each of the EBTs and BOs of each of the knowledge map.
4. Use the generated CRC cards to generate 5–7 significant scenarios from different domain for each of the generated CRC cards in 3.

14 Future Work and Conclusions

Change is the law of life. And those who look only to the past or present are certain to miss the future.

John F. Kennedy
Zimbardo and Boyd 2008

The work accomplished through this book fulfills numerous requirements that were enumerated during the beginning of this venture. We had enlisted and planned to answer the following three main questions:

1. How do we classify, develop, and utilize analysis and design patterns together toward solving a problem resolution?
2. What is the *behind-the-scene* language that guides the process of sewing or gelling of patterns as a whole?
3. How can we overcome challenges other than pattern composition problems (patterns traceability) that can hinder and impede the development of a system of patterns?

Throughout this book, we have provided a number of answers and some practical suggestions to follow a clear-cut process that arises from these answers. The major backbone to answer these questions was the integration of two approaches: software stability concepts and the pattern languages (Coplien and Schmidt 1995; Fayad 2002a, 2002b; Fayad and Altman 2001; Schmidt, Fayad, and Johnson 1996).

While conducting detailed research on this topic, we could analyze how numerous drawbacks and shortages of current software approaches that deal with software patterns especially in software patterns' composition, traceability, generality, and so on hindered the quality of built systems in one way or another (e.g., design trade-offs, loss of generality). In order to overcome these drawbacks and shortages, we have also provided a standard way for conceiving, building, and deploying systems by using a topology of software patterns. This topology is known as knowledge maps. The knowledge map will serve as the road map or supporting technique to guide software practitioners to delve into the rationale, business rules, and context of application of a set of problem domains and come up with a high-quality software system.

Throughout this book, we were constantly writing and highlighting that the essence of knowledge maps is twofold: a clear methodology and a precise visual representation. For the methodology approach, we have provided a set of guidelines, heuristics, and quality factors that will ease the process of creating knowledge maps, along with the realization and documentation of building blocks. However, for visual representation, we have provided the visual gadgets or symbols that convey how the knowledge maps and their enclosed elements look like, and in what manner they interact with other enclosed elements or other knowledge maps. In conjunction, both methodology and visual representation serve as the road maps for building systems from software patterns in a cost-effective manner. In addition to this, this road map will also allow the creation of synergies between managers and technical staff, especially when creating systems in terms of goals and capabilities. As a result, these synergies will provide the ways and means for reducing existing communication gaps between the managerial and technical staff.

While using knowledge maps, we can expect great team dynamics between managers and technical staff. They will be able to create an environment where the initial clashing of ideas that might occur because of own beliefs and experience is immediately detected and recognized for immediate action and finding suitable solutions. This environment will also allow managers and technical staff to focus on the merit of the problem and not on irrelevant trifles and details, like implementation details. At the same time, it will also create a common language for communicating ideas between manager and the technical staff. As a result, the process for creating software solutions will become immensely thrilling and interesting, because the solutions will be able to broaden the scope and horizon of discussions. Furthermore, as some of the knowledge maps' architectural processes are based on a trial and error approach, managers and technical staff might lose their apprehension in synthesizing ideas, because it becomes clear that no idea is bad, but it is only part of the process of getting it right.

We are currently working on the proof of concept of knowledge maps by developing a knowledge maps framework, where the completion of the knowledge maps framework fulfills what we are aiming in terms of knowledge maps provisioning and usage. The next section will describe more about the future work and goals related to knowledge maps that we would try to achieve and accomplish.

14.1 FUTURE WORK

Implementing knowledge maps as a tool for creating cost-effective software solutions will not only enhance the levels of communication and interaction among different software teams (e.g., managers and programmers), but also make the process of creating software enjoyable, exciting, and extremely interesting. These useful benefits, however, not only apply to the software realm; it has also opened doors to other areas of research and development, such as education and management. Therefore, the following list will only represent an initial draft of the work that will be carried out to enhance the usage of knowledge maps.

1. *IDE for knowledge maps management.* Several projects can assist in streamlining the utilization of the knowledge maps in software development, like the development of an integrated development environment (IDE) for knowledge maps management, the standardization and formalization of the process of building knowledge maps or topology of patterns, the development of facilities to automate the process of testing remote knowledge (remote knowledge maps), and the generation of software architectures. The completion and combination of these projects will provide us additional tools to create cost-effective and robust software systems. The additional tools might include the automation of dynamic analysis of software products to achieve a reduction in the software development life cycle, automation of intuitive verification and validation techniques to assure software quality, and the automation of the process of integrating numerous knowledge maps.
2. *Proper utilization of existing technologies.* To create a complete IDE for knowledge maps management, additional technologies such as Resource Description Framework, JDOM, and XML schema could be added or supplemented to our

current work. The addition of these technologies will also streamline knowledge map's framework. These technologies can be used, for example, to make knowledge maps portable, reusable, and representative to the information they conveyed regardless of the environment of deployment.

3. *Standardize and formalize knowledge maps building process.* Another possible project would be the standardization and formalization of the process of building knowledge maps by using formal languages, such as Object Z and Z++ (Object-Z; Z++ Language Syntax Chart). The utilization of formal languages will empower knowledge maps processes with validity, integrity, efficiency, and authority to be properly used by different software practitioners (with different levels of expertise).
4. *Automation of remote knowledge's testing and software architectures generation.* To automate the testing of remote knowledge and the generation of software architectures, many more dynamic methods for information gathering and pattern identification/definition, intuitive verification and validation techniques, and interaction with backend databases could be integrated in our knowledge map's overall framework. The practical utilization of these methods will empower the process of building knowledge maps with a dynamic analysis, integration with other knowledge maps, software development life-cycle reduction, and a faster return of investment (ROI).

SUMMARY

The work report that we have presented in this book would bring significant contributions to the domain of software engineering, especially in the domain of software patterns. These contributions range from software pattern identification, evaluation, and definition heuristics, knowledge maps formation, evaluation, management, and deployment processes to the initial framework that will assist the development of software by using the software stability concepts approach (Fayad 2002a, 2000b; Fayad and Altman 2001).

The other main aspects of this work were related to our personal contributions to knowledge map users and some of them were suggesting an initial framework for developing knowledge maps, especially the part that deals with the implementation and testing of enduring business theme, business objects, industrial objects, and hooks and providing a set of stable analysis and design patterns that creates the core of the knowledge maps approach. The documentations of these patterns, as well as the source code of this framework, are also included in this book

In general, the main benefit of using our approach is to bring to the software practitioners the ways and means for answering three important questions: How can we classify, develop, and utilize analysis and design patterns together toward problem resolution? What is the *behind-the-scenes* language that guides the sewing of patterns as a whole? How can we overcome different challenges, other than patterns composition problems (e.g., patterns traceability), that can hinder the development of a system of patterns? The backbone to answering these questions and later creating knowledge maps was introducing a subtle combination of traditional pattern languages techniques and the software stability concepts approach. This delicate combination that gives birth to the creation of knowledge maps not only provides an answer to the aforementioned questions, but also validates and verifies each one of the elements that will be generated from it.

Toward forming a future perspective, we are also planning to develop and enhance the knowledge maps approach's building blocks, by means of implementing tools that will ease the process of building, evaluating, and deploying stable software patterns.

REVIEW QUESTIONS

1. What are the advantages of using knowledge maps?
2. Knowledge maps foster team dynamics between managers and the technical staff. Explain.
3. Provide ways of enhancing knowledge map's usage.

EXERCISES

1. Research ways of enhancing knowledge maps' usage (other than pointers given in this chapter).
2. Research existing applications and discuss problems with them. Explain whether those problems can be fixed by using knowledge maps.

PROJECTS

Consider the following patterns:

1. Discovery Stable Analysis Pattern
2. Knowledge Stable Analysis Pattern
3. AnyMap Stable Design Pattern
4. AnyContext Stable Design Pattern

and answer the questions below:

1. Describe and document the business rules, and how you can extend them for two of scenarios of each pattern.
2. Define the business rules, business policies, business facts, in relation to each of the pattern.
3. Illustrate the business rules that derived from each of the pattern.

Appendix A
Pattern Documentation Templates

A.1 PATTERN DOCUMENTATION—DETAILED TEMPLATE (PREFERRED)

- *Name*. Presents the name of the presented pattern.
    - Provide short definition of the term (Name).
    - Compare the name of the patterns with other selective name and conclude with the right selection of the name.
    - Why did you choose that specific name?
    - Justify the name (such as why use *Any*... as a prefix for business object [BO] only).
- *Known as*. Lists all the terms that are similar to the name of the pattern. Two possible sources that one can use to fill this section are as follows: (1) similar patterns that are proposed in the literature and (2) other names that you may find relative to the developed pattern. In some cases, several names might make sense, so you can keep a list of few of these names under this section.
- *Discuss the following cases briefly*.
    - Names match the pattern name: Just list similar names and why?
    - Names match with doubts: List them, describe, and indicate doubts, and why?
    - Names do not match, but people think they match the pattern name: List them, describe, and show why they do not match.
- *Context*. Gives possible scenarios for the situations, in which the pattern may recur. It is important in this section that you motivate the problem you solve in an attractive way. For example, if I were writing a pattern about trust, I would flush the trust in the context of e-commerce. Keep this section short, yet exciting. (This section somewhat serves as an introduction in conventional paper.)
    - Describe the boundaries.
    - List basic scenario—Context.
    - Show by good examples, where the pattern can be applied.
    - For example, *account*... would have ownership and handler context, can be applied to banking Internet providers, private clubs, and so on.
- *Problem*. Presents the problem the pattern concentrating on. This is one of the hardest parts in the pattern writing. Do not try to write it quite well in the first iteration, and most probably, you will not be able to! The problem should focus on the core purpose of the pattern and should be able to answer the question: In what situation, I may benefit from your pattern? Try and keep this section as short as possible; otherwise, the reader may get confused.

- *Length: 1/4 to 1/3 Pgs.*
  - Has to be about a specific problems and descriptions = actual requirements of the pattern (functional and nonfunctional requirements of the pattern described in the template).
  - It must be within the domain. There are two basic domain analyses/design and own fields of existence.
  - Discuss all the elements of the goal of the pattern.
  - You may create a list of the subgoals for requirements of the pattern.
- *Challenges and constraints.* Illustrates the challenges and the constraints that the pattern needs to resolve. You may create two subsections: (1) challenges and (2) constraints. In particular, in this section, you try to say that this is not a trivial problem and that trivial solution may not work. Be clear and brief. One major mistake in writing this section is that you mix the problem statement with the forces themselves. After writing this section, try to read the problem statement again and make sure that they are not the same! It always happens!
  - Describe some of the challenges that must be overcome by the pattern.
  - Describe the constraints related to the pattern, such as multiplicities, limits, and range.
  - Make sure to list the challenges and constraints as bullets.
- *Solution.*
  - *Pattern structure and participants.* Gives the class diagram of the pattern (enduring business theme [EBT] or BO). It also introduces briefly each class and its role. Associations, aggregations, dependencies, and specializations should be included in the class diagram. Association classes, constraints, interfaces, tagged values, and notes must be included in the class diagram and also include the hooks (show each of the BOs connections to industrial objects [IOs]). A *full description of the class diagram should be included with the final submission.*
  - *CRC cards.* Summarizes the responsibility and collaboration of each participant (class). Each participant should have only one well-defined responsibility in its CRC card. Participants with more than one responsibility should be presented with more than one CRC card, when each CRC card will handle one of these responsibilities. Refer to Appendix B—CRC card layout.
  - *Behavior model (whenever is possible).* If the abstraction of the pattern prevents you from writing an appropriate behavior model, then you can flush the dynamics of the pattern later on within the example section.
    - **Description**
      - Describe the constraints related to the pattern such as multiplicities, limits, and range.
      - Describe some of the challenges that must be overcome by the pattern.
      - *Note.* Not all IO and BO may have inheritance.
    - **Detail Models**
      - Describe the model, role story, such as scenarios, and how they play together.

- **Participants**
  - Each name, and its short description, and how it behaves within the model, such as classes and patterns in the patterns.
- **CRC Cards**

- *Consequences.* How does the pattern (EBT or BO) support its objectives or goals? What is the trade-off and results of using the pattern? It is also important to highlight the things that the pattern does not cover and reason about why you choose to exclude them. Another point that I found useful in this section is to highlight other components that may arise from using the proposed patterns. For example, in AnyAccount pattern, we can say that using this pattern for banking systems will require the integration of entries and logs to keep track of the accounts. However, this does not mean that the pattern is incomplete, but this is the nature of patterns anyway, they need to be used with other components.
  - List and briefly describe the good (the benefits) of this pattern.
  - List and briefly describe the bad (side effects) of the pattern with suggested solutions.
- *Applicability with illustrated examples.* Provides two clear and detailed case studies for applying the pattern in different contexts. The following subelements represent the required details in one case.
  - *Case studies.* Shows the scenario of two cases studied from different contexts.
  - *Class diagram.* Presents the EBTs, BOs, and IOs.
  - *Use case template.* Gives detailed description for a complete use case. It includes test cases for the EBT and all the BOs—Abstraction of actors; roles; classes; class types, such as EBT, BOs, and IOs; attributes; and operations. Refer to Appendix B—Use case template.
  - *Behavior diagram.* Maps the above use case into a sequence diagram.
    - Show two to three distinct scenarios.
    - Description of the problem statement of the particular problem.
    - Describe the model—Class diagram.
    - Use case description with test cases (do not need to do use case diagrams).
    - Sequence diagram/use case.
- *Related patterns and measurability.* Shows other patterns that usually interact with the described pattern and those who are included within the described pattern. Related patterns can be classified as *related analysis or/and related design patterns.* Related patterns usually share common forces and rationale. In addition, it is possible that you might give some insights of other patterns that can or need be used with the proposed patterns; for example, in the case of AnyAccount pattern, we might point out to the AnyEntry pattern as a complementary pattern. There are rooms for contrasting and comparing the existing patterns with the documented pattern. This section also provides a few metrics for measuring several things related to the pattern structure, such as complexity and size, cyclomatic complexity, lack of cohesion, and coupling between object classes.

- This section is divided into two parts:
  - *Related pattern.* Two approaches:
    - Search for an existing traditional pattern on the same topic. Compare with traditional existing pattern models with reference to ours.
    - If existing patterns do not exist, select a single definition of the name of our pattern, develop a traditional model class diagram, and describe it briefly.
  - *Measurability.*
    - Measurability compares our pattern to other models on the number of behaviors and number of classes. Justification of why the numbers of behavior or classes are so high or low.
    - You may compare and comment on other quality factors, such as reuse, extensibility, integration, scalability, and applicability.
    - Two approaches: Compare the traditional with stability models in two of the following approaches:
      - *Quantitative measurability.*

        Number of behaviors or operations per class
        Number of attributes per class
        Number of associations
        Number of inheritance
        Number of aggregations
        Number of interactions per class
        Number of EBTs versus number of requirements classes in traditional model
        Number of classes
        Documentation—Number of pages
        Number of IOs
        Number of applications
        Estimation metrics
        Measurement metrics
      - *Qualitative measurability.*

        Scalability
        Maintainability
        Documentation
        Expressiveness
        Adaptability
        Configurability
        Reuse
        Extensibility
        Arrangement and rearrangement
- *Modeling issues, criteria, and constraints.* There are a number of modeling issues, criteria, and constraints that you need to address, in such a way as to explain them, and make sure that the model satisfies all the modeling criteria and constraints.
- *Modeling Issues Are as Follows*:
  - *Abstraction.* Describe the abstraction process of this pattern, list, and discuss briefly the abstractions within this pattern.
    - Show the abstractions that are required for the patterns (EBT, BOs, and IOs).
    - Elaborate on the abstraction of why EBTs and BOs are selected?

  - Show examples of unselected EBTs, and why?
  - Show examples of unselected BOs, and why?
- *Static models.* Illustrate and describe one or two of the static models of this pattern and list and discuss briefly the complete story of the pattern models using actual objects.
  - Determine the sample model that you are planning to use: CRC cards, class diagram, component diagram, and so on—Show the model.
  - Tell a complete story of the pattern models using objects.
  - Repeat a complete story with other objects.
- *Dynamic models.* Illustrate and describe one or two dynamic models of this pattern and list and discuss briefly the behavior of the pattern through the selected dynamic models.
  - Determine the sample model that you are planning to use: Interaction diagram or state transition diagram—Show the model.
- *Modeling essentials.* Examine the pattern using the modeling essentials and list and discuss briefly the outcome of this examination.
  - List or reference to the model essentials, and use them as criteria to examine the pattern.
  - Elaborate on how to examine the model of the pattern by using the model essential criteria.
  - Briefly describe the outcome.
- *Concurrent development.* Show the role of the concurrent development of developing this pattern.
  - Describe and show with illustration the concurrent development of this pattern.
- *Modeling heuristics.* Examine the pattern by using the modeling heuristics and list and discuss briefly the outcome of this examination.
  - List or reference to the modeling heuristic, and use them as criteria to examine the pattern.
  - Elaborate on how to examine the model of the pattern using the modeling heuristics.
  - Briefly describe the outcome.
  - Modeling heuristics
    - No dangling
    - No star
    - No tree
    - No sequence
    - General enough to be reused in different applications

- *Design and implementation issues.* For each EBT, discuss and elaborate on the important issues required for linking the analysis phase to the design phase, and for each BO, discuss the important issues required for linking the design phase to the implementation phase, for example, hooks. Describe the design issues (EBT), for example, hooking issues. Alternatively, discuss the implementation issues (BO), for example, why using relationship rather than inheritance, hooking, and hot spots problems. Show segments of code here.
  - *Here Is a List of Analysis Issues*
    - Divide and conquer
    - Understanding

  - Simplicity
  - One unique base that is suitable to many applications
  - Goals
  - Fitting with business modeling
  - Requirement specification models
  - Packaging
  - Components
  - Type (TOP) (A)
  - Actors/roles
  - Responsibility and collaborations
  - Generic and reusable models
- *Design Issues (EBT)*
  - For example, hooking issues.
  - Implementation issues (BO)
  - For example, why using aggregation or delegation rather than inheritance.
  - For example, hooking and hot spots problems.
  - Can show code here.
- *Here Is a Sample List of Design and Implementation Issues*
  - Framework models (D)
  - Classes (TOP) (D)
  - Collaborations (D)
  - Refinement (D)
  - Generic and reusable designs (D)
  - Precision (I)
  - Hooks (I)
  - Pluggable parts (I)
  - Navigation (I)
  - Object identity (I)
  - Object state (I)
  - Associations/aggregations (I)
  - Collections (I)
  - Static invariants (I)
  - Boolean operators (I)
  - Collection operators (I)
  - Dictionary (D) (I)
  - Behavior models (D) (I)
  - Pre–Post-conditions specify actions (I)
  - Joint actions (use cases) (D)
  - Localized actions (I)
  - Action parameters (I)
  - Actions and effects (I)
  - Concurrent actions (I)
  - Collaborations (I)
  - Interaction diagrams (D)
  - Sequence diagrams with actions (D) (I)
  - Pattern 1: Continuity
  - Pattern 2: Performance

  - Pattern 3: Reuse
  - Pattern 4: Flexibility
  - Pattern 5: Orthogonal abstractions
  - Pattern 6: Refinement
  - Pattern 7: Deliverables
  - Pattern 8: Recursive refinement
  - Package (D) (I)
- *Here Is a List of Java Patterns*
  - *Fundamental design patterns*
    - Delegation (when not to use inheritance)
    - Proxy
  - *Creational patterns*
    - Abstract factory
    - Builder
    - Factory method
    - Object pool
    - Prototype
    - Singleton
  - *Partitioning patterns*
    - Composite
    - Filter
    - Layered initialization
  - *Structural patterns*
    - Adaptor
    - Bridge
    - Cache management
    - Decorator
    - Dynamic linkage
    - Façade
    - Flyweight
    - Iterator
    - Virtual proxy
  - *Behavioral patterns*
    - Chain of responsibility
    - Command
    - Little language/interpreter
    - Mediator
    - Null object
    - Observer
    - Snapshot
    - State
    - Strategy
    - Template method
    - Visitor
- *Testability*. Describes the test cases, test scenarios, testing patterns, and so on. (This is a very important point, but sometimes it is very hard to write for an isolated pattern, I am not sure what is the best way to write this part!) You can use three ways

to document testability: (1) test procedures and test cases within classes members of the patterns; (2) propose testing patterns that are useful for this pattern and other existing patterns; and (3) check, if the pattern fit with as many scenarios as possible, without changing the core design.

- Mention to people to try to find scenarios within the context that cannot work with this pattern.
- Show how you can test the requirements and the design artifacts within use cases.
- Can also use exhaustive testing of behaviors (may require more pages) by using testing patterns.

- *Formalization using Z++, object Z, or object-constraints language (OCL) (optional).* Describes the pattern structure by using the formal language (Z++ or Object Z), BNF, EBNF, and/or XML.
- *Business issues.* Cover one or more of the following issues.
  - *Business rules.* Describe and document the business rules, and how you can extend them in the context and scenarios that are listed.
    - Define the business rules, business policies, business facts, in relation to the pattern.
    - Illustrate the business rules that are derived from the pattern.
  - Check the following links:
    - http://en.wikipedia.org/wiki/Business_rules
    - http://www.businessrulesgroup.org/bra.shtml
    - http://www.businessrulesgroup.org/first_paper/br01c0.htm—pdf format file
    - http://www.businessrulesgroup.org/brmanifesto.htm
  - Define the business rules in relation to the pattern.
  - Illustrate the business rules that are derived from the pattern.
- Business models: Issues
  - *Business model design and innovation*
  - *Business model samples* (http://en.wikipedia.org/wiki/Business_model):
    - Subscription business model
    - Razor and blades business model (bait and hook)
    - Pyramid scheme business model
    - Multilevel marketing business model
    - Network effects business model
    - Monopolistic business model
    - Cutting out the middleman model
    - Auction business model
    - Online auction business model
    - Bricks and clicks business model
    - Loyalty business models
    - Collective business models
    - Industrialization of services business model
    - Servitization of products business model
    - Low-cost carrier business model
    - Online content business model
    - Premium business model
    - Direct sales model

  - Professional open-source model
  - Various distribution business models
  - Describe the pattern. If it is part of or it is a business model.
  - Describe the direct impacts of the pattern on the business model.
  - Describe the indirect impacts of the pattern on the business model.
- Describe the same for the following business issues:
  - *Business standards*
    - Vertical standards versus horizontal standards
  - *Business integration*
    - Data integration
    - People integration
    - Tools integration
  - *Business processes or workflow*. Here are some of the business processes issues. Business process management (BPM) is a systematic approach to improving those processes, which are given as follows:
    - Business process modeling and design
    - Business process improvement
    - Continuous business process improvement
    - Business process categories: Management processes, operational processes, and supporting processes
    - Business process ROI
    - Business process rules
    - Business process mapping
  - e-Business
    - e-Commerce
    - e-Business models (http://en.wikipedia.org/wiki/E-Business)
      e-shops
      e-commerce
      e-procurement
      e-malls
      e-auctions
      Virtual communities
      Collaboration platforms
      Third-party market places
      Value-chain integrators
      Value-chain service providers
      Information brokerage
      Telecommunication
    - *e-business categories* (http://en.wikipedia.org/wiki/E-Business)
      Business-to-business (B2B)
      Business-to-consumer (B2C)
      Business-to-employee (B2E)
      Business-to-government (B2G)
      Government-to-business (G2B)
      Government-to-government (G2G)
      Government-to-citizen (G2C)

      Consumer-to-consumer (C2C)
      Consumer-to-business (C2B)
    - *Web applications*
  - *Business patterns*
    - Business modeling with UML
    - Business knowledge map
  - *Business strategies*
    - Business strategy modeling
    - Business strategy frameworks
    - Strategic management
    - Strategic analysis
    - Strategy implementation
    - Strategy global business
  - Business performance management (BPM)
    - Methodologies
    - BPM framework
    - BPM knowledge map
    - Assessment and indication
  - *Business transformation*
  - *EBTs*
  - *Security and Privacy*
- *Known usage.* Give examples of the use of the pattern within existing systems or examples of known applications that may benefit from the proposed pattern. Mention some projects that used it.
- *Tips and heuristics.* List and briefly describe all the lessons learned, tips, and heuristics from the utilization of this pattern, if any.
  - What did you discover?
  - Why did you included or excluded different classes?
  - Are there any tips on usage such as scaling, adaptability, flexibility?

A.2 PATTERN DOCUMENTATION—SHORT TEMPLATE

This template consists of five fields:

- *Name.* Presents the name of the presented pattern.
- *Context.* Gives possible scenarios for the situations, in which the pattern may recur.
- *Problem.* Presents the problem the pattern concentrating on.
- *Solution and participants.* This section is the same as what is given in the full template, expect that we do not need to have two class diagrams. Usually, it is sufficient to present the simple version of the stable object model.

Appendix B
Other Templates

B.1 FAYAD'S CLASS RESPONSIBILITY AND COLLABORATION CARD LAYOUT

Class/Pattern Name (Class/Pattern Role) (Class Type)

Responsibility	**Collaboration**	
	Clients	**Server**
A single responsibility for this class/pattern should be listed here briefly. Unique Within context	A list of all the classes/patterns that have a relationship with the named class/pattern. Two or more clients	A list of all the servers that named class/pattern provides. 5–12 operations (services)

Attributes: Class/pattern seven or more attributes

Source: Fayad, M. E., H. S. Hamza, and H. A. Sanchez. "A Pattern for an Effective Class Responsibility Collaborator (CRC) Cards." Paper presented at the 2003 IEEE International Conference on Information Reuse and Integration, Las Vegas, NV, October 2003.

B.2 FAYAD'S USE CASE TEMPLATE

Use Case Id	Insert a number of the use case in the sequence.
Use Case Title	Insert a use case name that starts with a verb.
AnyActor and/or Any/Party	*Roles*
List AnyActor's or AnyParty's Type[a]	Insert corresponding roles.

- *AnyActor has four different types.* Human, hardware, software, and creatures such as animals, trees, and animated characters.
- *AnyParty has four types as well.* Human, organizations, countries, and political parties.

Class Name	Type[a]	Attributes	Operations

- Each class is classified as an EBT, BO, or IO.

Use Case Description

Describes the first step in the scenario.
Describes the second step in the scenario.
Describes likewise subsequent steps in the scenario.

- The use case should be at least five or more steps, and it should be written with stability in mind, with test cases of EBT and BO names, attributes, and operations only.
- Use case description must be numbered.
- Each use case may contain 6–12+ steps.

Alternatives

(As an example) Insert an alternative scenario for step 2 in the original sequence.
Repeat the alternative, as many time as required, for any of the use case steps

Source: Fayad, M. E., D. Naney, and A. D. Pace. "Should Novice System Developers Use Use Cases to Develop Core Requirements?" White Paper at University of Nebraska, Lincoln, October 2000.
BO, business object; EBT, enduring business theme; IO, industrial object.

B.3 FAYAD'S STABLE ANALYSIS/DESIGN/ARCHITECTURAL PATTERNS (LAYOUT)

B.3.1 Stable Analysis Pattern or Stable Design Pattern Layout (1 EBT and 2–14 BOs)

EBT	BOs

B.3.2 Stable Architectural Patterns Layout (2–5 EBTs Where 3 Is the Most Common)

EBTs	**BOs**
EBT1	
EBT2	
....	

Sources: Fayad, M. E., and S. Wu. "Merging Multiple Conventional Models in One Stable Model." *Communications of the ACM*, 45, No. 9 (2002): 102–106. Hamza, H., and M. E. Fayad. "A Pattern Language for Building Stable Analysis Patterns." *In the Proceedings of 9th Conference on Pattern Languages of Programs 2002 (PLoP02)*, Monticello, IL, September 2002.

B.4 Fayad's Stable Analysis/Design Pattern Applications Layout [Hamza and Fayad 2002]

B.4.1 Stable Analysis Pattern or Stable Design Pattern Applications Layout

EBT	**BOs**	**IOs**

B.4.2 Stable Architectural Pattern Applications Layout

EBTs	BOs	IOs
EBT1		
EBT2		
....		

Source: Hamza, H., and M. E. Fayad. "A Pattern Language for Building Stable Analysis Patterns." *In the proceedings of 9th Conference on Pattern Languages of Programs 2002 (PLoP02)*, Monticello, IL, September 2002.

B.5 FAYAD'S KNOWLEDGE MAP TEMPLATE

Knowledge Map of Domain/Subject/Topic

1. *Knowledge Map Name.*
2. *Knowledge Map Nickname.*
3. *Knowledge Map Domain/Subject/Topic Description.*
4. *EBTs/Goals.* Name the EBTs of the *Domain/Subject/Topic* and provide a short description of each EBT and organize your answer in Table B.1.

TABLE B.1
EBTs of *Domain/Subject/Topic*

EBTs/Goals	Description

TABLE B.2
BOs of *Domain/Subject/Topic*

BOs/Capabilities	Description

TABLE B.3
Knowledge Map of *Domain/Subject/Topic*

EBTs	BOs

5. *BOs/Capabilities.* Name the BOs of the *Domain/Subject/Topic* and provide a short description of each BO and organize your answer in Table B.2.
6. *Knowledge Map (Core Knowledge*).* Map each EBT to its BOs of the *Domain/Subject/Topic* and organize your answer in Table B.3.

Source: Sanchez, H. A. *Building Systems Using Patterns: Creating Knowledge Maps.* Masters Thesis. San Jose State University, San Jose, CA, May 2006.

* Core knowledge of a domain/subject/topic = EBTs and BOs of a domain/subject/topic.

Appendix C
Stable Patterns Catalog

C.1 STABLE ANALYSIS PATTERNS

SAP Name	Pattern Description	Chapter No.
Discovery	*Discovery* is defined as the act of finding or discovering something. It could be a disease, a drug, or hidden patterns. For example, the main goal in data mining is to discover hidden patterns and knowledge from the data, which is available widely. The purpose is to model a discovery pattern that can be used in any application. Discovery of hidden patterns, trends, associations, anomalies, and statistically significant structures and events in data has a great impact on the formulation of strategies that can be employed to get better insights about the market and increase productivity.	4
Knowledge	Knowledge can be gained through experience or studies. It represents a collection of facts, rules, tips, or lessons learned with respect to anything that must be synthesized to create knowledge. Sometimes, it might not be possible to obtain complete knowledge about a subject and it results in partial knowledge. As a result, this partial knowledge needs to be used to solve a problem. The knowledge pattern will be used to represent knowledge synthesis and acquisition.	5

SAP, stable analysis pattern.

C.2 STABLE DESIGN PATTERNS

SDP Name	Pattern Description	Chapter No.
AnyMap	A map is a very commonly recurring concept that we encounter in our everyday life. Almost everything we do involves a map in one way or another.	7
AnyContext	Context is essential in communicating the correct information. The objective of being in context is to deliver the relevant information to the stakeholders based on the environment and current interactions. It helps in passing the message across the board in a clever and effective manner. The objective of pattern is to generalize the idea of context, so that one can use it as a basis for initiating interactions.	8

SDP, stable design pattern.

Appendix D
Sample Requirements

D.1 OCEAN RESOURCES MANAGEMENT SYSTEM

D.1.1 Introduction

To improve, retain, enhance, and sustain the ecological, cultural, economic, and social benefits, which we derive from ocean resources. The opportunity to enjoy ocean's plentiful bounty is very essential to our health and well-being. However, there is an insufficient planning associated between the government and nonprofit organizations, who are working tirelessly in this aspect. If we want our future generations to play and make merry in oceans and enjoy the comforts of healthy ocean life, we would have to change and transform our current practices, laws, regulations, and community–government interactions. It is necessary to alter the ways and mode in which we handle the ocean resources; otherwise, they would become extinct one day, and our future generation would be cursing us for all possible unlikely scenarios. Hence, it is necessary to implement a system that collaborates the activities of government organization, nonprofit sectors, and various other communities in preserving the ocean resources and helps in its better administration and management. It makes sense to completely remove the sector-based approaches that are currently associated in handling or managing ocean resources. At present, the management efforts are extremely fragmented with a number of gaps and overlaps in implementation, just because of this inefficient approach.

Seamless integration of efforts by the government and nonprofit organizations working toward preserving ocean resources.

Ocean zoning is a concept considered as a means to guide human uses of the ocean, to optimize utilization of marine resources, and to provide protection of marine ecosystems.

D.1.2 Description of Domain

The main domain of the system is *Natural Resources Conservation*. This is a very vast field, wherein all the small blocks of the society are trying to contribute their parts. As everything is sector based, the thing that is happening has no mutual coordination and integration among the various communities that are working together. Due to this lack of this perceived integration, we generally waste a large amount of resources on a number of unwanted and wasteful purposes. Replenishment of natural resources is actually a big challenge and ocean resources are no exceptions. Hence, how do we conserve our available ocean resources, how to best utilize them, so that it helps the country financially, and how best to maintain them, so that they are in perfect synchronization with the environment and we enjoy its soothing nature in future, are the key milestones to be considered here while developing this system. Thus, this forms the basis of domain of the system.

Soil erosion, waste management, and ocean ecosystem protection are some of the few domains to be considered, while designing this system. Working with integrity and coordination forms the foundation of domains to be included in the system.

D.1.3 Block Diagram

Ocean resource management system, as shown in Figure D.1, has various departments and sectors in it, and the detailed requirements of each department have been mentioned as under. All the different departments intercommunicate with each other, and they have to coordinate with each other, in order to achieve the final goal.

D.1.4 Description of the Program That Is Wanted

The system should be able to satisfy the following high-level goals/functions:

1. Integrate relationship between land and sea.
   a. Water quality monitoring.
   b. Protection of beaches and coastal communities from shoreline erosion and other natural hazards.
   c. Maintenance and appropriate use of environmental infrastructure.

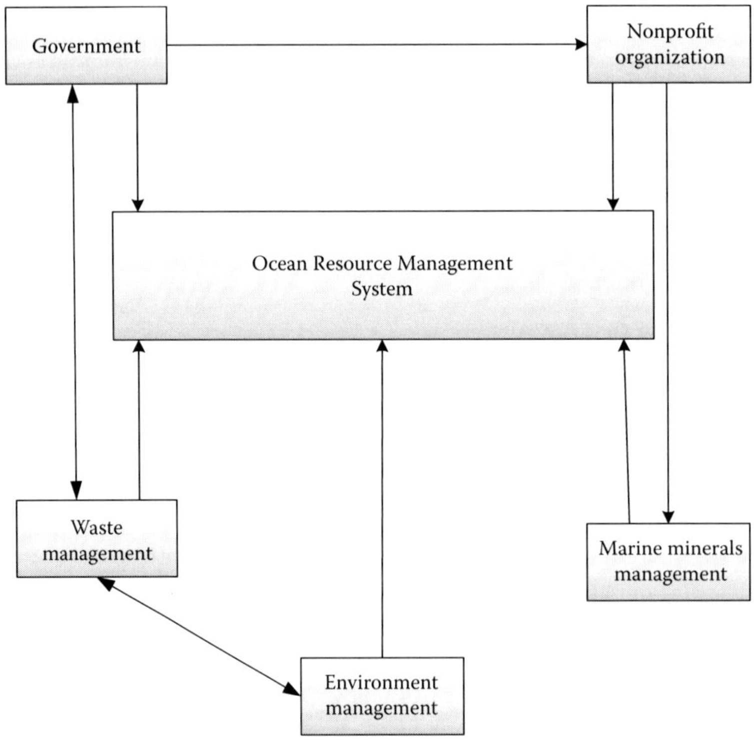

FIGURE D.1 Ocean resource management system block diagram.

2. Preserving the ocean heritage.
   a. Improve the quality of the ocean resources for traditional, commercial, and recreational purposes.
   b. Enhance public access and appropriate coastal-dependent uses of shoreline.
   c. Promote appropriate and responsible ocean recreation for tourists and residents.
   d. Encourage ocean science and technology, with safeguards for ocean resource protection.
3. Promoting collaboration.
   a. Permit integrated and place-based approaches to the management of the ocean resources.
   b. Institutionalize integrated ocean resources management.

D.1.5 Detailed Requirements

The detailed requirements consist of two different types of requirements: (1) functional requirements and (2) nonfunctional requirements.

D.1.5.1 Functional Requirements

1. Reduce soil erosion emanating from upland forest ecosystems and conversation lands.
   a. Uprooting of plants, loss of native forest species, weeds, and other manmade factors are mainly responsible for soil erosion that ends up in streams, and eventually to ocean waters. The system needs to address this issue.
2. Reduce pollutant loads from residential, agricultural, and commercial land uses in priority watersheds.
   a. Construction activities expose soil, which are washable into streams that lead to ocean water contamination, because of improper sediment control.
   b. Agricultural activities near the beaches involve the use of pesticides and insecticides that can enter the ocean water and make it hazardous for human beings.
3. Implementation of comprehensive and integrated shoreline policy.
   a. This should address the impacts of chronic and episodic coastal hazards.
   b. Site-specific management techniques should be developed for the beaches, which allow the natural erosion to occur with minimum impact on the ocean resources.
4. Encourage appropriate coastal-dependent development that reduces a number of risks from coastal erosion and other hazards in priority coastal areas.
   a. Coastal hazard assessment should be carried out, in order to incorporate future uncertainties and imponderables, like faster rates of erosion and high level of flooding.
5. Inspect and maintain sewer collection systems, including the detection of leaks.
   a. Upgrade and maintain the sewage system to minimize impact during flood conditions.
6. Reduce illegal stormwater discharges to the wastewater system.
   a. Conduct public education campaign explaining the impacts of illegal stormwater discharges.
   b. Develop new rules and regulations establishing penalties for noncompliance.
7. Minimize the introduction and spread of marine alien and invasive species.
   a. Develop risk-based approach to identify species and areas with highest potential for economic damage.

8. Establish wastewater-discharge restricted zones and conditions for commercial vessels plying in archipelagic waters.
   a. Enforce laws on wastewater discharge, with a close monitoring and enforcement plan.
9. Provide appropriate waste management infrastructure and facilities to support commercial and recreational marine facilities.
   a. Provide temporary pump-out facilities.
   b. Provide adequate solid waste management activities.
   c. Increase frequency of inspection of marine sanitation devices.
10. Strengthen and expand marine protected area management.
    a. Develop place-based marine protected area plans for priority areas.
    b. Identify limits of acceptable change.
11. Develop ecosystem-based approaches for nearshore fisheries management practices.
    a. Identify, protect, and restore fish habitat for nearshore fish stocks.
12. Establish and institutionalize new approaches for restoring, operating, and preserving ancient coastal fishponds and salt ponds.
    a. Provide support and incentives to the communities and individuals to facilitate restoration process.
13. Improve enforcement capacity and voluntary compliance with existing rules and regulations for ocean resource protection.
    a. Conduct education/research campaigns, community-based partnerships to judge ocean resource management issues.
14. Enhance the conservation of marine protected species, unique habitats, and biological diversity.
    a. Develop educational materials that are responsible for providing awareness to the people and support the efforts to improve the marine water quality.
15. Enhance and restore existing public shoreline areas and scenic vistas.
    a. Develop enhancement and restoration plans for the purpose stated above.
    b. Develop interagency agreements.
    c. Provide funding.
16. Establish new shorelines area for public and appropriate coastal dependent uses.
    a. Establish criteria for identifying priority coastal areas for public acquisition.
17. Develop community-based frameworks and practices for identifying and mitigating ocean recreational use conflicts.
    a. Work with existing/new advisory groups to develop tool for resource protection and conflict management.
18. Promote responsible and sustainable ocean-based tourism.
    a. Establish performance standards to ensure responsible commercial ocean-based tourism.
19. Promote alternative ocean energy sources.
    a. Conduct the analysis of the impact of nonocean energy resources on ocean.
20. Plan and develop sustainable commercial aquaculture in coastal areas and ocean water.
    a. Establish a database to locate coastal and ocean aquaculture projects in environmentally suitable sites.
21. Expand ocean science and technology.
    a. Facilitate appropriate research and innovation in marine technologies.

22. Develop standardized tools for ocean resource management.
    a. Investigate how limit of Acceptable change, can be used in resource management.
23. Develop legislative and administrative proposals to improve management of ocean resources.
24. Build the required capacity for community participation for preserving the ocean resources.
25. Monitor and evaluate ocean resource management plan implementation.
    a. Establish multisector Ocean Resource Management Plan and monitoring group.
    b. Establish public advisory group, in order to help the above group for assessment.

D.1.5.2 Nonfunctional Requirements

1. Lifetime.
    a. When is the present project supposed to be completed?
    b. After how many years should the project be reviewed again for any changes?
2. Cost.
    a. What is the estimated cost behind the project?
3. Reconfigurable.
    a. Can the project change as environmental/human requirement changes?
4. Scalability.
    a. Can we add a new functionality to project without disturbing its original implementation?
5. Robustness.
    a. Is the project versatile enough to provide support for all the beaches in the country?

D.1.6 Use Case and User Context

The following section provides the names, brief descriptions, and the actors' names and their corresponding roles' names of 10 different use cases.

D.1.6.1 Use Case 1

Use Case Name. Improve coastal water quality.

Use Case Description. For improving the coastal water quality, different departments of this system must concentrate on reducing land-based sources of pollution and restoring natural habitats.

Actors. Tourists, Fishermen, Exporters/Importers, Residents, Sea life

D.1.6.2 Use Case 2

Use Case Name. Protecting ocean resources from coastal hazards.

Use Case Description. Develop a comprehensive and integrated shoreline policy and guideline for the coastal hazards, which addresses the impacts of chronic and episodic coastal dangers. Also, develop shoreline management plan with specific measure of erosion and other coastal areas.

Actors. Residents, researchers, ecosystem protector, sea life

D.1.6.3 Use Case 3

Use Case Name. Maintain environmental infrastructure.

Use Case Description. Maintain sewer collection systems by including the detection of leaks, which also reduce illegal stormwater discharges to the wastewater system. Also, improve and ensure maintenance and appropriate use of environmental infrastructure.

Actors. Ecosystem protectors, residents, researches, tourists, sea life

D.1.6.4 Use Case 4

Use Case Name. Reduce marine sources of pollution.

Use Case Description. Establish wastewater discharge restricted zones and conditions for commercial vessels plying in archipelagic waters. Also, provide appropriate waste management infrastructures to support commercial and recreational marine facilities.

Actors. Fishermen, residents, tourists, exporters/importers, sea life

D.1.6.5 Use Case 5

Use Case Name. Improve health of coastal and ocean resources.

Use Case Description. Develop ecosystem-based approaches for fisheries management and also establish and institutionalize approaches for restoring, operating, and preserving ancient coastal fishponds and salt ponds. Enhance the conservation of marine protected species, unique habitats, and biological diversity.

Actors. Resident, ecosystem protectors, exporters/importers, sea life

D.1.6.6 Use Case 6

Use Case Name. Enhancing public access.

Use Case Description. Enhance and restore existing public shoreline areas and scenic vistas. Also, establish new shoreline areas for public and appropriate coastal-dependent uses.

Actors. Residents, tourists, sea life

D.1.6.7 Use Case 7

Use Case Name. Preserve ocean resources.

Use Case Description. Encourage cutting edge and appropriate ocean science and technology measures with safeguards for ocean resource protection.

Actors. Researchers, fishermen, residents, ecosystem protectors, sea life

D.1.6.8 Use Case 8

Use Case Name. Manage ocean resources.

Use Case Description. Apply integrated and place-based approaches to the management of ocean resources. Develop standardized tools and build additional capacity for community participation in ocean resources management.

Actors. Fishermen, exporters/importers, sea life

D.1.6.9 Use Case 9

Use Case Name. Integrate the beaches and surrounding area.

Use Case Description. Maintain the beaches for the tourism and provide the different kinds of facilities for the tourists.

Actors. Tourists, residents, sea life

D.1.6.10 Use Case 10

Use Case Name. Create awareness.

Use Case Description. Local residents need to be educated, so that they understand and comprehend the importance of ocean resources. Slideshows, seminars, and workshops must be organized by the native education and research centers and nonprofit organizations, to help people realize the essentials of ocean resources.

Actors. Residents, fishermen, researchers, tourists, sea life

Sources

http://coastalmanagement.noaa.gov/.

Johannes, R. E. and Hickey, F. R.; UNESCO (2004). Evolution of village-based marine resource management in Vanuatu between 1993 and 2001, Paris, France: UNESCO.

Vierros, M., Tawake, A., Hickey, F., Tiraa, A., and Noa, R. (2010). Traditional Marine Management Areas of the Pacific in the Context of National and International Law and Policy. Darwin, Australia: United Nations University–Traditional Knowledge Initiative. UNU-IAS, 2010.

D.2 DENGUE FEVER PREVENTION AND OUTBREAK MANAGEMENT SYSTEM

D.2.1 Introduction

Dengue fever is a serious mosquito-borne viral disease. It spreads very quickly and can be fatal; its fatality rate is approximately 5% in most of the countries around the world. It has not been possible to eradicate this fever completely forever. Vaccines are in the process of development for this epidemic. Efficacy trials in human volunteers are still not completed. An effective vaccine may not be available to the public for the next 5–10 years.

Our project focuses specially on finding measures to suppress the spread of this disease and promote the development of vaccines for the disease. The dengue prevention and outbreak management system will help assist authorities to prevent a dengue epidemic and guide health authorities, in case of an outbreak of this deadly disease.

D.2.2 Description of Domain

First outbreak of dengue fever occurred between 1779 and 1780 in the regions of Africa, Asia, and North America. Dengue fever, which was very typical in tropical and subtropical continents, has now spread over the entire globe soon after World War II.

Dengue is an infectious disease, characterized by frequent bouts of severe pain in the eyes, head, and extremities, later accompanied by catarrhal symptoms. This disease occurs because of a mosquito bite. One of these four closely related virus stereotypes (DEN-1, DEN-2, DEN-3, and DEN-4) causes dengue and dengue hemorrhagic fever (DHF). Infections with one of these viruses build immunity for only that stereotype.

The *Dengue Prevention and Outbreak Management System* focuses on controlling the epidemic at the slightest appearance of epidemic signs. This project also aims at finding out the main reasons behind the dramatic and lightening spread of dengue. Because no vaccines are available yet, one can prevent the spread of dengue only by increasing awareness among people and educating them about the symptoms of dengue and possible preventive measures. During an outbreak, officials can work closely with citizens to prevent breeding of mosquitoes and isolate patients, and the scientists can study different viruses found in the patient's body. They can also prepare better medications for cure.

D.2.3 Description of the Program That Is Required

This system must comprise of a hand of assistance from the local government for supporting the health infrastructure, health authorities, and a management team to overlook and monitor operations of this system.

Dengue Prevention and Outbreak Management System can be classified into two parts—one being measures that are taken to control the epidemic from spreading further and the other being measures taken to prevent an outbreak.

D.2.4 Detailed Requirements

The system must fulfill the following requirements:

Functions for controlling the epidemic

<R10> Medical facilities should monitor number of cases of dengue per unit of area.

Medical facilities can determine and assess about the outbreak of dengue. They can monitor dengue patients arriving for treatment and later group them by their location and addresses.

<R20> Notify local government.

Medical facilities must notify local government about an outbreak and about possible reasons for the same. Local government broadcasts the occurrence of outbreak to the public.

<R30> Set up facilities to treat patients.

The system should set up facilities far from urban areas to quarantine, isolate, house patients affected, and treat them.

<R40> Management of funds.

The system must be able to manage an easy flow of funds to manage and cater this facility.

Funds for providing medical needs to patients
Funds for running the facility
Funds for sanitizing affected areas

<R50> Suppress the spread of epidemic by effectively controlling the spread of mosquitoes and their breeding activities.

<R60> Medical facilities must work closely with scientists.

In the worst-case scenarios, medicines may not prove as effective as they should. Medical facilities must work with scientists to create new treatment to save lives.

<R70> Additional treatment.

In case of ineffective treatments, local government can coordinate control efforts among medical facilities, local and outside scientific communities, and other authorities, for a quick and effective treatment.

<R80> Medical facilities must also work with local government to create new treatment measures.

<R90> Medical facilities can update necessary preventive care information shared by local government with public.

Functions for preventing the epidemic

<R100> Identify problems.

The system must make a sincere attempt to understand the problems behind the emergence of this disease.

<R110> Create awareness.

Educate public about the causes and symptoms for dengue.

<R120> Promote development of vaccines.

Vaccines are under development for dengue; the system can help this process by initiating volunteers for human trials.

Encourage and provide facilities and amenities for research regarding these vaccines.

<R130> Information related to both preventive and control measures are maintained in an efficient database.

D.2.5 Use Cases and User Context

The users of this system will comprise of government officials, doctors, scientists, patients, and other citizens. The main purpose of this system is to collect and provide information for better management of dengue prevention and its outbreak to save lives.

1. *Horizon-UC1.* Nondengue patient seeking diagnosis.
   a. *Description.* A citizen has a fever. He/she registers with City Hospital for diagnosis. He/she is diagnosed as a nondengue patient and given medication for other diseases.
   b. *Actors/Roles.* Citizen, City Hospital primary physician
2. *Horizon-UC2.* Dengue patient seeking diagnosis.
   a. *Description.* A citizen has a fever. He/she registers with City Hospital for diagnosis. He/she is diagnosed as having dengue fever. He/she is put on dengue treatment and monitored very closely.
   b. *Actors/Roles.* Citizen, City Hospital primary physician
3. *Horizon-UC3.* Dengue patient recovers.
   a. *Description.* After *N* number of days, the condition of the dengue patient is found to recover and improve. Further testing concludes that he/she is cured completely and now can go home. He/she will not cause any danger to his/her community.
   b. *Actors/Roles.* Dengue patient, City Hospital physician
4. *Horizon-UC4.* Treatment not effective.
   a. *Description.* After *N* number of days, the condition of the dengue patient is still worsening. City Hospital doctors feel that the patient is not responding well to the treatment. City Hospital specialist is then consulted. He/she concludes that this may be because of a new variant of virus. Medical scientist collects samples and starts working on a new treatment.
   b. *Actors/Roles.* Dengue patient, City Hospital physician, City Hospital specialist, Local medical scientist

5. *Horizon-UC5.* New treatment is successful.
   a. *Description.* After *N* number of days, the condition of the dengue patient is still worsening. City Hospital consults with a local Medical Scientist. He/she creates a new treatment. Patient is cured and goes home.
   b. *Actors/Roles.* Dengue patient, City Hospital physician, City Hospital specialist, Local medical scientist
6. *Horizon-UC6.* New treatment fails.
   a. *Description.* After *N* number of days, the condition of dengue patient is still worsening. City Hospital consults with a local Medical Scientist. He/she creates a new treatment. However, it is not effective and the patient later dies.
   b. *Actors/Roles.* Dengue patient, City Hospital physician, City Hospital specialist, Local medical scientist
7. *Horizon-UC7.* Many dengue patients a day and normal case.
   a. *Description.* City Hospital provides services to citizens from different areas. It receives approximately the same number of dengue patients every day. This is considered as a normal condition.
   b. *Actors/Roles.* Dengue patients, City Hospital physicians
8. *Horizon-UC8.* Above-average number of patients and advisory issued.
   a. *Description.* City Hospital provides services to citizens from different areas. It is observed that more than average numbers of dengue patients are arriving from different regions of the city. City Hospital now notifies local government authorities about this sudden rise in the number of patients. Local government issues advisory to its citizens, reminding them about precautionary measures. Employees are sent to those regions to find any obvious reasons, such as standing water and unhygienic areas. Clean up ponds and still water bodies.
   b. *Actors/Roles.* Dengue patients, City Hospital physicians, Local government, and Citizens
9. *Horizon-UC9.* Above-average number of patients from an area, advisory issued, and cleanup actions are undertaken.
   a. *Description.* City Hospital provides services to citizens from different areas. It is observed that more than average number of dengue patients are arriving from specific areas of the city. The number is very high per given region. City Hospital now notifies local government authorities about possible outbreak in these regions. Local government issues an advisory to its citizens, reminding them about precautionary measures. Employees are then sent to those regions to find any visible reasons such as standing water and unhygienic areas. Clean up ponds and still water bodies.
   b. *Actors/Roles.* Dengue patients, City Hospital physicians, Local government, Citizens
10. *Horizon-UC10.* Above-average number of patients from many areas of the city and outbreak declared.
    a. *Description.* City Hospital provides services to citizens from different areas. It is observed that more than average numbers of dengue patients are arriving from many areas of the city. The number is very high per given region. City Hospital now notifies local government authorities about possible outbreak in these regions. Local government declares an outbreak. Outbreak control action plan is executed and carried out.
    b. *Actors/Roles.* Dengue patients, City Hospital physicians, Local government, Citizens

11. *Horizon-UC11.* Outbreak declared. Treatment is effective. Situation is under control.
    a. *Description.* Government declares an outbreak condition. Many dengue patients are being treated at City Hospital. Treatment is proving effective and the patient's condition is improving. Situation is under perfect control.
    b. *Actors/Roles.* Dengue patients, City Hospital physicians, Local government, Citizens
12. *Horizon-UC12.* Outbreak declared. Treatment is not effective. Situation is not under control.
    a. *Description.* Government declares an outbreak condition. Many dengue patients are being treated at City Hospital. Treatment is not proving effective and patient's condition is still worsening. Many patients are dying. External scientific help is sought. Federal government, other local governments, and scientist community are consulted for new treatment immediately.
    b. *Actors/Roles.* Dengue patients, City Hospital physicians, Local government, Other governments, Scientist community
13. *Horizon-UC13.* Outbreak declared. Many patients are dying. Biohazardous condition.
    a. *Description.* Government declares an outbreak condition. Many dengue patients are still dying. Government isolates patients from other citizens. Dead bodies are disposed of with great care and concern.
    b. *Actors/Roles.* Dengue patients, City Hospital physicians, Dead patients
14. *Horizon-UC14.* Outbreak declared. New treatment is proving effective. Situation is coming under control.
    a. *Description.* Government declares an outbreak condition. Many dengue patients are being treated at City Hospital. Treatment is not proving effective and patient's condition is still worsening. Many patients are dying. External scientific help is sought. Federal government, other local governments, and scientist community are consulted for new treatment immediately. New treatment is available and is proving quite effective.
    b. *Actors/Roles.* Dengue patients, City Hospital physicians, Local government, Other governments, Scientist community
15. *Horizon-UC15.* Local government arranges awareness camp.
    a. *Description.* Local government with the help from City Hospital arranges awareness camps in the city to educate citizens about dengue and preventive measures.
    b. *Actors/Roles.* Citizens, Local government, City Hospital
16. *Horizon-UC16.* Local government broadcasts updates in preventive measures.
    a. *Description.* Scientists find new preventive measures and notify the local government. Government broadcasts these new measures to its citizens.
    b. *Actors/Roles.* Citizens, Local government, Scientists

Sources

Halstead, S. B. (2008). *Dengue*. London: Imperial College Press. pp. 180, 429.

Wiwanitkit, V. (2010). "Unusual mode of transmission of dengue." *Journal of Infection in Developing Countries* 4 (1): 51–4.

Wiwanitkit, V. (2010). "Dengue fever: Diagnosis and treatment." *Expert Review of Anti-Infective Therapy* 8 (7): 841–5.

D.3 ORGANIZING CRICKET WORLD CUP

Cricket is a very popular game in many parts of the world. Every year, a number of tournaments are organized among different teams, and at different locations around the world. ICC Cricket World Cup is the universal form of championship in one-day cricket, and it is held once every four years. All test playing teams and qualifying teams compete against one another to acquire global superiority in cricket. Apart from entertainment, this event also serves to foster and enhance goodwill and better relationship among the teams, brings huge economical benefits to the nation where it is being held, and generates huge revenue for the organizers. The main purpose of this project is to depict the activities involved in organizing a world cup.

D.3.1 Domain Description

The cricket world cup is scheduled once every four years, but the preparations start well in advance. The International Cricket Council (ICC), which foresees all of the global activities in cricket, selects a country to host this tournament, after taking the political and economical issues into consideration. Normally, the president of a country or a very famous dignitary officially declares open the event. This tournament lasts for about two months, during which various teams from all over the world participate and play in it. Event organizers assign various teams to take care of the venues, player security, media interaction, and so on. Millions of viewers all over the world enjoy watching these games. Once the final match is played, there will be a winning team, which gets to keep the world cup for the next four years. Figure D.2 shows a cricket stadium. Figure D.3 shows the media coverage and scoreboard.

D.3.2 Description of the Program That Is Wanted

Organization of world cup cricket tournament includes the following functions and requirements:

- Deciding on the country with proper geographic locations, in which the game will be played.
- Getting the sponsors, who will provide the funding to organize various activities.

FIGURE D.2 Cricket stadium.

FIGURE D.3 Media coverage and scorecard.

- Selecting the media, who will promote this tournament.
- Constructing and updating of stadium and pavilions.
- Inviting qualified teams and arrange for their trips, food, and accommodation.
- Establishing various security systems and monitoring them during the game play.
- Using the new technology systems like advanced software, which will be useful for showing various statistical data on screen and scoreboard and for some other visual analysis and recording.
- Building the media hype around the world cup, by using media and by selling tickets online to attract people.
- Arranging for a grand opening ceremony and closing presentation in one of the main stadium.
- Arranging warm-up matches between various international teams.
- Making a whole game plan (fixtures and results).
- Deciding about various rules and regulations for the entire event and for every action and situation.
- Managing the various computing systems that count the score during the games and show player's statistics, his records, and other details and team ranking.
- Arranging for the movement of players and equipment by hiring Event Tour and Travel Partner.

D.3.3 Detailed Requirements

ICC schedules the World Cup Cricket match every four years. ICC will always decide that in which country the tournament will be held. To award this event, ICC will take into account various deciding criteria like economy of the country and availability of stadium and hotels. The hosting country must also meet all eligibility conditions like minimum financial requirements. The country that likes to host the tournament must ensure that it has a group of corporate sponsor/s. Hosting a world cup tournament is a big challenge and a difficult task, as the host country needs to meet many difficult conditions and requirements.

Besides the issue of economy, many other things are a matter of big concern like the security aspects to ensure players and official safety. Before awarding the contract, information about stadium, geographical locality, and tourism is also seriously considered. After confirming the country and the tournament sponsors, it is important that ICC will choose a good event manager/organizer, who can conduct a good and satisfying tour for everyone.

Event organization group will create game plans and hosting schedules and manage all other aspects of the tournament. The game plan will include drawing a board containing four or five groups, each of them consisting of four cricket-playing countries. After draws are entered, there will be group matches to eliminate failing teams and select winning teams. By the end of group playing sessions, all the emerging countries will be asked to play for quarter-, semi-, and final matches. The event organizer will thus conduct the tourney very successfully.

It is also important to choose different media like television, news, websites, and advertising companies to create a favorable opinion of the world cup tournaments and to attract international tourists. Each participating country will choose its own team and members, coach, and medical personnel. The country that hosts the tournament must decide on the stadium on which the games will be played, and the main condition to choose a particular stadium is the availability of enough hotel rooms, transportation, and tourism spots.

A hospitality group, who provides accommodation, food, entertainment, and transportation facilities, treats all playing countries with good hospitality. Stadium manager will be assigned to take care of the construction aspects of stadium and the estimated capacity crowd that attends to watch the game. He or she will take care of all aspects and activities that are connected with the management of stadium. It may include arranging for TV telecasting of the event, audience services, arranging for computers, and a full-pledged media center.

A pitch curator is appointed to prepare pitches on which the cricket will be played; he or she will also make sure that the turf and the stadium areas are kept clean and managed well. A manager will also be appointed and assigned the duty of arranging seats for dignitaries, VIP's, and manage refreshment facilities.

Stadium manager will also import new gadgets, tools, instruments, and software that will be used in security and to broadcast the live game show. They must have high-resolution cameras and video editing teams, to provide a good-quality viewing on the TV. These software systems must also enable the third umpire to take proper and unbiased decisions. The software system must also be able to maintain and show a series of statistical records of the game and the team, whereas the scorecard for a given game should also be maintained by the software itself. As the event is of an important and critical nature, all software systems that manage the show must be perfect, authentic, and completely stable.

Security and safety of players, teams, dignitaries, and the stadium is also very important, and care must be taken to ensure their safety throughout the tenure of the tournament. Both the government and stadium manager will involve in the security-related issues. Governmental security agencies will take care of the security of team members and the dignitaries, when they arrive at the stadium, and while they are in the country, maintain law and order if something bad happens in the city where the game is played.

Organizers also choose one group for evaluating experiences, rules and regulation of the game, and to make new rules and legislations for the present tournament. Umpires, referees, and other key members will also be selected by the ICC to participate and manage the tournament. ICC will also look after the ranking of teams and in maintaining the player's ethics and integrity clauses. However, tournament organizers and event sponsors will ensure that all teams, winners, losers, and semifinalists get their share of prize money. Once the tournament is finished, they will also organize a grand closing ceremony to sign off the event.

D.3.4 Use Cases and User Context

The following section provides the names, brief descriptions, and the actors' names and their corresponding roles' names of 10 different use cases.

D.3.4.1 Use Case 1

Use Case Name. Selects the location (country).
Description
Selecting the location (country) is the first step of World Cup Cricket.
Selection is made according to the current economy, tourism, and political issues of the country.
ICC makes some tender and the country should meet the all financial requirements according to the tender.
Actors/Roles. ICC (International Cricket Control), cricket board of selected country

D.3.4.2 Use Case 2

Use Case Name. Select sponsors.
Description
Sponsors are the main part for the successful events. There will be many sponsors.
The sponsors will give the financial support during the whole event.
The event organizers will select sponsors by studying the history and their relationship with the hosts, such as credits, rules, and service.
Actors/Roles. Event organizers, sponsors

D.3.4.3 Use Case 3

Use Case Name. Schedule games (fixtures).
Description
Only qualified top 16 teams are selected for the game.
Those teams are divided into four groups.
The team will play each other in round-robin format progressively, and finally, four best teams will advance to the semifinal and the winners of that will then compete in the final.
The date and the stadium will be selected for each game, by taking into account factors like weather and stadium condition.
The umpires (referees) will also be selected by the ICC for each game.
Actors/Roles. ICC, event organizers

D.3.4.4 Use Case 4

Use Case Name. Promote events.
Description
Promote the events to attract the viewers.

Promoters show advertisement for world cup through various electronic media like TV and Internet.

Sponsors will also promote the event through their advertisements.

They will promote the event until the final game to keep the interest alive.

Actors/Roles. Promoters, sponsors, media

D.3.4.5 Use Case 5

Use Case Name. Evaluate, construct, and upgrade the stadiums.

Description

Event organizer will evaluate all the stadiums and their conditions and then make some suggestion to stadium manager to improve and renovate.

Stadium manager will implement the changes as suggested by the organizer.

The changes can be in increasing the capacity of the stadium, improving the ground condition, infrastructure, and so on.

Actors/Roles. Stadium manager, event organizer

D.3.4.6 Use Case 6

Use Case Name. Set up the security.

Description

Stadium manager and special security squad will handle security set up during and before the game.

During the stadium renovation, security squad sets up special security cameras in all over the stadiums.

There will be a special security arrangement in players and VIP section.

During the game, the squad will also take care of the crowd to prevent any misdemeanors.

Actors/Roles. Stadium manager, security squad, security tools and software

D.3.4.7 Use Case 7

Use Case Name. Host the teams.

Description

Invited team should be provided good accommodation and food.

All kind of travel and transportation should be handled.

Taking care of the all team member's security during their staying and traveling.

Extra entertainment like tour and cultural festival should be managed.

Actors/Roles. Hosting team, team members, security squad

D.3.4.8 Use Case 8

Use Case Name. Telecast the game.

Description

Event organizer will select the channels, which will broadcast the game in different countries.

In addition, it will be broadcasted by other media like radio, cell phone, and through Internet.

Good commentators, high-resolution cameras, and sound arrangement should be used by the telecasting team.

There should be well-organized arrangement for live video editing.

Actors/Roles. Telecasting team, commentators, video editing teams, camera team, media, audience

D.3.4.9 Use Case 9

Use Case Name. Sell the tickets.

Description

There will be different categories of tickets depending upon stadiums and seats.

Tickets can be bought through Internet, phone or personally, or some agencies that are assigned by world cup organizers.

Media is used to let people know about ticket availability.

Tickets are categorized into different types like general, family, and VIP, and each ticket consists of seat number, gate number, and person number.

Actors/Roles. Stadium manager, media, internet, audience

D.3.4.10 Use Case 10

Use case Name. Provide software support and update centralized data.

Description

Update the scoreboard during the game.

Track the records of each team and player and show their records on screen, when they achieve some significant milestones.

Update the records in centralized database after each game.

Some statistical software like *wagon* is used for showing statistic information of team and players, through graphs and charts.

Some special software like *Hawkeye* is used for tracking the ball direction and path to make clear decision.

Actors/Roles. Software team, software and hardware applications

D.3.4.11 Use Case 11

Use Case Name. Maintain the rules and regulations.

Description

ICC makes some decision whether to add new rules and regulation to *ICC rule book*.

These rules and regulations are for everyone, who is going to be the part of the event, such as players, event managers, and stadium managers.

Umpires should follow and execute all the regulation mentioned in the ICC rulebook.

New technologies as *Hawkeye* and Ultra slow motion should be used to make perfect decision.

In case of any critical situation, when the whole match cannot be completed or match abandoned, the ultimate decision should be made by both umpires and ICC representatives.

Actors/Roles. ICC, umpires (referees)

D.3.4.12 Use Case 12

Use Case Name. Ceremonies held during the world cup.

Description

There should be a grand ceremony to attract and create passion for the game.
VIP dignitaries like the president, prime minister, and ICC representative of the host country participates in opening ceremony.
At the end of each game, award should be given to the outstanding performance.
At the end of the world cup cricket final, the awards like *Man of the Series* and *Best Team* would be presented.

Actors/Roles. ICC, event organizers, dignitaries, sponsors, stadium manager, audience, players

Sources

Browning, M. *A Complete History of World Cup Cricket 1975-1999.* East Roseville, N.S.W.; London: Simon & Schuster, 2000.
http://cricketworldcup.indya.com/.

D.4 POLLUTION MANAGEMENT

Today, the most pressing and critical issue that is engulfing our world is the growing environmental degradation and pollution. Due to inconsiderate activities and by uncontrolled exploitation of nature by humans on the pretext of industrialization, modernization, and urbanization, considerable environmental imbalance has resulted in the creation of unusual climatic conditions, extinction of land and marine species, and an imbalance in the ecosystems. We are facing real-time environmental crises. The need of the hour is to spread awareness about the environmental issues and formulate workable policies and solutions to control pollution.

Pollution management also helps us to study about the different agents and effects of the pollutants on the environment. The system helps to spread the awareness and actions, which has to be taken in order to reduce the alarming rate of pollution.

D.4.1 Description of Domain

Since the dawn of our civilization, we as humans have tried to alter the basic structure of the planet Earth, so that we can achieve all the material comforts. This indiscriminate use without wisdom has led us to the point, where we will need to start thinking of reversing the effects of our past; else, our future generation would be definitely doomed. We have successfully ransacked the planet in devious ways to get burning fuels and raw materials, which have been the cause of extinction of an unthinkable amount of plants and animals, and have multiplied our population to that of a plague.

Air and water pollution are the major contributors to the ever-growing environmental degradation. Around 80% of the pollution observed in seas and oceans is the direct result of land-based activities. Toxic air pollution damages our natural environment and jeopardizes public health. Once released into the air, toxins build up in the environment and work their way up the food chain, eventually ending up in the food that we eat. Eating contaminated food and breathing contaminated air can cause cancer, birth defects, and other serious health problems. The lifestyle choices that we make, the products that we use, and the efforts that we undertake to save our ecosystem greatly affect the quality of our environment.

D.4.2 Block Diagram

Despite all of the damage that we have caused, most of it is quite reversible. We can restore habitats, clean rivers, replenish the topsoil, and replant forests. However, these activities do not relieve the worst symptoms of the already occurred damage. We still have to fix the main source of these problems.

To protect the environment from the adverse effects of pollution, nations worldwide have enacted a number of legislations to regulate various types of pollutions, as well as to mitigate the adverse effects of uncontrolled pollution. Nevertheless, there is a long way ahead and we need to put in dedicated and concerted efforts, before we can make our planet a safe place to live once again.

D.4.3 Description of the Program That Is Wanted

The pollution management is to be built and managed mainly to control and reduce the pollution. This system also gives a real-life feel of what is actually happening at a smaller scale.

It achieves this objective in the following way:

- The pollution management maintains the list of various types of pollutions that are affecting the environment and their adverse effects on human health, as well as nature, animals, and birds.
- Data from research studies are used to assess the risk of the various pollutants and depict it graphically. Based on the risk models available, it is able to predict the severity of the health hazards for the various pollutants.
- The system also maintains the data and information collected from various governmental and nongovernmental agencies to keep track on the incidents of pollution and efforts expended in keeping pollution under real check.
- The system helps toward creating awareness by educating people.

D.4.4 Detailed Requirements

The detailed requirements consist of two different types of requirements: (1) functional requirements and (2) nonfunctional requirements.

D.4.4.1 Functional Requirements

1. Study different types of pollutions that are contaminating our planet Earth.
   The chief pollutants in the air are carbon dioxide, nitrogen dioxide, and hydrocarbons. By inhaling such an unhealthy air, the health of human and animal gets affected. Water is contaminated from variety of sources like discharge of industrial waste, sewer, and oil spills.
2. Understand the effect of pollutants.
   Air pollution results in ozone layer depletion, global warming, and acid rain. The contaminated drinking water poses serious health problems in the form of dysentery, typhoid, cholera, and diarrhea. Oil spills and leaks have created very serious adverse effects on the marine life.
3. Control and monitor damages caused by pollution.

By studying and understanding the damages caused by pollution, we can come up with many ways to mitigate them.

4. Devise ways to reduce pollution.
   Analyzing data collected from research to come up with solutions to reduce pollution. Using recyclables, disposables, renewable sources of energy over fossil fuels, conserving energy by use of other modes of transport and fuel-efficient vehicles are some ways to reduce pollution.
5. Spread awareness.
   Awareness can be created among public, through community education, road shows, advertisement, volunteers, and schools.
6. Reach out to organizations.
   Develop relations with both governmental and nongovernmental organizations to provide funds for research activities and creating awareness.

D.4.4.2 Nonfunctional Requirements

1. *Data requirement*. Data collected must be correct, authentic, and verifiable. They must be in the form that is easily readable by the system and can be analyzed.
2. *Scalability*. Pollution management must be able to interact with other systems.
3. *Ease of use*. The system should also be easy and simple to use by anybody, who is interested in contributing toward controlling of pollution. It should be such that even a nonprofessional can use the system to see what the system does.

D.4.5 Use Cases and User Context

The following section provides the names, brief descriptions, and the actors' names and their corresponding roles' names of 10 different use cases.

D.4.5.1 Use Case 1

Use Case Name. Contaminate water.

Brief Description. Clean water is an essential ingredient for the well-being of all living things. Factories dump chemical by-products into the water without treating them. Another culprit is the sewage treatment plant, which spews human wastes into the water bodies. Polluted water is responsible for various waterborne diseases like cholera and dysentery. It also affects the birds, animals, and other aquatic animals and disrupts their life cycle.

Users. Person (polluter, company policy maker, industrialist), creature (animal, bird)

D.4.5.2 Use Case 2

Use Case Name. Degrade soil.

Brief Description. Soil pollution occurs due to introduction of harmful substances, pesticides, and chemicals into the soil, which adversely degrades the quality of soil. This, in turn, affects the normal process of soil reuse and adversely affects the health of humans, animals, birds, and other living organisms.

Users. Person (farmer), creature (animal, bird, pathogen, earthworm)

D.4.5.3 Use Case 3

Use Case Name. Identify the effects of noise pollution.

Brief Description. Noise pollution results in hearing loss, cardiovascular impacts, high blood pressure, and psychological effects in human beings. High noise levels may also interfere with the natural cycles of animals, including feeding behavior, breeding rituals, and migration paths. The most significant impact of noise to animal life is the systematic reduction of usable habitat.

Users. Person (polluter, company policy maker, industrialist), creature (animal, bird), manufacturer (aircraft manufacturer, motor vehicle manufacturer)

D.4.5.4 Use Case 4

Use Case Name. Identify and manage adverse effects of air pollution on humans.

Brief Description. Air pollution of any sort can cause numerous ill effects, which can have either acute or chronic effects on health of humans. Initiate a study to identify these adverse effects on health and then find solutions to reduce and prevent them.

Users. Person (volunteer, researcher, scientist, environmentalist)

D.4.5.5 Use Case 5

Use Case Name. Research the factors causing pollution.

Brief Description. Environmental Protection Agency (EPA) carries out studies to identify factors that are causing pollution. People are subjected to health hazards in the form of contaminated drinking water, polluted air, and urban stress factor, such as noise. Government organizations and private companies fund these research activities.

Users. Person (concerned individual, researcher, scientist, environmentalist)

D.4.5.6 Use Case 6

Use Case Name. Predict the severity of health hazards.

Brief Description. The risk assessors collect data, analyze, and synthesize scientific data to produce hazard identification and exposure assessment portion of the risk assessment. This group includes scientists and statistician. The risk managers integrate the risk assessment to predict the severity. External experts and the public contribute to the development of site-specific risk assessments.

Users. Risk assessor (scientist, statistician), decision maker (risk manager), person (expert, concerned individual)

D.4.5.7 Use Case 7

Use Case Name. Minimize fossil fuels usage to control pollution.

Brief Description. Make use of renewable sources of energy like solar energy, wind energy, geothermal energy, water energy, and biofuel in homes and factories in order to conserve fossil fuels. Use of fuel-efficient vehicles like hybrid cars would also help in reducing pollution. Considering mass-transit, use of carpools, riding bicycles, or walking too can help in keeping pollution under control.

Users. Person (consumer, buyer, commuter)

D.4.5.8 Use Case 8

Use Case Name. Plant trees.

Brief Description. Trees are carbon sinks and help reduce the carbon dioxide in the atmosphere by converting them into oxygen during the process of photosynthesis. Trees can be planted by individuals or by organizations to reverse the effects of deforestation.

Users. Person (volunteer, concerned individual)

D.4.5.9 Use Case 9

Use Case Name. Use of recyclables.

Brief Description. Recycling is the reprocessing of materials into new products. It is an easy way to save energy and conserve resources. Whenever possible, make use of items packaged in recyclable materials or in those that can be recycled. Recycling of plastic bottles, paper, and cans would help to conserve energy and it is environment friendly too.

Users. Person (consumer), manufacturer (recyclable packaging manufacturer)

D.4.5.10 Use Case 10

Use Case Name. Regulate use of plastic bags.

Brief Description. One of the biggest contributors to soil contamination is plastic bags that are predominantly offered at retail shops. By offering some sales tax break to consumers, who are not using plastic bags, will reduce the use of plastic bags significantly.

Users. Person (consumer, distributor), manufacturer (plastic bag manufacturer)

D.4.5.11 Use Case 11

Use Case Name. Avoid disposables.

Brief Description. Disposable products like cups, plates, and plastic wrap ends up in litter, which wash into storm drains clogging the streams and create flooding problems. So, one must make use of the reusable alternatives that are cheaper to disposable products.

Users. Person (consumer), manufacturer (reusable products manufacturer)

D.4.5.12 Use Case 12

Use Case Name. Create Awareness through education.

Brief Description. The objective is to educate people to prevent the worsening of environment due to air, solid, and water pollutants through anti-litter campaigns, usage of renewable energy for healthy living, and recycling of inorganic materials.

Users. Person (concerned individual, resident, environmentalist, media)

D.4.5.13 Use Case 13

Use Case Name. Enact laws.

Brief Description. Laws and regulations are the major tools in protecting the environment. The concerned individuals write to their congressional representatives, who vote on the pollution control, fix budgets for enforcements of safety regulations and the preservation of forests and wildlife. To put these laws into effect, congress authorizes certain government agencies to create and enforce regulations.

Users. Person (concerned individual, senator, regulator)

D.4.5.14 Use Case 14

Use Case Name. Reuse.

Brief Description. The items that students discard at the end of the school year are collected in volunteer run events. Student volunteers print flyers to promote the event. The donors place the items in collection boxes. This is later sold and the profits are given to charity.

Users. Person (volunteer, donor, buyer)

D.4.5.15 Use Case 15

Use Case Name. Refurbish.

Brief Description. Volunteers salvage used computers by repairing, upgrading, and later donating them to nonprofit organizations, schools, low-income families, and people with disabilities. The volunteers work with the recipient organizations to set up the equipment.

Users. Person (volunteer, donor, recipient)

D.4.6 Interfaces

Following is the list of interfaces required for pollution management:

1. *EPA.* A government organization works to protect human health and the environment.
2. *Environmental management system.* A set of processes and practices that enable an organization that can reduce environmental impacts and help increase operating efficiency.

Sources

Hales, D. (2012). "An Invitation to Health." 15th Edn., Stamford, CT: Cengage Learning, January.

Hill, M. K. (2010). *Understanding Environmental Pollution*, 3rd Edn. Cambridge, UK: Cambridge University Press, May 24.

"Spill in China Underlines Environmental Concerns," *The New York Times*, March 2, 2013.

D.5 NATURAL DISASTER TRACKING SYSTEM

D.5.1 Goals/Purposes

- To build a successful tracking system that tracks natural disasters with a fair degree of accuracy and precision.
- To minimize the loss of life in the event of natural disasters.
- To minimize the costs that are associated with natural disasters, by involving different agencies.

D.5.2 Motivations

It is a well-known fact that natural disasters can strike all countries, both developed and developing. They may become the essential reasons for causing massive destruction, creating human sufferings, and producing harmful impacts on national economies. Due to the different climatic conditions that are present in different parts of the globe, various types of natural disasters like floods, droughts, earthquakes, cyclones, landslides, and volcanoes strike according to the susceptibility of the given area.

D.5.3 Brief Description

A natural disaster tracking system (NDTS), as shown in Figure D.4, aids scientists, media, government agencies, and the general public to be more aware and prepared to face the difficulties of catastrophic events.

D.5.4 Challenges

- To gather all the necessary data and to interpret them in real time.
- To accurately estimate time and place of natural disasters.
- To communicate critical information with external entities.

D.5.5 Accomplishments

- Reduce loss of life.
- Better evacuation plans.
- Better informed populace.

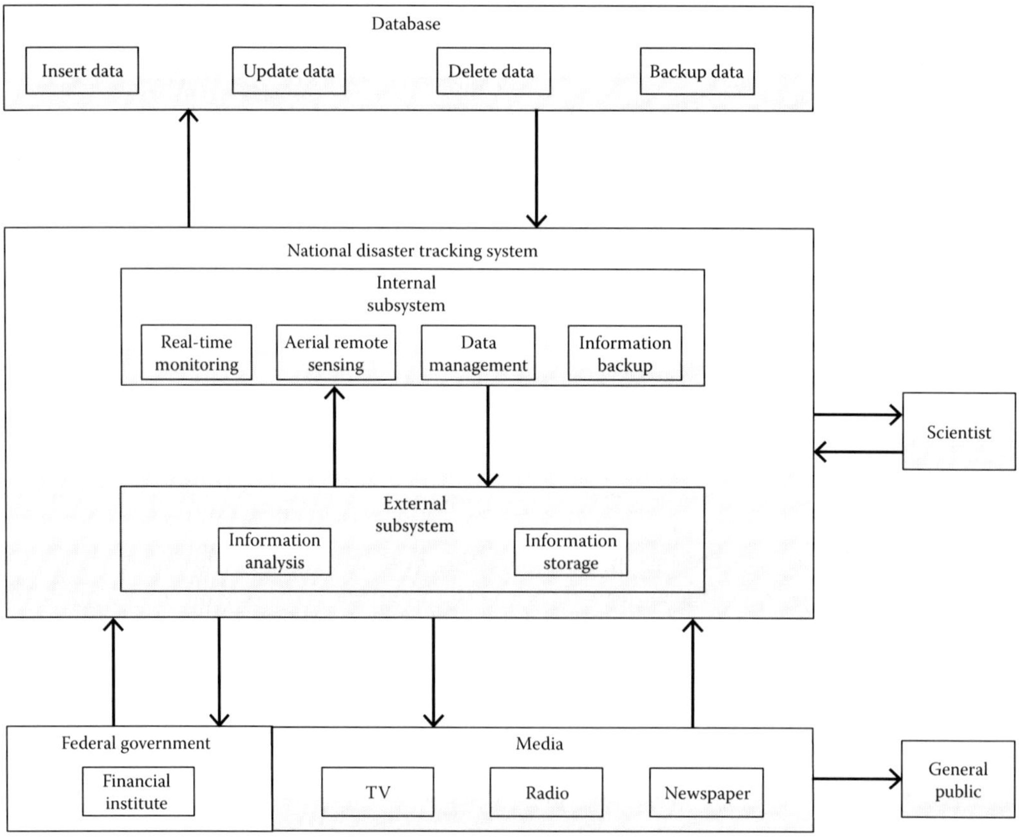

FIGURE D.4 Nature disaster tracking system's block diagram.

D.5.6 Project Results

- Successfully track all natural disasters.
- Successfully involve all external entities by including scientists, media, government agencies, and the public.

D.5.7 Description of the Domain

Some of them have been explained below diagrammatically:

A *volcano* is mainly an opening (or rupture) in the earth's surface or crust, which allows hot, molten rock, ash, and gases to escape from deep below the surface. Volcanic activity involving the extrusion of rock tends to form mountains or features like mountains over a period.

Hurricane, tropical cyclone, and typhoon are different names for the same phenomenon, that is, a cyclonic storm system that forms over the oceans. It is caused by evaporated water that comes off the ocean and becomes a storm, see Figure D.5.

D.5.8 Block Diagram

The block diagram (see Figure D.4) shows an overview of an NDTS and its mode of interaction with the outside entities, when a large natural disaster occurs. On the top level, the NDTS communicates with an external database system, which is in charge of storing, deleting, and updating sensitive information that are obtained from the NDTS. The NDTS is further divided into internal and external subsystems. The internal subsystem is in charge

FIGURE D.5 Hurricane.

of the core activities of the NDTS, such as data management, local data backup, aerial remote sensing, and real-time monitoring. The external subsystem is in charge of information/data analysis and information storage that are obtained from the internal subsystem. On the lower level of the diagram, there is real-time information exchange with the government and federal agencies, and the media to inform the public, should the possibility of a disaster becomes eminent.

D.5.9 Use Cases and User Context

D.5.9.1 Use Case: Gather Satellite Information

ID. UC 01.

Brief Description

Communication is established with the satellite.
Images and data are gathered by the satellite, as per request.
Images are transferred to NDTS.
System stores the updated information.
NDTS process the information received by the satellite.

Actors. Scientists, NDTS, satellite

D.5.9.2 Use Case: Communicate with Media

ID. UC 02.

Brief Description

Effective communication with the local and national television and radio.
Verification by NDTS.
Communication with newsgroups.
Newsgroups inform general public.

Actors. Newsgroups, NDTS, external tracking system, government agencies

D.5.9.3 Use Case: Manage and Model Data

ID. UC 03.

Brief Description

Provides detailed information about the solar winds, sea magnitude level, and global climate.
Saving and loading of updated information.
Predictions based on data processing.
Updating status of information like forecasts.
Raw information sent to and received by external tracking system.

Actors. Database administrator, scientists, external tracking system

D.5.9.4 Use Case: Backup and Transfer Information

ID. UC 04.

Brief Description

Backup critical information and electronically transfer it to other locations in the country, with a low risk of an impact.

Validation of data.

Database administrator archives the data. Data transferred electronically in case of a possible urgent situation.

Validation of backed up data.

Actors. Database administrator, External tracking system

D.5.9.5 Use Case: Monitor in Real Time

ID. UC 05.

Brief Description

Monitoring the planet for its changes on a real-time basis to ensure that during unpredicted event, critical information is always available.

Acquire all information from latest point of interests.

Update and store all new information and details.

Generate new models on the basis of the newly gathered information.

Update front-ends.

Actors. Scientists, NDTS, external tracking system, NDTS employees

D.5.9.6 Use Case: Gather Remote Aerial Sensor Data

ID. UC 06.

Brief Description

Allow us to gather data much faster than ground-based observation. It will help in obtaining photographs from sensors, in order to predict the early set in of a disaster.

Request information from aerial transmitters.

Information along with images sent to NDTS.

Processed information verified and stored.

Actors. Scientists, NDTS, external tracking system

D.5.9.7 Use Case: Correlate Information

ID. UC 07.

Brief Description

Correlation of information to obtain better and enhanced data and communicating this information with external systems.

Save the updated information.

Correlate the information to external systems to allow constant updates.

Validation of correlated information by external systems.

Actors. Scientists, newsgroups, government agencies

D.5.9.8 Use Case: Allocate and Manage Resources

ID. UC 08.

Brief Description

Properly allocate resources and investments to acquire maximum system throughput.

Allocation of resources.

Depending upon the allocated resources, system performance is maximized.

Actors. System administrator, NDTS, government agencies

D.5.9.9 Use Case: Track Natural Disasters

ID. UC 09.

Brief Description

Provides public awareness to decrease the damage repair cost in a radical manner.
Tracking the occurrence of a natural disaster.
Analyzing and predicting the data.
Examining the predictions to track disasters.

Actors. Natural disasters, external tracking system

D.5.10 Detailed Requirements

The detailed requirements consist of two different types of requirements: (1) functional requirements and (2) nonfunctional requirements.

D.5.10.1 Nonfunctional Requirements

In the United States, the direct cost of carrying out natural damage repair is over 20 billion dollars per year. One of the main reasons for such an enormous expense is the perceived inability to correlate and communicate the information or details that are gathered from the research and forecasting system to the public, federal government, and emergency services. As a result, more harm is done, even though the severity of natural disasters has not increased dramatically for over a decade.

Because it has become extremely necessary to increase public awareness, our novel proposal for implementing a natural disaster system would not only decrease the damage repair cost in a dramatic manner, but also reduce the total number of causalities by at least 10%–20%. Our system would also be able to help decision makers to allocate resources and investments to acquire maximum system throughput and results.

D.5.10.2 Functional Requirements

The NDTS has to have the following criteria:

1. In the event of an upcoming natural disaster, the system shall be able to communicate effectively with the local and national television and radio media. This is particularly very useful and beneficial because local and national media could effectively communicate the upcoming incidence of danger to the public, so that further actions such as evacuation plan could be easily accommodated.
2. The NDTS shall also incorporate the satellite orbiting system. The satellite orbiting system provides scientists and researchers with useful information, statistics, and images that could precisely foretell an upcoming natural disaster with extremely precise probability.
3. The NDTS shall also incorporate the hydrologic prediction system (HPS). The HPS is a real-time data management and modeling system that provides detailed information about the solar winds, sea magnitude level, and global climate.
4. In the event of an emergency, the NDTS shall also be able to back up critical information and electronically transfer it to other locations in the country with low risk of an impact.
5. Our system would be more sophisticated and multidisciplinary in providing analysis of disasters, before their occurrence compared to other existing tracking systems.

6. The capacity of our system also extends to monitoring the planet for its changes on a real-time basis, to ensure that during unpredicted events, all the critical information required is available, when it is most needed.
7. Our system will also have the capability of remote sensing, which will allow us to gather data much faster than ground-based observation. Aerial remote sensing will also help in obtaining photographs from sensors, in order to predict the early set in of a disaster.

D.5.11 Interface

The NDTS is required to interface with various devices and sensors that are used in the field to gather necessary data. This interface is very critical to help predict and locate natural disasters; therefore, it needs to be reliable, useful, and performance oriented.

The NDTS is also required to interface with the media, scientists, and general users. It is therefore critical to have an easy-to-use interface, which brings the interpreted, but also raw data to them. A web-enabled interface would be able to reach all these users and would therefore be needed.

The NDTS also needs to communicate with other tracking systems, and again a web interface seems to be the most logical solution. It is critical to be connected and sharing information.

Sources

Davis, L. (2008). "Natural Disasters." New York, NY: Infobase Publishing, 2008.

Knabb, R. D.; Rhome, J. R.; Brown, D. P. (2012). National Hurricane Center (December 20, 2005) (PDF). Hurricane Katrina: August 23–30, 2005 (Tropical Cyclone Report). United States National Oceanic and Atmospheric Administration's National Weather Service. Retrieved December 10.

Lockwood, J. P.; Hazlett, R. W. (2010). Volcanoes: Global Perspectives. 2010, p. 552.

Miles, M. G.; Grainger, R. G.; Highwood, E. J. (2004). "The significance of volcanic eruption strength and frequency for climate." Quarterly Journal of the Royal Meteorological Society 130: 2361–2376.

NSTA Press (2007). "Earthquakes, Volcanoes, and Tsunamis." Resources for Environmental Literacy. Retrieved April 22, 2014.

Sheikh, P. A. (October 18, 2005). "The Impact of Hurricane Katrina on Biological Resources." Congressional Research Service. Archived from the original on June 24, 2008. Retrieved April 14, 2010.

D.6 GLOBAL WARMING CONTROL SYSTEM

The global warming control system (GWCS) (see Figure D.6) represents the intensive effects of global warming on the world climatic system and those control measures that are necessary to reduce global warming, along with possible ways to deal with all predicted consequences. Indeed, the manifestations of global warming have significantly intensified over the past few years, most evident in globally documented changes or shifts in weather patterns, leading to drought in some areas and typhoons of immense magnitude in other places. Of late, an increasing number of scientists have voiced out their opinions, that left unchecked or unattended, the ill effects of global warming could potentially be

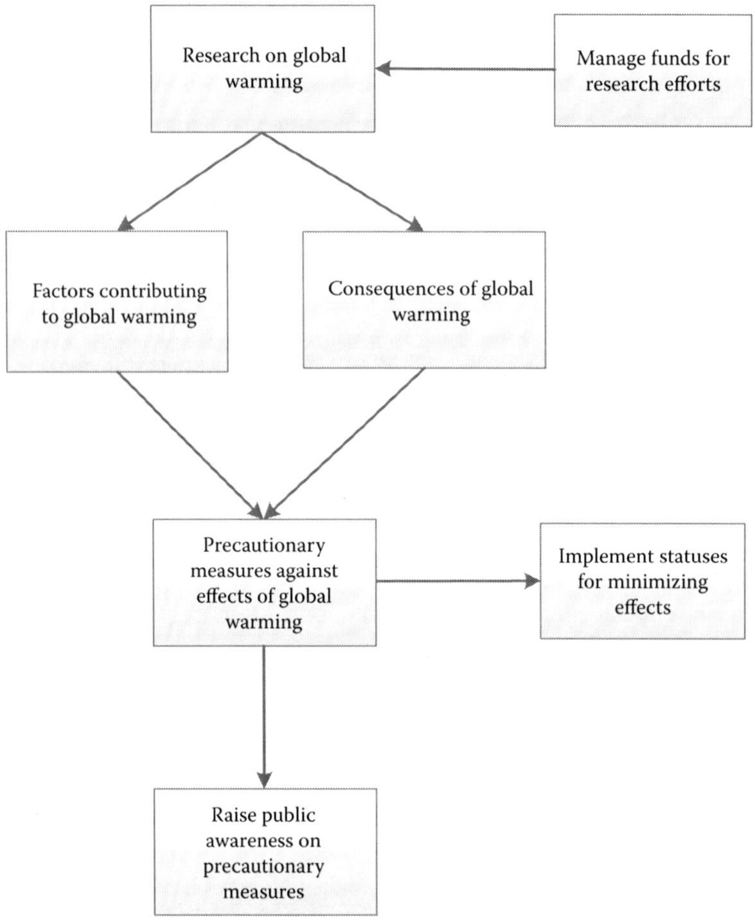

FIGURE D.6 Block diagram on overall coverage of the global warming control system.

catastrophically disastrous. Therefore, the relevance of studying the factors that are affecting global warming could not perhaps be overstated.

The study of global warming and its damages shows the increased frequency/intensity of extremely fickle weather situations like floods, droughts, heat waves, hurricanes, and tornados. Other consequences also include higher or lower agricultural yields, increased glacial retreat, reduced summer stream flows, species extinctions, and increases in the ranges of disease vectors. Simply studying the effects, however, should not preclude the analysis of the causes that have given rise to the global warming issue. More importantly, taking both cause and effect into consideration would allow all concerned parties to propose a set of appropriate preventive measures to mitigate the escalation of global warming to even more alarming levels. Moreover, making a concerted effort to educate and involve the citizenry in every country in the world, about the causes and effects of global warming, may spur and initiate them to make their own contributions, such as buying hybrid vehicles or patronizing environmentally friendly energy resources. In addressing the important global warming issue, every contribution really counts, no matter how seemingly small in scale.

In fact, many organizations and institutions are operating all over the world, which are all focused on studying the factors that are contributing to global warming and subsequently controlling its effects. For example, government officials have proposed a number of laws that will require certain industries to control their greenhouse gas emissions to an internationally acceptable level. Nonprofit environmental organizations have also been launching information campaigns in various areas around the world, through different media, in an attempt to increase local public awareness on the possible preventive measures that may be taken at present.

GWCS is an attempt to illustrate the process, whereby different persons, organizations, and institutions attempt and try to understand the factors and parameters that are necessary to reduce or reverse the effects of global warming. Necessary precautions or measures, dealing with the avoidance of the predicted dire or serious consequences of global warming, are also taken into consideration.

D.6.1 Description of the Domain

Global warming is the pronounced increase in the average temperature of the earth's atmosphere and oceans. The human factors that affect global warming may include the increased amount of carbon dioxide (CO_2) and green house noxious gases released from burning of fuels, the generation of industrial wastes, and the carbon emission of home appliances. Researchers have predicted that the temperature may increase between 2.5°F and 10.5°F in the next 10 years or so. By itself, temperature increase is not entirely alarming; however, taking into consideration the disruption on weather patterns, this abnormal temperature increase may cause around the world many disastrous consequences, should the effects of global warming be allowed to escalate unmitigated.

Reaching a common consensus on several approaches of dealing with the realities of the global warming issues is an important step in reducing the factors contributing to the rapid escalation of global warming effects. To control global warming, concerned individuals from various institutions and organizations like government officials and volunteers from nonprofit organizations should understand various factors that lead to global warming. Later, they should propose precautionary measures that are necessary to control the bad effects of global warming in order to minimize its dangerous consequences, as well as promote awareness among the public, on the contributions ordinary citizens can make regarding this issue. All concerned persons across the globe must make a concerted effort to participate in activities, like patronizing more environmentally friendly products.

Further research efforts that build on existing studies about the factors contributing to global warming should also be undertaken regarding this critical issue. Although much has been uncovered in recent years about the factors contributing to global warming, as well its projected effects in the near future, a lot still needs to be done to bring a greater sense of clarity on the pending issue. To this end, defining a comprehensive and cohesive model to encompass the various factors that are related to global warming is thought to be a crucial step in bringing into clear focus what has already been done and the many tasks that still lie ahead.

D.6.2 Description of the Program That Is Required

Here we hope to develop the following subsystems within the GWCS:

- The GWCS brings together the organizations and institutions that are dealing with both studying the factors and parameters affecting global warming and controlling its effects. The system must contain many provisions to enlist new organizations with the same level of thrust and enthusiasm, whenever and wherever possible. Collectively, these organizations work together to create greater awareness within the society, regarding the effects of global warming, support ongoing research efforts in relation to global warning causes and effects, and also encourage the introduction of preventive measures aimed at reducing the effects of global warming, wherever appropriate.

- The GWCS also maintains a solid knowledge base of various human and non-human factors that are uncovered during the research processes, which also ultimately contribute to the escalation of the effects of global warming, such that this escalation is never left unchecked.
- These organizations and institutions collaborate in finding means and methodologies for controlling the human factors that are known to cause global warming.
- Further, the GWCS also consolidates the results from various, past research studies to identify the possible hidden, invisible factors that potentially lie unnoticed, which may result to disastrous effects later on.
- The GWCS also deals with predicting the consequences of global warming and the actions that need to be undertaken, should critical conditions arise because of the escalation of global warming.
- The GWCS also deals with collecting and managing funds from various organizations and institutions for disbursement to the appropriate researchers for their global warming related areas of study.
- The GWCS catalogs the laws and provisions proposed and approved by government officials, aimed at curtailing the greenhouse emissions both in the commercial and household arenas.
- The GWCS is also involved with the various awareness campaigns that are launched by volunteers from environmental organizations, to contribute to educating the public on the factors, effects, and preventive measures related to global warming.

D.6.3 Detailed Requirements

The detailed requirements consist of two different types of requirements: (1) functional requirements and (2) nonfunctional requirements.

D.6.3.1 Functional Requirements

This section discusses the functional requirements of each of the subsystems requirements of GWCS.

- *Consolidate organizations and institutions working to reduce global warming.* The GWCS maintains a comprehensive list of various voluntary and government organizations that are working together to reduce the effects of global warming. Any new organization with similar goals and objects may be included at any time.
- *Finance management.* The GWCS is also responsible for managing funds sourced from various organizations and institutions. Funds may be disbursed for research efforts that are related to global warming causes, along with finding ways to control global warming.

- *Study factors affecting global warming*. The researcher provides various human and nonhuman factors that affect global warming. This also involves disseminating new findings and advanced research into previously uncovered factors.
- *Monitor damages caused by global warming*. The various adverse effects of global warming such as extinction of species, new global warming-related diseases found, disasters induced by climate change, and other factors are efficiently and properly documented. The system also monitors the ozone layer, which is always checked to determine the nature of global warming and the latest extent of its effects.
- *Literate people on global warming*. The volunteers and other concerned parties from organizations and institutions involve themselves in making the public aware of the various factors leading to the escalation of global warming and the necessary precautionary measures that should be undertaken. In particular, disaster management in the face of emergencies induced by major changes to the climate should also be brought into emphasis.
- *Provide environmental policies governing global warming*. The government officials, who are also effectively the policy makers, will also participate in the policy-making by implementing statutes for mitigating the causes of global warming, such as measures for reducing greenhouse gas emissions. Examples are energy taxes on usage of appliances emitting green house gases and a CO_2 tax. Incentives such as industry–government agreements on energy efficiency in appliances, and the reduction on vehicular emissions are some of the other proactive programs that are being followed in some countries.

D.6.3.2 Nonfunctional Requirements

This section discusses the nonfunctional requirements or the quality factors of GWCS.

- Understandability.
  The basic understanding of global warming and its effects on the world is necessary in order to implement this project
- Data requirements.
  The GWCS also requires information and data regarding the various factors that are responsible for causing global warming.

D.6.4 Use Cases

D.6.4.1 Use Case 1

Use Case Name. Consolidate organizations and institutions.

Brief Description. Submit any concerned parties within an environmental organization or institution, in both the private and public sectors, for inclusion into the GWCS. These organizations and institutions are thus provided a common avenue or platform to share and access information regarding global warming.

Users. Person (environmental organization volunteer, government official, media reporter)

D.6.4.2 Use Case 2

Use Case Name. Allocate funds for research.

Brief Description. The GWCS manages funds collected from various organizations and institutions and disburses the funds to the endorsed research study being conducted that is related to global warming.

Users. Person (treasurer, researcher)

D.6.4.3 Use Case 3

Use Case Name. Inspect factors affecting global warming.

Brief Description. Conducts a research study to identify the human and nonhuman factors that are affecting global warming in order to discover the possible consequences arising from these factors. Furthermore, based on the factors identified, prepare future recommendations and suggestions for managing the effects resulting from the factors uncovered.

Users. Person (scientist, engineer, technical writer)

D.6.4.4 Use Case 4

Use Case Name. Manage environmental consequences.

Brief Description. Regularly monitor the effects that arise from the prevalence of the different factors that contribute to the global warming. Activities may involve tracking potential climate changes, ozone layer depletion, reduction in the polar ice caps, and so on in order to mitigate the potential damages.

Users. Person (scientist, engineer, meteorologist)

D.6.4.5 Use Case 5

Use Case Name. Maintain statutes governing global warming factors.

Brief Description. Propose laws and policies for mitigating the prevalence of factors affecting global warming. This may cover cases such as regulating green house gas emissions by certain industries and introducing incentives for manufacturing and purchasing energy-efficient appliances and vehicles. Compliance with these laws and policies must constantly be monitored and enforced, and appropriate legal action should be taken whenever necessary.

Users. Person (government official, law enforcer, manufacturing corporation employee, shopper)

D.6.4.6 Use Case 6

Use Case Name. Educate the public regarding preventive measures.

Brief Description. To mitigate the growing consequences of the effects of global warming, governmental and private institutions and volunteer-initiated efforts must focus on building awareness among local residents. This may be done through disseminating practical, energy-saving ways and means that they can undertake, in order to save not only money, but also protecting the environment during the process. Such ways may include using energy-efficient appliances, recycling, patronizing alternative energy resources, or contributing to funds for research initiatives concerning global warming. The media may also contribute through making the information available by using various means such as radio, television, and the Internet.

Users. Person (resident, environmental organization volunteer, government official, media reporter)

D.6.5 Interfaces

- Air pollution system
- Meteorological forecasting system
- EPA

Sources

Bello, D. (4 September 2009). "Global Warming Reverses Long-Term Arctic Cooling." *Scientific American*. Retrieved June 8, 2011.

Meehl, G. A. et al. (18 March 2005). "How Much More Global Warming and Sea Level Rise." *Science* 307 (5716): 1769–1772.

Weart, S. (2008). "The Carbon Dioxide Greenhouse Effect." *The Discovery of Global Warming*. American Institute of Physics. Retrieved April 21, 2009.

D.7 CIRCUS

I love to walk in rain because nobody knows I am crying.

These momentous and gripping words, filled with extreme gloom overridden by unbridled ecstasy, originated from one of the greatest comedians ever to live, Charlie Chaplin. There are too many people in this world like him who try to drive away the gloom with their art and relinquish the life of the common man with happiness, joy, and enjoyment.

Our project *CIRCUS* is also inspired from such an abstract art. The objective behind this project is to model the performances in a circus show (Figure D.7).

"Circus is a place where animal can see humans acting fool." The idea behind this quote is circus is a platform where skilled performers like jugglers, acrobats, clowns, and animals present their skills to entertain the audience.

By the end of this project, one will know what it costs to bring a smile on the face of hundreds and thousands of people.

D.7.1 Description of the Domain

Circus, a Latin word derived from the Greek word *Kirkos*, *Circle, and Ring* alludes to a place where a group of performers like acrobats, clowns, trained animals, tight rope walkers, jugglers, and other stunt artists demonstrate their skills. Here, horses, ponies, and elephants are allowed to see men, women, and children acting the fool. It is typically a circular arena with tiered seating. Most circuses have to move from one place to another, in which case they usually take place in a huge tent.

Circus is a source of entertainment since ancient Roman civilization, which is considered to be the era when circus was originated. In early roman circuses, exhilarating performers fought to the death for freedom; gallant equestrians and chariot races amused the Roman people. Even in the present day, circuses remain the most perennial form of entertainment.

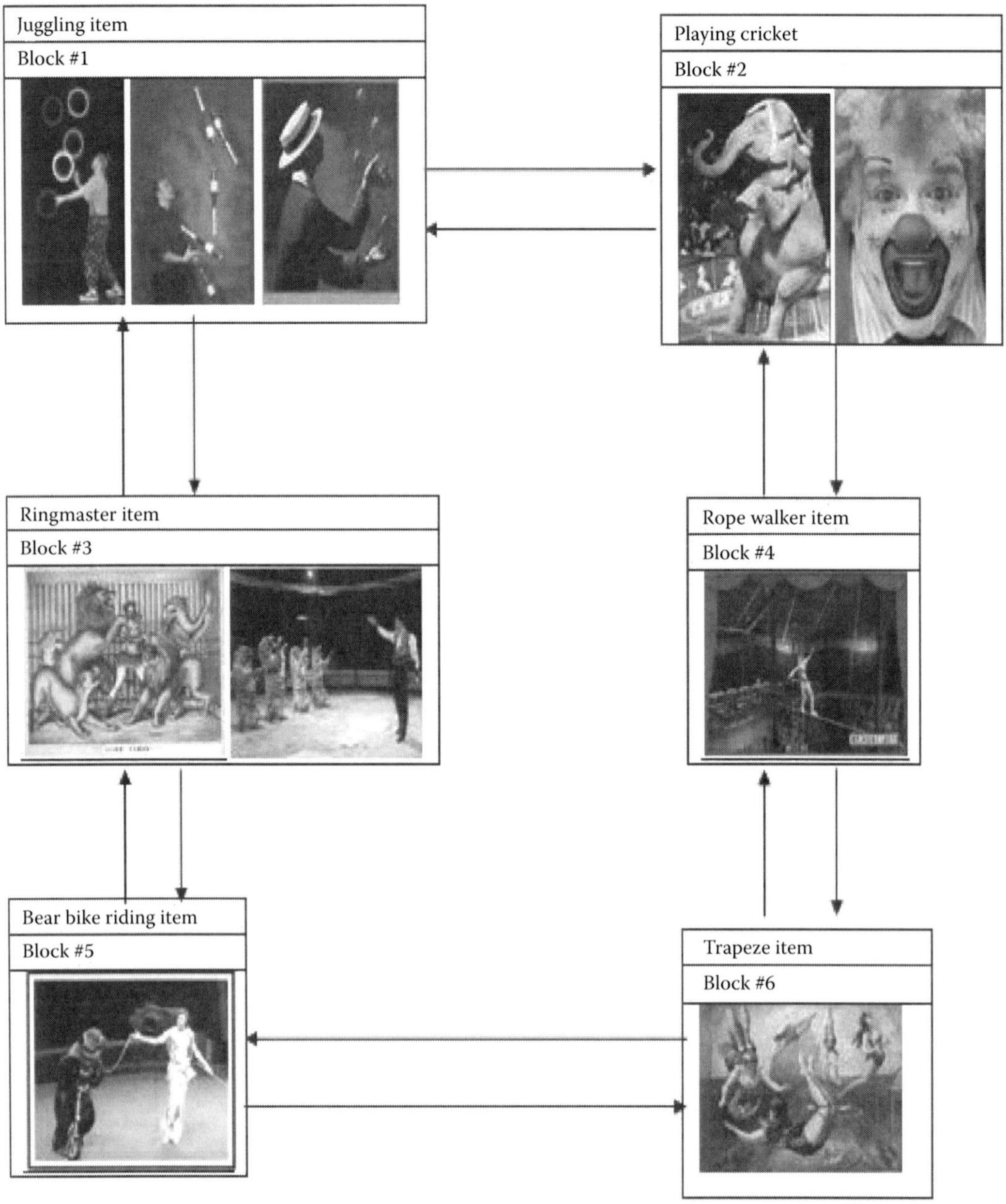

FIGURE D.7 Circus block diagram.

Apart from entertainment being the chief motto, circus is an epitome of art that helps in boosting and exhibiting different genre of talents.

D.7.2 Description of the Program That Is Wanted

Our goal is to build a system that is a model of various shows (*items* called in the circus jargon) in the circus. The model should illustrate all the activities and interactions that happen between the performers and the system related to following items:

1. The elephant show
2. Rope walker show
3. Trapeze show
4. Ringmaster show
5. Bear riding a bike show
6. Juggler show

D.7.2.1 Detailed Requirements

Following are the requirements to model the different items that are shown in Figure D.7.

- *The elephant show.* This is an interesting item, which is highly amusing among all other circus items. A trained elephant is required to play cricket with the clowns. Elephant should know how to hold a bat with its trunk, and it should also know how to hit the ball using the bat into the audience. Some clowns are also required to throw the ball to the elephant. They are also responsible for making foolish gestures to entertain the audience. Clown should recollect the ball from the audience. A commentator is also required to comment during the item.
- *Rope walker show.* For this item, an experienced acrobat is required to walk on the rope tied between distant poles. The acrobat is required to use a stick to balance himself on the rope. A security net is required to be tied below the rope to avoid any damage caused to acrobat in case he falls from the rope.
- *Trapeze show.* For this item, expert trapeze artists are required to swing on the trapeze. They are required to balance themselves on the trapeze. They should leave the trapeze while they are on a swing and should perform rolls, somersaults, and twists in the air and again hold the other trapeze. A security net is required to be tied below the trapeze to avoid any damage caused to the performers in case they fall from the trapeze.
- *Ringmaster show.* For this item, a trained ringmaster and lions are required. Ringmaster should instruct the lions to climb on the table and make gestures to amuse the audience. Ringmaster should also control and protect himself from the lions. Lions are required to jump through the ring of fire.
- *Bear riding a bike show.* For this item, a trained bear and a bear instructor are required. The bear should ride a bike according to the instructions of the instructor. The instructor should hold the rope tied to the bear. The instructor should instruct the bear in a way that the bear rides the bike in a circular manner.
- *Juggler show.* For this item, an experienced juggler is required. Juggler needs to juggle glass bottles, balls, rings, and sticks. He should also balance the bottles on his head, nose, and chest.

A host for all the items is required, who should comment during performance to create enthusiasm among the audience. The audience is required to appreciate the performances and motivate the performers.

D.7.3 Use Cases and User Context

The following section provides the names, brief descriptions, and the actors' names and their corresponding roles' names of eleven different use cases.

D.7.3.1 Use Case 1

Use Case Name. Playing Cricket.

Description

Elephant will hold the bat from its trunk.

Clown will throw the ball to the elephant.

Elephant will hit the ball with the bat into the audience.

Audience will return the ball to the clowns.

Actors/Roles. Creature (elephant), human (clowns, audience)

D.7.3.2 Use Case 2

Use Case Name. Performing rope balancing.

Description

Acrobat will climb on the platform with the help of hanging stairs.

Acrobat will go from one end of the platform to another end on the rope.

Acrobat will wave his hands to the audience.

Acrobat will come down through the hanging stairs.

Actors/Roles. Human (acrobat)

D.7.3.3 Use Case 2.1

Use Case Name. Going from one end of the platform to another end.

Description

Acrobat will pick the stick in his hands.

Acrobat will walk on the rope.

Acrobat will maintain his balance by using the stick.

Acrobat will concentrate on his work and ignore the surrounding noise.

Actors/Roles. Human (acrobat)

D.7.3.4 Use Case 3

Use Case Name. Performing the trapeze show.

Description

Trapeze artists will climb on their respective platforms using hanging stairs.

Trapeze artists will swing on the trapeze to reach other platform.

Trapeze artists will perform the swinging activity several times.

Trapeze artists will jump on the security net to get down and finish the act.

Actors/Roles. Human (trapeze artist)

D.7.3.5 Use Case 3.1

Use Case Name. Swinging from one platform to another.

Description

Trapeze artist will hold the trapeze.

Trapeze artist will start swinging.

Trapeze artist will leave the trapeze, perform rolls, somersaults, and twists in the air and hold the other trapeze back.

Trapeze artist will again come back to the platform.

Actors/Roles. Human (trapeze artist)

D.7.3.6 Use Case 4

Use Case Name. Presenting ring master show.

Description

Ringmaster will instruct the lions to come out of the cage.
Lions will come out of the cage.
Ringmaster will control the displeased lions.
Lions will climb on the respective tables.
Lions will make gestures to the audience.
Lions and ringmaster will carry out *ring of fire*.

Actors/Roles. Human (ringmaster), creatures (lions)

D.7.3.7 Use Case 4.1

Use Case Name. Controlling the displeased lions.

Description

Displeased Lions will attack the ringmaster.
Ringmaster in turn will use hunter to defend himself.
Lions will roar on the ringmaster after beaten up.

Actors/Roles. Human (ringmaster), creatures (lions)

D.7.3.8 Use Case 4.2

Use Case Name. Performing *ring of fire*.

Description

Ringmaster will instruct the lion to climb on the platform.
Lion will climb on the platform.
Ringmaster will hold the ring of fire.
Ringmaster will direct the lion to jump through the ring of fire.
Lion will jump through the ring of fire.

Actors/Roles. Human (ringmaster), creatures (lions)

D.7.3.9 Use Case 5

Use Case Name. Performing the bike ride item.

Description

Bear instructor will hold the belt tied with the bear.
Bear instructor will instruct the bear to climb on the bike.
Bear will climb on the bike.
Bear will start riding the bike.
Bear instructor will give directions to the bear to ride the bike in circular manner.

Actors/Roles. Human (bear instructor), creature (bear)

D.7.3.10 Use Case 6

Use Case Name. Juggling.

Description

Juggler will pick the items among bottles, sticks, and balls.
Juggler will juggle the item picked.
Juggler will balance the item on his head, nose, or chest.

Actors/Roles. Human (juggler)

D.7.3.11 Use Case 7

Use Case Name. Hosting the show.

Description

Host will announce the upcoming item.
Host will comment during the item.
Host will ask audience to applaud on a performance.
Audience will clap on the performance.
Host will thank the performers and audience.

Actors/Roles. Human (host), human (audience)

Sources

Speaight, G. (1980). "A History of the Circus," The Tantivy Press, London.

Stoddart, H. (2000). "Rings of Desire: Circus History and Representation," Manchester University Press, Manchester.

D.8 JURASSIC PARK

D.8.1 Goals/Purposes

To model Jurassic Park by incorporating scenarios with interaction between dinosaur and human beings.

D.8.2 Motivation

- Jurassic Park is a hard-hitting drama of two different worlds coming together: one that existed almost 65 million years ago, and the other of the present time.
- It resurrects the greatest and most powerful creatures our planet has ever known, which ruled the world for 165 million years. It extends the limits to human imagination, thereby resulting in extraordinary things happening to ordinary people. This powerful experience was certainly worth modeling.

D.8.3 Description of Domain

Jurassic Park was created on Isla Nubar, an island in the Pacific Ocean off the coast of Costa Rica. It consisted of a research lab for the study and cloning of dinosaurs, the storage of scientific data, the visitor center, and the central control, which monitors all the activities on the island. The medical center provided care for injured and diseased dinosaurs. The major part of the island was intended to be a showroom for cloned dinosaurs. There were a number of enclosures with dinosaurs in separate regions. Programmed excursion routes were being designed to allow visitors to navigate different regions of the park.

- Entry to the Jurassic Park (see Figure D.8).
- Attack by the Tyrannosaurus on the children's car (see Figure D.9).
- Attack on the visiting experts (see Figure D.10).

FIGURE D.8 Entry to the Jurassic Park.

FIGURE D.9 Attack by the Tyrannosaurus on the children's car.

D.8.4 Jurassic Park Block Diagram

D.8.4.1 Description of the Program That Is Wanted

Our aim is to model most of the significant events that occur in the movie right from the creation of the dinosaurs, the attempt to harbor a controlled environment, where humans can see the dinosaurs in human life, and later the failure of the entire system, which leads to the break out of the dinosaurs, deaths, and injuries of many of the people in the park. We

FIGURE D.10 Attack on the visiting experts.

will bring out maximum moments of interaction between the dinosaurs and humans, which were shown in the movie.

D.8.4.2 Detailed Requirements

Following are the requirements to model the different subsystems that are shown in Figure D.11.

1. There was an incident of attack on an InGen employee, while releasing a Velociraptor, leading to CEO John Hammond being pressured by investors for a safety inspection (see Figure D.11).
2. For safety inspection, CEO John Hammond invites paleontologist Alan Grant, paleobotanist Ellie Sattler, chaos theorist Ian Malcolm, and his investor's attorney Donald Gennaro to perform the inspection.
3. The group studies the different enclosures for different species of dinosaurs present in the Park.
4. The group takes a vehicular tour of the park, along with the two children (CEO John Hammond's grand children), namely, Tim and Lex, who just arrive. In the midst of the tour, there occurs a tropical storm. Most InGen employees leave except for Hammond, game warden Robert Muldoon, chief engineer Ray Arnold, and leading computer programmer Dennis Nedry.
5. Bribed by a rival geneticist, Nedry takes an opportunity to shut down the park's security system, so that he can steal dinosaur embryos and deliver them to an auxiliary dock. As a result, the Tyrannosaurus breaks through the deactivated electric fence surrounding its pen, devouring Gennaro, attacking Tim and Lex hiding in the car, and wounding Malcolm. Nedry crashes his Jeep and, while trying to winch it, is killed by a Dilophosaurus. The children and Grant only narrowly avoid being killed and eaten.
6. While hiking to safety the next morning, they discover hatched eggs, which mean that the dinosaurs are actually breeding. Grant realizes that the frog DNA is responsible: some species of frog are known to spontaneously change sex in a single-sex environment.

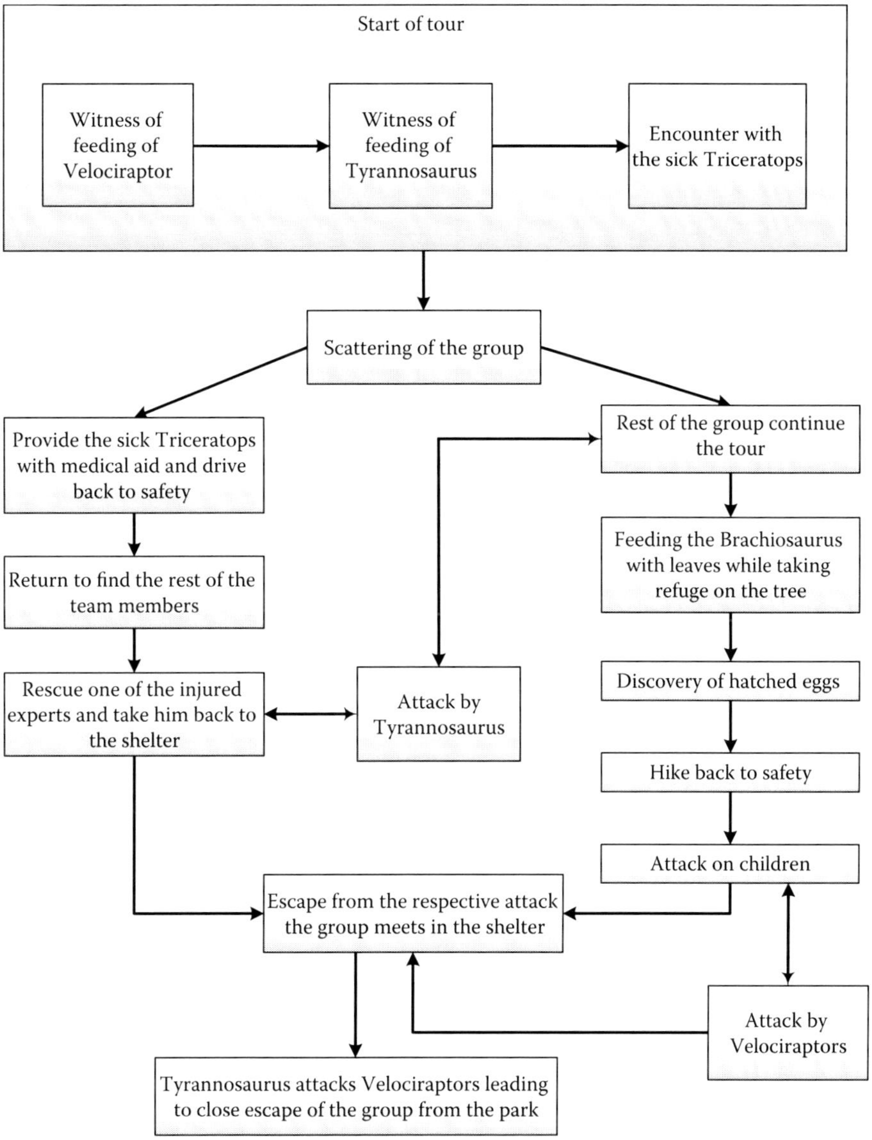

FIGURE D.11 The block diagram of the Jurassic Park.

7. Arnold tries to hack Nedry's computer to turn the power back on but fails, so he does a full system restart, which requires the circuit breakers to be manually reset from the utility shed. When he does not return, Ellie and Muldoon follow and discover the raptors have escaped, the shutdown having cut off power to the electric fences around their pen. Muldoon realizes that they are near and tells Ellie to go to the utility shed herself and turn the power back on while he tries to hunt them down. A lurking raptor attacks and kills Muldoon, while Ellie escapes from another, after discovering Arnold's remains in the maintenance shed.
8. After managing to turn on the power and escaping the raptor, she meets Grant, and they both go back to Malcolm and Hammond in the emergency bunkers. Lex

and Tim narrowly escape from two of the raptors in the kitchen (locking one in the freezer), and Lex is finally able to restore the Park's computer systems in order to call Hammond to request a helicopter rescue of the survivors. Grant and Ellie hold off a raptor that was trying to open the door to the computer room, until the power is restored and the electromagnetic locks begin working.

9. Help from an unlikely source—a Tyrannosaurus suddenly appears and kills both raptors, saving Grant, Ellie, Lex, and Tim in the process.
10. The four then climb into Hammond and Malcolm's Jeep and leave. Grant says he will not endorse the park, a choice with which Hammond concurs. As all fly away in the helicopter, the children fall asleep beside Grant, who contemplatively watches the birds flying nearby, the surviving relatives of the dinosaurs they escaped.

D.8.5 Use Cases

D.8.5.1 Use Case 1: Attack by Velociraptor, While Releasing

Release into special built enclosure.
Attacks the employee.
Lawsuit from the family of the injured.
Other employees try to rescue the victim.

Actors/Roles. Human (employee), creature (velociraptor)

D.8.5.2 Use Case 2: Tour at the Park

Board the vehicle for tour.
Witness dinosaurs being given prey for their food (Velociraptor).
Encounter a sick Triceratops.
Provide medical aid to the sick dinosaur.
Get stranded due to vehicle failure.

Actors/Roles. Human (experts, children, attorney expert), creature (triceratops, velociraptor, cow, goat)

D.8.5.3 Use Case 3: Attempt to Steal Dinosaur Embryo

Shuts down park security system.
Steal dinosaur embryos.
Escape from the research lab, drive to auxiliary dock.
Meet with car accident.
Killed by a Dilophosaurus.

Actors/Roles. Human (computer expert [Nedry]), creature (Dilophosaurus)

D.8.5.4 Use Case 4: Attack Due to Park's Shut Down Security System

Tyrannosaurus break through the park's deactivated electric fence surrounding its pen. Tyrannosaurus devour Attorney expert.
Tyrannosaurus smash children's vehicle.
Tyrannosaurus injures one of the expert.
Children and visiting paleontologist escape from the Tyrannosaurus.

Actors/Roles. Human (experts, children), creature (Tyrannosaurus)

D.8.5.5 Use Case 5: Visiting Paleontologist and Children Hike to Safety

Children and visiting paleontologist hike to safety.
Children feed a Brachiosaurus a branch of leaves.
Visiting paleontologist discovers hatched egg of dinosaurs.
Children flee from a flock of running dinosaurs.
Children witness a Tyrannosaurus hunting down a flock of dinosaurs.

Actors/Roles. Human (children, visiting paleontologist), creature (Brachiosaurus, Tyrannosaurus)

D.8.5.6 Use Case 6: Attempt to Restart Computer and Power

Security guard and visiting paleobotonist run to the maintenance shed to restart system.
Raptor kills security guard.
Visiting paleobotonist follows plumbing lines and finds the control unit.
Visiting paleobotonist turns the power back on.
Raptor attacks visiting paleobotonist.
Visiting paleobotonist escapes to emergency bunkers.

Actors/Roles. Human (children, visiting paleontologist, security guard), creature (Raptor)

D.8.5.7 Use Case 7: Children and Visiting Paleontologist Reach the Shelter

Children and visiting paleontologist encounter the security perimeter fence.
Children encounter a dinosaur approaching them.
Children and visiting paleontologist climb across fence.
Visiting paleobotanist turns on the perimeter fence voltage.
Children run to the shelter.

Actors/Roles. Human (children, visiting paleontologist, security guard), creature (Raptor)

D.8.5.8 Use Case 8: Children Are Attacked by Raptors

Children eat at the tables at the shelter.
Raptors enter the shelter.
Children escape to the kitchen area.
Raptors charge after the children.
Children lock one Raptor in the freezer.
Children escape out of the room.

Actors/Roles. Human (children), creature (Raptor)

D.8.5.9 Use Case 9: Raptors Attack the Control Room

Children, visiting paleontologist, and visiting paleobotanist escape to the computer room. Raptor attempts to open the door to the room.
Grand and visiting paleobotanist hold of the Raptor.
Visiting paleontologist repairs the computer system and the electromagnetic locks begin working.
Raptor breaks into the room through the window.
Children, visiting paleontologist, and visiting paleobotanist climb up the ceiling crawlspace.

Actors/Roles. Human (children, visiting paleontologist, security guard), creature (Raptor)

D.8.5.10 Use Case 10: Escape from the Island

Children, visiting paleontologist, and visiting paleobotanist reach the visitors center.

Raptors surround children, visiting paleontologist, and visiting paleobotanist.

Tyrannosaurus attacks the Raptors.

Children, visiting paleontologist, and visiting paleobotanist escape out of the building.

Children, visiting paleontologist, and visiting paleobotanist leave through a jeep.

Children, visiting paleontologist, and visiting paleobotanist fly away in the helicopter.

Actors/Roles. Human (children, visiting paleontologist, security guard), creature (Raptor)

Sources

Corliss, R. (1993). "Behind the Magic of Jurassic Park." *TIME*, April 26.

"Jurassic Park." Box Office Mojo. Retrieved May 6, 2013.

White, J. (2013). "Jurassic Park Joins The Billion Dollar Club." *Empire*. August 23.

References

SECTION I

Baader, F., D. Calvanese, D. L. McGuinness, D. Nardi, and P. F. Patel-Schneider. eds. *The Description Logic Handbook: Theory, Implementation, and Applications*. Cambridge, UK: Cambridge University Press, 2003.

Barwise, J. ed. *Handbook of Mathematical Logic*. North Holland, the Netherlands, Elsevier, 2006.

Besnard, P. *An Introduction to Default Logic*. Berlin; Heidelberg, Germany: Springer, 1989.

Boddu, R., L. Guo, S. Mukhopadhyay, and B. Cukic. "RETNA: From Requirements to Testing in a Natural Way." Paper presented at the IEEE International Conference on Requirements Engineering, Kyoto, Japan, September, 2004.

Chandrasekaran, B., and J. R. Josephson, and V. R. Benjamins. What Are Ontologies, and Why Do We Need Them? *IEEE Intelligent Systems*, January/February 1999.

Fayad, M. E. "Accomplishing Software Stability." *Communications of the ACM* 45, no. 1 (2002a): 111–115.

Fayad, M. E. "How to Deal with Software Stability." *Communications of the ACM* 45, no. 4 (2002b): 109–112

Fayad, M. E., and A. Altman. "Introduction to Software Stability." *Communications of the ACM* 44, no. 9 (2001): 95–98.

Fayad, M. E., D. C. Schmidt, and R. E. Johnson. *Building Application Frameworks: Object-Oriented Foundations of Framework Design*. New York, NY: Wiley, 1999.

Froehlich, G., H. J. Hoover, L. Liu, and P. Sorenson. "Hooking into Object-Oriented Application Frameworks." Proceedings of the International Conference on Software Engineering, Boston, MA, pp. 491–501, May 1997.

Gruber, T. R. "A Translation Approach to Portable Ontologies." *Knowledge Acquisition* 5, no. 2 (1993): 199–220.

Gruber, T. R. "Toward Principles for the Design of Ontologies Used for Knowledge Sharing." *International Journal of Human-Computer Studies* 43, no. 4–5 (1995): 907–928.

Sanchez, H. A. "Building Systems Using Patterns: Creating Knowledge Maps." Master's Thesis, San Jose State University, San Jose, CA, May 2006.

Shtivastava, P. The Hook Facility, MS Project Report, San Jose State University, San Jose, CA, May 2005.

CHAPTER 1

Appleton, B. "Patterns and Software: Essential Concepts and Terminology." *Object Magazine Online* 3, no. 5 (1997): 20–25. http://www.cmcrossroads.com/bradapp/docs/patterns-intro.html.

Buschmann, F. *Pattern-Oriented Software Architectures: A System of Patterns*. New York, NY: Wiley, 1996.

Chen, Y., H. S. Hamza, and M. E. Fayad. "A Framework for Developing Design Models with Analysis and Design Patterns." Paper presented at the 2005 IEEE International Conference on Information Reuse and Integration, Las Vegas, NV, August 15–17, 2005, pp. 592–596.

Coplien, J. *Software Patterns*. New York, NY: SIGS, 1996.

Fayad, M. E. "Accomplishing Software Stability." *Communications of the ACM* 45, no. 1 (2002a): 111–115.

Fayad, M. E. "How to Deal with Software Stability." *Communications of the ACM* 45, no. 4 (2002b): 109–112.

Fayad, M. E. *Stable Analysis Patterns for Software and Systems*, Auerbach Publications, 2015a.

Fayad, M. E. *Stable Design Patterns for Software and Systems*, Auerbach Publications, 2015b.

Fayad, M. E. *Software Architectures on Demand*, Auerbach Publications, 2015c.
Fayad, M. E., and Altman, A. "Introduction to Software Stability." *Communications of the ACM* 44, no. 9 (2001): 95–98.
Fayad, M. E., H. S. Hamza, and H. A. Sanchez. "Towards Scalable and Adaptable Software Architectures." Paper presented at the IEEE International Conference on Information Reuse and Integration, Las Vegas, NV, 2005.
Fayad, M. E., and H. Kilaru. "Any Information Hiding: A Stable Design Pattern." Paper presented at the 2005 IEEE International Conference on Information Reuse and Integration, Las Vegas, NV, 2005, 108–115.
Fincher, S. "What Is a Pattern Language?" Paper presented at the Chi'99, ACM SIGCHI Conference on Human Factors in Computing Systems in Pittsburgh, PA, May 15–20, 1999.
Fowler, M. *Analysis Patterns*. Reading, MA: Addison-Wesley Professional, 1997.
Gamma, E., R. Helm, R. Johnson, and J. Vlissides. *Design Patterns: Elements of Reusable Object-Oriented Software*. 1st edn. Reading, MA: Addison-Wesley Professional, 1995.
Hamza, H. "A Foundation for Building Stable Analysis Patterns." Master's Thesis Report, University of Nebraska, Lincoln, NE, 2002.
Hamza, H., and M. E. Fayad. "A Pattern Language for Building Stable Analysis Patterns." Paper presented at the 9th Pattern Languages of Programs Conference, Monticello, IL, September 2002.
Hamza, H., and M. E. Fayad. "Applying Analysis Patterns through Analogy: Problems and Solutions." *Journal of Object Technology* 3, no. 4 (2004): 197–208.
Mahdy, A., and M. E. Fayad. "A Software Stability Model Pattern." Paper presented at the Proceedings of the 9th Conference on Pattern Language of Programs, Allerton Park, Monticello, IL, September 8–12, 2002.
Niccolò, M. *Catholic Encyclopedia*. New York, NY: Robert Appleton Company, 1913.
Salingaros, N. A. "The Structure of Pattern Languages." *Architectural Research Quarterly*, vol. 4, Cambridge University Press, pp. 149–161, 2000. http://www.math.utsa.edu/sphere/salingar/StructurePattern.html.
Schmidt, D. C., M. E. Fayad, and R. E. Johnson. "Software Patterns." *Communications of the ACM* 39, no. 10 (1996): 37–39.
Wu, S., H. Hamza, and M. E. Fayad. "Implementing Pattern Languages Using Stability Concepts." Paper presented at ChiliPLoP, Carefree, AZ, March 2003.

CHAPTER 2

Abbot, J. "Program Design by Informal English Descriptions." *Communications of the ACM* 26, no. 11 (1983): 882–94.
Berard, E. V. "Abstraction, Encapsulation, and Information Hiding." White paper, Berard Software Engineering, Inc., Baithersburg, MD, 1991.
Fayad, M. E. "Accomplishing Software Stability." *Communications of the ACM* 45, no. 1 (2002a): 111–115.
Fayad, M. E. "How to Deal with Software Stability." *Communications of the ACM* 45, no. 4 (2002b): 109–112.
Fayad, M. E., and A. Altman. "Introduction to Software Stability." *Communications of the ACM* 44, no. 9 (2001): 95–98.
Fayad, M. E., H. S. Hamza, and H. A. Sanchez. "Towards Scalable and Adaptable Software Architectures." Paper presented at the IEEE International Conference on Information Reuse and Integration, Las Vegas, NV, August 15–17, 2005.
Gruber, T. "Ontology." in *Encyclopedia of Database Systems*, L. Liu and M. T. Özsu (Eds.), Springer-Verlag, 2009.
Gruber, T. R. "Toward Principles for the Design of Ontologies Used for Knowledge Sharing." *International Journal of Human-Computer Studies* 43, no. 4–5 (1995): 907–928.

Gruber, T. R. “A Translation Approach to Portable Ontologies.” *Knowledge Acquisition* 5, no. 2 (1993): 199–220.
Hamza, H., and M. E. Fayad. “A Pattern Language for Building Stable Analysis Patterns.” Paper presented at the 9th Pattern Languages of Programs Conference, Monticello, IL, September 8–12, 2002.
Hamza, H., and M. E. Fayad. “Engineering and Reusing Stable Atomic Knowledge (SAK) Patterns.” Paper presented at the IEEE International Conference on Information Reuse and Integration, Las Vegas, NV, October 27–29, 2003.
Josef, A. “Artist and Teacher, Dies.” *New York Times*. March 26, 1976. p. 33. Retrieved March 21, 2008.
Mahdy, A., and M. E. Fayad. “A Software Stability Model Pattern.” Paper presented at the Proceedings of the 9th Conference on Pattern Language of Programs, Allerton Park, Monticello, IL, September 8–12, 2002.
Sanchez, H. A. “Laying the Foundations for Branding as a Stable Analysis Pattern.” Paper presented at the 19th European Conference on Object-Oriented Programming, 2005.
Sanchez, H. A., B. Lai, and M. E. Fayad. “The Sampling Analysis Pattern.” Paper presented at the Workshop on Timeless and Stable Architectures, IRI’03 Conference, Las Vegas, NV, October 27–29, 2003.

SECTION II

Lapouchnian, A. “Goal-Oriented Requirements Engineering: An Overview of the Current Research Department of Computer Science.” White Paper, University of Toronto, Ontario, Canada, June 28, 2005.

CHAPTER 3

Abbot, J. “Program Design by Informal English Descriptions.” *Communications of the ACM* 26, no. 11 (1983): 882–94.
Anton, A. I. “Goal-Based Requirements Analysis.” Paper presented at the Proceedings of the IEEE International Conference on Requirements Engineering, Colorado Springs, CO, April 15–18, 1996.
Anton, A. I., and C. Potts. “The Use of Goals to Surface Requirements for Evolving Systems.” Paper presented at the 20th International Conference on Software Engineering, Kyoto, Japan, 1998.
Cline, M., and M. Girou. “Enduring Business Themes.” *Communications of the ACM* 43, no. 5 (2000): 101–6.
Fayad, M. E. “Accomplishing Software Stability.” *Communications of the ACM* 45, no. 1 (2002a).
Fayad, M. E. “How to Deal with Software Stability.” *Communications of the ACM* 45, no. 4 (2002b): 109–112.
Fayad, M. E., and A. Altman. “Introduction to Software Stability.” *Communications of the ACM* 44, no. 9 (2001): 95–98.
Fayad, M. E., H. S. Hamza, and H. A. Sanchez. “Towards Scalable and Adaptable Software Architectures.” Paper presented at the IEEE International Conference on Information Reuse and Integration, Las Vegas, NV, August 15–17, 2005.
Fayad, M. E., and S. Telu. “The Learning Stable Analysis Pattern.” Paper presented at the IEEE International Conference on Information Reuse and Integration, Las Vegas, NV, August 15–17, 2005.
Hamza, H., and M. E. Fayad. “A Pattern Language for Building Stable Analysis Patterns.” Paper presented at the 9th Pattern Languages of Programs Conference, Monticello, IL, September 8–12, 2002.
Hamza, H., and M. E. Fayad. “Engineering and Reusing Stable Atomic Knowledge (SAK) Patterns.” Paper presented at the IEEE International Conference on Information Reuse and Integration, Las Vegas, NV, October 27–29, 2003.

Khadpe, P. "Pattern Language for Data Mining." Master's Thesis Report, San Jose State University, San Jose, CA, May 2005.

van Lamsweerde, A. "Goal-Oriented Requirements Engineering: A Guided Tour." Paper presented at the Proceedings of the 5th IEEE International Symposium on Requirements Engineering, Toronto, Ontario, Canada, August 27–31, 2001.

Woodley, M. S., Digital Project Planning & Management Basics, The Library of Congress and the Association for Library Collections & Technical Services, Washington, DC, April 2008.

CHAPTER 4

Bowen, J. *Formal Specification and Documentation using Z: A Case Study Approach.* International Thomson Computer Press, London, 1996.

Brodersen, K. H., C. S. Ong, K. E. Stephan, and J. M. Buhmann. The balanced accuracy and its posterior distribution. Proceedings of the 20th International Conference on Pattern Recognition, pp. 3121–3124, 2010.

Chen, J., G. B. Call, E. Beyer et al. "Discovery-Based Science Education: Functional Genomic Dissection in *Drosophila*, Undergraduate Researchers," *PLoS Biology* 3, no. 2 (2005): P207.

Clark, M. A. *Getting Set for e-Discovery*. EDDix LLC, 2008.

Cohen, A. I. and D. J. Lender. *Electronic Discovery: Law and Practice*, 2011.

Darling, D. J., *The Universal Book of Astronomy from the Andromeda Galaxy to the Zone of Avoidance*, Hoboken, NJ: Wiley, 2004.

Darling, D. J., "Formation of Planetary Systems," *The Internet Encyclopedia of Science*, accessed September 23, 2007.

Davies, J., and J. Woodcock. *Using Z: Specification, Refinement, and Proof.* Prentice Hall International Series in Computer Science, London, 1996.

Email Archiving: A Proactive Approach to e-Discovery, Proofpoint, Inc., Sunnyvale, CA, Archive White paper, July 2008.

Fayad, M. E., and M. Laitinen. *Transition to Object-Oriented Software Development*. New York, NY: Wiley, 1998.

Fayad, M. E. *Stable Analysis Patterns for Software and Systems*, Auerbach Publications, 2015a.

Fayad, M. E. *Stable Design Patterns for Software and Systems*, Auerbach Publications, 2015b.

Fayad, M. E., and S. Wu. "Merging Multiple Conventional Models into One Stable Model." *Communications of the ACM* 45, no. 9 (2002): 102–106.

Grand, M. *Patterns in Java I—A Catalog of Reusable Design Patterns Illustrated with UML*. Wiley, New York, NY, 1998.

The Growing Importance of e-Discovery on Your Business, White paper, Black Diamond, Washington, DC: Osterman Research, sponsored by Google, June 2008.

Hamza, H., and M. E. Fayad. "A Pattern Language for Building Stable Analysis Patterns." Paper presented at the 9th Conference on Pattern Language of Programs, Monticello, IL, September 8–12, 2002.

Hamza, H., and M. E. Fayad. "Stable Analysis Patterns." Paper presented at the Proceedings of the 4th ACS/IEEE International Conference on Computer Systems and Applications, Dubai/Sharjah, UAE, March 8–11, 2006.

Herbert, F., *The Worlds of Frank Herbert*. London, UK: New English Library, 1970.

Jacky, J. *The Way of Z: Practical Programming with Formal Methods*. Cambridge University Press, Cambridge, Angleterre, 1997.

Jacobs, J. M., J. N. Adkins, W. J. Qian, T. Liu, Y. Shen, D. G. Camp, and R. D. Smith. "Utilizing Human Blood Plasma for Proteomic Biomarker Discovery." *Journal of Proteome Research* 4, no. 4 (2005): 1073–1085.

Kyckelhahn, T. and T. H. Cohen. Civil Rights Complaints in U.S. District Courts, 1990–2006, U.S. Department of Justice, Office of Justice Programs, NCJ 222989, August 2008.

Linoff, G. S., and Berry, M. J. A. *Data Mining Techniques for Marketing, Sales, and Customer Relationship Management*, 3rd Edition, Wiley, New York, NY, April 2011.

Mobasher, B., N. Jain, E. Han, and J. Srivastava, "Web mining: Pattern discovery from world wide web transactions," Technical Report TR 96-050, University of Minnesota, Department of Computer Science, Minneapolis, MN,1996.

Norman, D. A., *The Design of Everyday Things*, New York, NY: Basic Books, 2002.

Ortega-Argiles, R., L. Potters, and M. Vivarell. "R&D and productivity: testing sectoral peculiarities using micro data." *Empirical Economics* 41, no. 3 (2011): 817–839.

Paul, S. M., D. S. Mytelka, C. T. Dunwiddie, C. C. Persinger, B. H. Munos, S. R. Lindborg, and A. L. Schacht. "How to improve R&D productivity: the pharmaceutical industry's grand challenge." *Nature Reviews Drug Discovery* 9, no. 3 (2010): 203–14.

Perry, M., and H. Kaminski. "SLA Negotiation System Design Based on Business Rules." Vol. 2, Proceedings of the IEEE International Conference on Services Computing, Honolulu, HI, July 7–11, pp. 609–612, 2008.

Podsiadlowski, P. "Planet Formation Scenarios." in *Planets around Pulsars*; Proceedings of the Conference, J. A. Phillips, J. E. Thorsest, and S. R. Kulkarni, eds. California Institute of Technology, Pasadena, CA. April 30–May 1, 1992, ASP Conference Series, 36, pp. 149–165, 1993.

Sain, N., and S. Tamrakar. "Web Usage Mining & Pre-Fetching Based on Hidden Markov Model & Fuzzy Clustering." *International Journal of Computer Science and Information Technologies*, vol. 3, no. 4 (2012): 4874–4877.

Spivey, J. M. *The Z Notation: A Reference Manual*. 2nd edn. Prentice Hall International Series in Computer Science, 1992.

Symantec Corporation. *What Is e-Discovery and Why Should IT Shops Care?* White paper, Symantec Corporation, March 2008, 10pp.

Taylor., J. R. *An Introduction to Error Analysis: The Study of Uncertainties in Physical Measurements*. University Science Books, pp. 128–129, 1999.

Various. In E. Casey, ed. *Handbook of Digital Forensics and Investigation*. Academic Press, 2009, p. 567.

Warren, J. B. "Drug discovery: Lessons from evolution." *British Journal of Clinical Pharmacology* 71 (2011): 497–503.

Yang, J. S. H., Y. H. Chin, and C. G. Chung. "Many-Sorted First-Order Logic Database Language." *The Computer Journal* 35, no. 2 (1992): 129–137.

CHAPTER 5

Bahrami, J. H., and S. Evans. "The Research Laboratory: Silicon Valley's Knowledge Ecosystem," in *Super-Flexibility for Knowledge Enterprises*. New York, NY: Springer, 2005.

Concise Oxford English Dictionary: Main edition, Oxford Dictionaries, August 2011.

Drucker, P. *The Age of Discontinuity: Guidelines to Our Changing Society*. New York, NY: Harper & Row, 1969.

Martin, P., and P. W. Eklund. "Knowledge retrieval and the World Wide Web," *IEEE Intelligent Systems*, 15, no. 3 (2000): 18–25.

Potter, S. "A Survey of Knowledge Acquisition from Natural Language." *Technology Maturity Assessment*. Retrieved July 9, 2014.

Powell, W. W., and K. Snellman. "The Knowledge Economy." *Annual Review of Sociology* 30, no. 1 (2004): 199–220.

Power, C. J. "Serving two masters: The student work dilemma." *Microbiology Australia* 24, no. 4 (2004).

Sanford, E. H. An Address To a Daughter. *Knowledge is the food of the mind*; and without knowledge the mind must languish. Ann Arbor, MI: The Gem of Science. vol. 1., no. 6., July 31, 1846.

Wolfram, S. *A New Kind of Science*. Champaign, IL: Wolfram Media, 2002.

Yao, Y., Y. Zeng, N. Zhong, and X. Huang. Knowledge Retrieval (KR). In Proceedings of the IEEE/WIC/ACM International Conference on Web Intelligence, IEEE Computer Society, Silicon Valley, CA, November 2–5, 729–735, 2007.

CHAPTER 6

Booch, G., J. Rumbaugh, and I. Jacobson. *The Unified Modeling Language User Guide*. 1st edn. Addison-Wesley Professional, Reading, MA, 1998.

Fayad, M. E. *Stable Design Patterns for Software and Systems*. Boca Raton, FL: Auerbach Publications, 2015.

Fayad, M. E. *Software System Engineering*, Lecture Notes, Required Course at Computer Engineering Department, College of Engineering, San Jose State University, San Jose, CA, 2002–2014.

Fayad, M. E., and A. Altman. "Introduction to Software Stability." *Communications of the ACM* 44, no. 9 (2001): 95–98.

Fayad, M. E., D. Hamu, and D. Brugali. "Enterprise Frameworks Characteristics, Criteria, and Challenges." *Communications of the ACM* 43, no. 10 (2000): 39–46.

Fayad, M. E., N. Islam, and H. Hamza. "The Stable Model-View-Mapping (MVM) Architectural Pattern." Paper presented at the Workshop on Reference Architectures and Patterns for Pervasive Computing, the 18th ACM Conference on Object-Oriented Programming, Systems, Languages, and Applications, Anaheim, CA, 2003.

Fayad, M. E., and H. Kilaru. "Any Information Hiding: A Stable Design Pattern." Paper presented at the 2005 IEEE International Conference on Information Reuse and Integration, Las Vegas, NV, 2005, 108–115.

Fayad, M. E., H. Sanchez, and H. Hamza. "A Pattern Language for CRC Cards." Paper presented at the 11th Conference on Pattern Language of Programs, Illinois, 2004.

Fayad, M. E., H. Sanchez, and H. Hamza. "Towards Scalable and Adaptable Software Architectures." Paper presented at the IEEE International Conference on Information Reuse and Integration, Las Vegas, NV, 2005.

Hamza, H., and M. E. Fayad. "A Pattern Language for Building Stable Analysis Patterns." Paper presented the 9th Pattern Languages of Programs Conference, Monticello, IL, 2002.

James, H. A., K. A. Hawick, and P. D. Coddington. "An Environment for Workflow Applications on Wide Area Distributed Systems." Technical Report DHPC-091, Distributed and High Performance Computing Group, Department of Computer Science, The University of Adelaide, Adelaide, Australia, 2000.

Lawrence, P. *Workflow Handbook*. John Wiley, New York, NY, 1997.

Odell, J. "Designing Agents: Using Life as a Metaphor." Springer Distributed Computing, July, 51–56, 1998.

Patel, D., J. Sutherland, and J. Miller. *Business Object Design and Implementation II*. Springer-Verlag, London, 1998.

Sanchez, H. A. "Laying the Foundations for Branding as a Stable Analysis Pattern." Paper presented at the 19th European Conference on Object-Oriented Programming, Glasgow, UK, July 25–29, 2005.

CHAPTER 7

Brandman, O., J. Cho, and N. Shivakumar. "Crawler-friendly web servers." *Proceedings of ACM SIGMETRICS Performance Evaluation Review* 28, no. 2 (2000).

Büttcher, S., C. L. A. Clarke, and G. V. Cormack. Information Retrieval: Implementing and Evaluating Search Engines. MIT Press, Cambridge, MA, 2010.

Fayad, M. E., and A. Altman. "Introduction to Software Stability." *Communications of the ACM* 44, no. 9 (2001): 95–98.

Fuller, R. B. "Dymaxion World." *LIFE* 41–55. March 1, 1943.

Griffiths, A. J. F., J. H. Miller, D. T. Suzuki, R. C. Lewontin, and W. M. Gelbart. Chapter 5. *An Introduction to Genetic Analysis*, 5th Edition, 1993.

Harvey, P. D. A. *The History of Topographical Maps: Symbols, Pictures and Surveys*, Thames and Hudson, 1980

Koen, P., G. Ajamian, R. Burkart et al. "Providing clarity and a common language to the 'fuzzy front end'." *Research Technology Management* 44, no. 2 (2001): 46–55.

Kraak, M. J., and F. Ormeling, *Cartography: Visualization of Spatial Data*, London, UK: Longman, 1996.

Monmonier, M. *Rhumb Lines and Map Wars: A Social History of the Mercator Projection*, Chicago, IL: The University of Chicago Press, 2004

Morville, P., and L. Rosenfeld, Information Architecture for the World Wide Web: Designing Large-scale Web Sites. O'Reilly & Associates, Inc. Sebastopol, CA, 3rd edition, December 2006.

Neligan, M. "Profile: Jean-Marie Messier". *BBC NEWS: Business* (BBC), July 2002. http://news.bbc.co.uk/1/hi/business/2078564.stm. Retrieved 2006-03-14.

Petchenik, B. B. "From Place to Space: The Psychological Achievement in Thematic Mapping." *The American Cartographer* 6, no. 1 (1979): 5–12.

Slocum, T. A., R. B. McMaster, F. C. Kessler, and H. H. Howard *Thematic Cartography and Geovisualization*, 3rd ed. Upper Saddle River, NJ: Pearson Prentice Hall, 2009.

Spink, A., D. Wolfram, Major B. J. Jansen, and T. Saracevic. "Searching the web: The public and their queries." *Journal of the American Society for Information Science and Technology* 52, no. 3 (2001): 226–234.

Wong, S. K. S. and C. Tong. "The influence of market orientation on new product success." *European Journal of Innovation Management* 15, no. 1 (2012): 99–121.

CHAPTER 8

Dey, A. K. "Understanding and Using Context." *Personal and Ubiquitous Computing* 5, no. 1 (2001): 4–7.

CHAPTER 9

Appleton, B. "Patterns and Software: Essential Concepts and Terminology." *Object Magazine Online* 3, no. 5 (1997): 20–25.

Buschmann, F. *Pattern-Oriented Software Architectures: A System of Patterns*. New York, NY: Wiley, 1996.

Cercone, N., and G. McCalla (eds.). *The Knowledge Frontier: Essays in the Representation of Knowledge*. Symbolic Computation/Artificial Intelligence. 1st edn. Springer-Verlag, New York, NY, 1987.

Connelly, S., J. Burmeister, A. MacDonald, and A. Hussey. "Extending and Evaluating a Pattern Language for Safety-Critical User Interfaces." *Conferences in Research and Practice in Information Technology Series: Sixth Australian Workshop on Safety Critical Systems and Software* 3 (2001): 39–49.

Coplien, J. *Software Patterns*. New York, NY: SIGS, 1996.

Devedzic, V. "Ontologies: Borrowing from Software Patterns." *Intelligence* 10, no. 3 (1999): 14–24.

Fayad, M. E. "Accomplishing Software Stability." *Communications of the ACM* 45, no. 1 (2002a): 111–115.

Fayad, M. E. "How to Deal with Software Stability." *Communications of the ACM* 45, no. 4 (2002b): 109–112.

Fayad, M. E. Software System Engineering, Lecture Notes, Required Course at Computer Engineering Department, College of Engineering, San Jose State University, San Jose, CA, 2002–2014.

Fayad, M. E., and A. Altman. "Introduction to Software Stability." *Communications of the ACM* 44, no. 9 (2001): 95–98.

Fayad, M. E., D. Hamu, and D. Brugali. "Enterprise Frameworks Characteristics, Criteria, and Challenges." *Communications of the ACM* 43, no. 10 (2000): 39–46.

Fayad, M. E., H. S. Hamza, and H. A. Sanchez. "Towards Scalable and Adaptable Software Architectures." Paper presented at the IEEE International Conference on Information Reuse and Integration, Las Vegas, NV, August 15–17, 2005.

Fayad, M. E., N. Islam, and H. Hamza. "The Stable Model-View-Mapping (MVM) Architectural Pattern." Paper present at the Workshop on Reference Architectures and Patterns for Pervasive Computing, The 18th ACM Conference on Object-Oriented Programming, Systems, Languages, and Applications, Anaheim, CA, November, 2003.

Fayad, M. E., A. Ranganath, and M. Pinto. "Towards Software Stability Engineering." Paper presented at the Sixth International Conference on the Unified Modeling Language, Workshop on Stable Analysis Patterns: A True Problem Understanding with UML, Workshop #8, San Francisco, CA, October 20–24, 2003.

Fayad, M. E., and S. Wu. "Merging Multiple Conventional Models in One Stable Model." *Communications of the ACM* 45, no. 9 (2002c): 102–106.

Fernandez, E. B. "Building Systems using Analysis Patterns." Paper presented at the Proceedings of the 3rd International Workshop on Software Architecture, Orlando, FL, November, 1998.

Fowler, M. *Analysis Patterns: Reusable Object Models*. Addison-Wesley, Reading, MA, 1997.

Gamma, E., R. Helm, R. Johnson, and J. Vlissides. *Design Patterns: Elements of Reusable Object-Oriented Software*, 1st edn. Reading, MA: Addison-Wesley Professional, 1995.

Hamza, H. "A Foundation for Building Stable Analysis Patterns." Master's Thesis Report, University of Nebraska, Lincoln, OR, 2002.

Hamza, H., and M. Fayad. "On the Traceability of Analysis Patterns." Paper presented at the ChiliPLoP, Carefree, AZ, March 11–14, 2003.

Judd, H. S. *Think Rich*. New York, NY: Delacorte Press, January 1978.

Laplante, P. A., and C. J. Neill. *Antipatterns: Identification, Refactoring, and Management*. CRC Press, Boca Raton, FL, 2006.

Mahdy, A., and M. E. Fayad. "A Software Stability Model Pattern." Paper presented at the Proceedings of the 9th Conference on Pattern Language of Programs, Allerton Park, Monticello, IL, September 8–12, 2002.

Manns, M. L., and L. Rising. "Introducing Patterns (or any New Idea) into Organizations." Paper presented at the Addendum to the 2000 Proceedings of the Conference on Object-Oriented Programming, Systems, Languages, and Applications, Minneapolis, MN, October 15–19, 2000.

Oestereich, B. *Developing Software with UML: Object-Oriented Analysis and Design in Practice*. 2nd edn. Boston, MA: The Addison-Wesley Object Technology Series, January, 1999.

Sanchez, H. A. "Laying the Foundations for Branding as a Stable Analysis Pattern." Paper presented at the 19th European Conference on Object-Oriented Programming, Glasgow, UK, July 25–29, 2005.

Schmidt, D. C., M. Fayad, and R. E. Johnson. "Software Patterns." *Communications of the ACM* 39, no. 10 (1996): 37–39.

Wu, S., H. Hamza, and M. Fayad. "Implementing Pattern Languages Using Stability Concepts." Paper presented at the ChiliPLoP, Carefree, AZ, March 11–14, 2003.

Yavari, E., and M. E. Fayad. "A Stable Software Model for MRI Visual Analyzer." Paper presented at the 18th Annual ACM SIGPLAN Conference on Object-Oriented Programming, Systems, Languages, and Applications Poster, Anaheim, CA, October 26–30, 2003.

CHAPTER 10

Bederson, B. B., and B. Shneiderman. *The Craft of Information Visualization: Readings and Reflections*, San Francisco, CA: Morgan Kaufmann, 2003, p. 120.

Bloch, J. *Effective Java: Programming Language Guide.* 2nd edn. Addison-Wesley, The Java™ Series, 2005.

Bruegge, B., and A. H. Dutoit. *Object Oriented Software Engineering: Using UML, Patterns, and Java.* 3rd edn. Prentice Hall, Englewood Cliffs, NJ, September, 2003.

Constantinides, C., T. H. Elrad, and M. E. Fayad. "A Framework Solution to Support Advanced Separation of Concerns." *Software: Practice and Experience* 32 (2002).

Dey, A. K. "Understanding and Using Context." *Personal and Ubiquitous Computing* 5, no. 1 (2001): 4–7.

Fayad, M. E. "Accomplishing Software Stability." *Communications of the ACM* 45, no. 1 (2002a): 111–115.

Fayad, M. E. "How to Deal with Software Stability." *Communications of the ACM* 45, no. 4 (2002b): 109–112.

Fayad, M. E., and A. Altman. "Introduction to Software Stability." *Communications of the ACM* 44, no. 9 (2001): 95–98.

Fayad, M. E., and A. Arun. "Identifying UML Elements That Can Be Used to Model Aspects." Paper presented at the UML'03—Workshop on Modeling Aspects Using UML, Workshop #4, San Francisco, CA, October 20–24, 2003.

Fayad, M. E., and M. P. Cline. "Aspects of Software Adaptability." *Communications of the ACM* 39, no. 10, (1996): 37–39.

Fayad, M. E., H. S. Hamza, and H. A. Sanchez. "Towards Scalable and Adaptable Software Architectures." Paper presented at the IEEE International Conference on Information Reuse and Integration, Las Vegas, NV, 2005.

Fayad, M. E., D. C. Schmidt, and R. Johnson. *Building Application Frameworks: Object-Oriented Foundations of Framework Design*. Wiley, New York, NY, 1998.

Fayad, M. E., and S. Wu. "Merging Multiple Conventional Models in One Stable Model." *Communications of the ACM* 45, no. 9 (2002).

Fayad, M. E. *Stable Design Patterns for Software and Systems.* Boca Raton, FL: Auerbach Publications, 2015.

Findler, R. B., and M. Felleisen. "Behavioral Interface Contracts for Java." Technical Report CS TR00-366, Department of Computer Science, Rice University, Houston, TX, 2000.

Lackner, M., A. Krall, and F. Puntigam. "Supporting Design by Contract in Java." *Journal of Object Technology (Special Issue: TOOLS USA 2002 Proceedings)* 1 no. 3 (2002): 57–76.

Monson-Haefel, R. *Enterprise Javabeans.* 2nd edn. O'Reilly Media, Sebastopol, CA, 2000.

Oestereich, B. *Developing Software with UML: Object-Oriented Analysis and Design in Practice.* 2nd edn. Boston, MA: The Addison-Wesley Object Technology Series, January, 1999.

CHAPTER 11

Bruegge, B., and A. H. Dutoit. *Object-Oriented Software Engineering: Using UML, Patterns, and Java.* 3rd edn. Prentice Hall, Englewood Cliffs, NJ, September, 2003.

Fayad, M. E., and M. Cline. "Aspects of Software Adaptability." *Communications of the ACM* 39, no. 10 (1996): 58–9.

Fayad, M. E., D. Hamu, and D. Brugali. "Enterprise Frameworks Characteristics, Criteria, and Challenges." *Communications of the ACM* 43, no. 10 (2000): 39–46.

Fayad, M. E., H. S. Hamza, and H. A. Sanchez. "Towards Scalable and Adaptable Software Architectures." Paper presented at the IEEE International Conference on Information Reuse and Integration, Las Vegas, NV, 2005.

Fayad, M. E., and R. S. Pradeep. "A Pattern Language for Performance Evaluation." Paper presented at the 4th Latin American Conference on Pattern Languages of Programming, Porto das Dunas, Ceará, Brazil, August 10–13, 2004.

Fayad, M. E., H. Sanchez, and H. Hamza. "A Pattern Language for CRC Cards." Paper presented at the 11th Conference on Pattern Language of Programs, Monticello, IL, 2004.

Jacobsen, E. E., B. B. Kristensen, and P. Nowack. "Characterizing Architecture as Abstractions over the Software Domain." Paper presented at the Position Paper for the 1st Working IFIP Conference on Software Architecture, San Antonio, TX, 1999.

Malan, R., and D. Bredemeyer. "Defining Non-Functional Requirements." Technical Report at Bredemeyer Consulting. Whitepaper, Bloomington, IN: Bredemeyer Consulting, August, 2001. http://www.bredemeyer.com.

Tran, E. "Verification/Validation/Certification." In Koopman, P. *Topics in Dependable Embedded Systems*. USA: Carnegie Mellon University, 1999. Retrieved January 1, 2008.

Tsai, W. T., Vishnuvajjala, R., and Zhang, D. Verification and validation of knowledge-based systems. *IEEE Transactions on Knowledge and Data Engineering* 11, no. 1 (1999): 202–212.

SECTION V

Buschmann, F. *Pattern-Oriented Software Architectures: A System of Patterns*. New York, NY: Wiley, 1996.

Fayad, M. E. "Accomplishing Software Stability." *Communications of the ACM* 45, no. 1 (2002a): 111–115.

Fayad, M. E. "How to Deal with Software Stability." *Communications of the ACM* 45, no. 4 (2002b): 109–112.

Fayad, M. E., and A. Altman. "Introduction to Software Stability." *Communications of the ACM* 44, no. 9 (2001): 95–98.

Gamma, E., R. Helm, R. Johnson, and J. Vlissides. *Design Patterns: Elements of Reusable Object-Oriented Software.* 1st edn. Reading, MA: Addison-Wesley Professional, November, 1994.

CHAPTER 12

Bloch, J. *Effective Java: Programming Language Guide.* 2nd edn. Addison Wesley, The Java™ Series, 2005.

Buschmann, F. *Pattern-Oriented Software Architectures: A System of Patterns.* New York, NY: Wiley, 1996.

Forman, I. R., and N. Forman. *Java Reflection in Action.* Manning Publications, Greenwich, CT, 2004.

Gamma, E., R. Helm, R. Johnson, and J. Vlissides. *Design Patterns: Elements of Reusable Object-Oriented Software.* 1st edn. Reading, MA: Addison-Wesley Professional, 1994.

Simmons, R. *Hardcore Java.* 1st edn. O'Reilly Media, Sebastopol, CA, 2004.

Stelting, S. A., and O. Maassen. *Applied Java Patterns.* 1st edn. Prentice Hall, Upper Saddle River, NJ, January, 2001.

CHAPTER 13

Abbott, A. *Chaos of Disciplines.* Chicago, IL: University of Chicago Press. 2001.

Ambler, S. W. *CRC Modeling: Bridging the Communication Gap between Developers and Users.* An AmbySoft Inc. White Paper, 1998.

Ambler, S. W. *The Object Primer: The Application Developer's Guide to Object Orientation and the UML*, 2nd edn. Cambridge; New York: Cambridge University Press; SIGS Books, May 2001.

Armitage, H. M. and S. Cameron. *Using Strategy Maps to Drive Performance*. The Society of Management Accountants of Canada, the American Institute of Certified Public Accountants, and The Chartered Institute of Management Accountants, Canada, 2006.

Armitage, U. "Can Navigation Aids Support Constructive Engagement with Hypermedia?" EC1V OHB, Centre for HCI Design, City University, Northampton Square, London.

Beck, K., and W. Cunningham. "A Laboratory for Teaching Object-Oriented Thinking." Paper presented at the OOPSLA'89, Conference Proceedings, New Orleans, LA, October 1–6, 1989.

Biddle, R., J. Noble, and E. Tempero. "Role-Play and Use Case Cards for Requirements Review." Paper presented at the Proceedings of the 20th Australian Conference on Information System, Melbourne, Australia, December, 2009.

Börstler, J., T. Johansson, and M. Nordström. "Teaching OO Concepts—A Case Study Using CRC-CARDS and BLUEJ." Paper presented at the 32nd ASEE/IEEE Frontiers in Education Conference, Boston, MA, November 6–9, 2002.

Coplien, J. O., and D. C. Schmidt (eds.). *Pattern Languages of Program Design*. Addison-Wesley, 1995.

Deming, W. E., *The New Economics for Industry, Government, and Education*. Boston, MA: MIT Press, p. 132, 1993.

Fayad, M. E. *Stable Design Patterns for Software and Systems*. Auerbach Publications, 2015.

Fayad, M. E. Software System Engineering, Lecture Notes, Computer Engineering Department, San Jose State University, San Jose, CA, 2002–2014.

Fayad, M. E., and H. Hamza. "Software Stability Background." Whitepaper at San Jose State University, San Jose, CA, 2003.

Fayad, M. E., H. S. Hamza, and H. A. Sanchez. "A Pattern for an Effective Class Responsibility Collaborator (CRC) Cards." Paper presented at the 2003 IEEE International Conference on Information Reuse and Integration, Las Vegas, NV, October 2003.

Fayad, M. E., H. A. Sanchez, and R. Goverdhan. "A Goal-Driven Software Development Life Cycle." Whitepaper at San Jose State University, San Jose, CA, 2005.

Fayad, M. E., V. Stanton, H. Sanchez, and H. Hamza. "A Closer Look at Class Responsibility Collaborator (CRC) Cards." Technical Report, Computer Engineering Department, San Jose State University, San Jose, CA, 2003.

Halbleib, H. "Software Design Using CRC Cards." *Real Time Magazine* 99-1, 1999: 28–32.

Hamza, H., and M. E. Fayad. "A Pattern Language for Building Stable Analysis Patterns." Paper presented at the 9th Pattern Languages of Programs Conference, Monticello, IL, September 8–12, 2002.

Kearsley, G., and B. Shneiderman. "Engagement Theory: A Framework for Technology-Based Teaching and Learning." *Educational Technology* 38, no. 5 (1998): 20–23.

Maciaszek, L. A. *Requirements Analysis and System Design—Developing Information System with UML*. Addison-Wesley Professional, Boston, MA, 2001.

Oleson, A. and J. Voss, eds. *The Organization of Knowledge in Modern America, 1860–1920*. Baltimore, MD: The Johns Hopkins University Press, pp. 285–312, 1979.

Oxford English Dictionary, 2nd ed. Oxford: Clarendon Press, 1989.

Pressman, R. S. *Software Engineering—A Practitioner's Approach*. McGraw-Hill Publishing Company, New York, NY, 2001.

Schmidt, D. C., M. E. Fayad, and R. E. Johnson. "Software Patterns." *Communications of the ACM* 39, no. 10 (1995): 65–74.

CHAPTER 14

Coplien, J. O., and D. C. Schmidt. eds. *Pattern Languages of Program Design*. Addison-Wesley Professional, 1995.

Fayad, M. E. "Accomplishing Software Stability." *Communications of the ACM* 45, no. 1 (2002a): 111–115.

Fayad, M. E. "How to Deal with Software Stability." *Communications of the ACM* 45, no. 4 (2002b): 109–112.

Fayad, M. E., and A. Altman. "Introduction to Software Stability." *Communications of the ACM* 44, no. 9 (2001): 95–98.

Schmidt, D. C., M. E. Fayad, and R. E. Johnson. "Software Patterns." *Communications of the ACM* 39, no. 10 (1996): 37–39.

Zimbardo, P., and J. Boyd. The Time Paradox – The New Psychology of Time That Will Change Your Life. New York, NY: Free Press. 2008, p. 135.

APPENDIX B

Fayad, M. E. "How to Deal with Software Stability." *Communications of the ACM* 45, no.4 (2002): 109–112.

Fayad, M. E. Software System Engineering, Lecture Notes, Required Course at Computer Engineering Department, College of Engineering, San Jose State University, San Jose, CA, 2002–2014.

Fayad, M. E., H. S. Hamza, and H. A. Sanchez. "A Pattern for an Effective Class Responsibility Collaborator (CRC) Cards." Paper presented at the 2003 IEEE International Conference on Information Reuse and Integration, Las Vegas, NV, October 2003.

Fayad, M. E., and S. Wu. "Merging Multiple Conventional Models in One Stable Model." *Communications of the ACM* 45, no. 9 (2002): 102–106.

Hamza, H., and M. E. Fayad. "A Pattern Language for Building Stable Analysis Patterns." Paper presented at the Proceedings of the 9th Conference on Pattern Languages of Programs 2002, Monticello, IL, September 2002.

Sanchez, H. A. Building Systems Using Patterns: Creating Knowledge Maps. Masters Thesis. San Jose State University, San Jose, CA, May 2006.

Index

Note: Locators followed by "*f*" and "*t*" denote figures and tables in the text

B

C

D